Nonsmooth Approach to Optimization Problems with Equilibrium Constraints

Nonconvex Optimization and Its Applications

Volume 28

The titles published in this series are listed at the end of this volume.

To Eva, Daniela, and Marlis

Contents

Preface

In the early fifties, applied mathematicians, engineers and economists started to pay close attention to the optimization problems in which another (lower-level) optimization problem arises as a side constraint. One of the motivating factors was the concept of the Stackelberg solution in game theory, together with its economic applications. Other problems have been encountered in the seventies in natural sciences and engineering. Many of them are of practical importance and have been extensively studied, mainly from the theoretical point of view. Later, applications to mechanics and network design have lead to an extension of the problem formulation: Constraints in form of variational inequalities and complementarity problems were also admitted. The term "generalized bilevel programming problems" was used at first but later, probably in Harker and Pang, 1988, a different terminology was introduced: *Mathematical programs with equilibrium constraints*, or simply, MPECs. In this book we adhere to MPEC terminology.

A large number of papers deals with MPECs but, to our knowledge, there is only one monograph (Luo et al., 1997). This monograph concentrates on optimality conditions and numerical methods. Our book is oriented similarly, but we focus on those MPECs which can be treated by the *implicit programming approach:* the equilibrium constraint locally defines a certain implicit function and allows to convert the problem into a mathematical program with a nonsmooth objective. Using the "nondifferentiable" calculus of Clarke, we derive necessary optimality conditions for this class of MPECs and propose a solution method based on the bundle technique of nonsmooth optimization. Therefore we speak of a *nonsmooth approach* to MPECs. These optimality conditions differ from the conditions in Luo et al., 1997 and our numerical method is not discussed there either. Besides Clarke's calculus, the main tool in our development is the stability and sensitivity theory due to Robinson. The numerical approach proved its efficiency on a series of tough nonacademic problems from the area of optimum shape design and economic modelling. These problems are of interest in their own and hence introduced in detail in the second part of the book.

The book is organized as follows:

In Chapter 1 we introduce MPEC and describe several motivating problems coming from different application areas. Further, we briefly survey several existing approaches both to optimality conditions and to a numerical solution of MPEC and give a short description of the nonsmooth approach. The chapter concludes with some simple conditions ensuring the existence of a solution.

Chapter 2 collects the basic results from nonsmooth analysis which are essential for the whole development. Further, some needful auxiliary results on projections and polyhedral set-valued maps are included. As all this is mostly standard material, some longer proofs are omitted.

Chapter 3 discusses two classes of methods which are basic for our numerical treatment of MPEC: The bundle approach from nonsmooth optimization and the nonsmooth Newton's method for the solution of Lipschitz equations.

In the sensitivity and stability studies of equilibrium problems, it is convenient to work with various equivalent formulations (variational inequalities, generalized equations, nonsmooth equations etc.). In the first part of Chapter 4 we present the formulations relevant for our framework. The second part of this chapter gives the basic existence and uniqueness results for the considered equilibrium problems.

In Chapter 5 we apply the essential parts from Robinson's stability theory to the studied types of equilibria. As a result, we get various criteria which ensure the existence and Lipschitz continuity of the implicit function mentioned above. This helps to decide, whether our approach can be applied to a given MPEC.

Chapter 6 deals with sensitivity issues. The behaviour of equilibria with respect to perturbations is described by directional derivatives and by generalized Jacobians of Clarke. Further, we verify an important property, called semismoothness, which relates directional derivatives to generalized Jacobians. This property is a crucial prerequisite for our numerical method.

Chapter 7 is of central importance. In the first part we establish necessary optimality conditions for equilibria governed by variational inequalities, nonlinear complementarity problems and implicit complementarity problems. Under certain assumptions we guarantee that these conditions are as "tight" as possible within the framework of Clarke's calculus. The second part is devoted to the numerical solution of MPEC. We analyze the conditions which guarantee the global convergence of our method to a stationary point and investigate a question, how to provide the subgradient information, required by the bundle-trust algorithm.

The aim of Part II—Applications is threefold: First we want to indicate the main sources of MPECs to which our nonsmooth approach can be applied; second we intend to show what has to be done in order to apply our numerical method to a given MPEC; third we would like to encourage the reader to use our approach.

Chapters 9–11 deal with several optimum design problems in nonsmooth mechanics. In Chapter 9 we introduce, analyze and solve three design problems for a membrane which is in contact with an obstacle. Chapter 10 studies design problems with elastic-plastic and masonry structures and in Chapter 11 contact problems with Coulomb friction are investigated. All problems appear in an infinite dimensional setting, then they are discretized by the finite element method and the resulting discrete MPECs are finally solved by the tools of Chapter 7. Each problem is accompanied with examples that show the efficiency of the numerical approach.

Chapter 12 presents two equilibrium problems from economic modelling. As a typical Stackelberg problem we consider the Stackelberg strategy for a producer in an oligopolistic market. This problem appears in MPEC form quite naturally. The second section deals with the computation of generalized Nash equilibria to which, after a simple reformulation, the proposed numerical method can also be successfully applied.

The book is closed with three appendices. The first one summarizes the tools needed for a successful solution of an MPEC. Reader only interested in the technique of solving his particular problem will find all the formulas in this Appendix A. Large part of the book deals with problems modelled by elliptic boundary value problems. From this reason we included Appendix B—a condensed elementary introduction to this discipline. Finally, Appendix C contains two technical results on complementarity problems.

This book addresses several types of readers. We hope it to be useful for applied mathematicians working in continuum mechanics, operations research or economic modelling. It could also be of interest to mathematical programmers working in sensitivity and stability. Finally, graduate and advanced undergraduate students may use this book to get an insight into an interesting and rapidly developing topic of mathematical programming. We have tried to keep Part II as self-contained as possible. Hence a reader, who is mainly interested in applications and ready to neglect the underlying mathematical theory, will have no major difficulties in understanding the text of this second part. Readers, familiar with Clarke's calculus, bundle methods and nonsmooth Newton variants may certainly skip Sections 2.2, 2.3 and the whole Chapter 3. The same hold for Sections 2.1, 2.4, the whole Chapter 5 and Section 6.1 for those, who are acquainted with the Robinson's stability and sensitivity theory.

This book owes a lot to F. H. Clarke and S. M. Robinson. Their contributions to nonsmooth and set-valued analysis with applications in optimization and variational inequalities definitely form the theoretical basis of this book. Further, we would like to mention many helpful discussions we had with S. Dempe, F. Facchinei, S. D. Flåm, J. Haslinger, J. Jarušek, D. Klatte, B. Kummer, J.-S. Pang, D. Ralph and S. Scholtes.

We thank R. Horst for welcoming our work into the series *Nonconvex Optimization and Its Application*.

Our thanks go also to I. Marešová who typed a great part of the manuscript.

Much of our work has been supported through the projects K1075601 and A1075707 of the Czech Academy of Sciences and the project 03ZO7BAY of the Federal Department of Education, Science, Research and Technology (BMBF), Germany. We express our gratitude for this support.

J.V.O., M.K., and J.Z.
Erlangen, Prague, March 1998

List of Notations

Special symbols

■	end of proof
$\triangle$	end of example

Scalars

$u^+ := \max(0, u)$	the nonnegative part of a scalar
$u^- := \max(0, -u)$	the nonpositive part of a scalar

Spaces

$\mathbb{R}^n$	real n-dimensional space
$\mathbb{R}$	the real line

Vectors

x^i	the ith components of a vector $x \in \mathbb{R}^n$
e_i	the ith vector of the canonical basis in $\mathbb{R}^n$
$\langle x, y \rangle := x^T y$	the standard inner product of vectors in $\mathbb{R}^n$
$\|x\| = \sqrt{\langle x, x \rangle}$	the Euclidean norm of a vector $x \in \mathbb{R}^n$
$x \geq y$	the (usual) partial ordering $x^i \geq y^i,\ i = 1, \ldots, n$
$x > y$	the strict ordering: $x^i > y^i,\ i = 1, \ldots, n$
$\min_c(x, y)$	the vector whose ith component is $\min(x^i, y^i)$
$\max_c(x, y)$	the vector whose ith component is $\max(x^i, y^i)$
$x^+ := \max_c(0, x)$	the nonnegative part of a vector x
$x^- := \max_c(0, -x)$	the nonpositive part of a vector x
$\mathbf{1} := (1, 1, \ldots, 1)^T$	

Matrices

A^{ij}	the (i,j)th element of a matrix A
A^i	the ith row of a matrix A
$\|A\|$	the Euclidean norm of a matrix A
$A_{I,J}$	the submatrix of an $m \times n$ matrix A with elements A^{ij}, $i \in I \subset \{1,2,\ldots,m\}$ and $j \in J \subset \{1,2,\ldots,n\}$
A_I	the submatrix of an $m \times n$ matrix A with rows A^i, $i \in I \subset \{1,\ldots,m\}$
$[A]_s$	the submatrix of an $m \times n$ matrix A with rows A^i, $i = 1,2,\ldots,s,\ s \leq m.$
E	the identity matrix of appropriate order
$\text{diag}\{a\}$	the diagonal matrix with diagonal elements equal to the components of the vector a
$\text{Ker}(A)$	the kernel of a matrix A
A/A_{11}	the Schur complement of the matrix A_{11} in the matrix A, cf. Lemma 5.6

Cones

Ω^*	the (positive) polar cone to a set Ω
Ω^{**}	the (positive) bipolar cone to a set Ω
$T_\Omega(x)$	the tangent cone to a convex set Ω at $x \in \text{cl}\,\Omega$, cf. Definition 2.5
$C_\Omega(x)$	the Clarke's tangent cone to a set Ω at $x \in \Omega$, cf. Definition 2.14
$N_\Omega(x)$	the normal cone to a convex set Ω at $x \in \text{cl}\,\Omega$, cf. Definition 2.6
$K_\Omega(x)$	the Clarke's normal cone to a set Ω at $x \in \Omega$, cf. Definition 2.15
$\mathbb{R}_+$	the set of nonnegative reals
$\mathbb{R}_-$	the set of nonpositive reals
$\mathbb{R}^n_+$	the nonnegative orthant of $\mathbb{R}^n$

Functions

$F[\mathcal{D} \to \mathbb{R}^n]$	a mapping with domain $\mathcal{D}$ and range in $\mathbb{R}^n$
$\mathcal{R}(F)$	the range of F
F^i	the ith component of a mapping F with range in $\mathbb{R}^n$
$F_1 \circ F_2$	the composite function
$\mathcal{J}F(x)$	the $m \times n$ Jacobi matrix (Jacobian) of a mapping $F[\mathbb{R}^n \to \mathbb{R}^m]$ $(m \geq 2)$ at $x \in \mathbb{R}^n$ with elements $\frac{\partial F^i(x)}{\partial x^j}$
$\nabla f(x)$	the gradient of a real-valued function f at x
$\mathcal{J}_x F(x, y)$	the partial Jacobian of F with respect to x
$\nabla_x f(x, y)$	the partial gradient of f with respect to x
$\mathcal{J} F_{I,J}(x)$	the submatrix of the $m \times n$ matrix $\mathcal{J}F(x)$ with elements $\frac{\partial F^i(x)}{\partial x^j}$, $i \in I \subset \{1, 2, \ldots, m\}$, $j \subset \{1, 2, \ldots, n\}$
$\mathcal{J} F_I(x)$	the submatrix of the $m \times n$ matrix $\mathcal{J}F(x)$ with rows $\left(\frac{\partial F^i(x)}{\partial x^1}, \frac{\partial F^i(x)}{\partial x^2}, \ldots, \frac{\partial F^i(x)}{\partial x^n}\right)$, $i \in I \subset \{1, 2, \ldots, m\}$
$\nabla^2 f(x)$	the Hessian matrix of a real-valued function f at x
$F'(x; h)$	the directional derivative of a mapping F at x in the direction h
$f^0(x; h)$	the Clarke's directional derivative of a real-valued function f at x in the direction h, cf. Definition 2.9
$\partial F(x)$	generalized Jacobian (gradient) of a (real-valued) function F at x, cf. Definition 2.12
F^{-1}	the inverse of F
$o(t)$	any function such that $\lim_{t \to 0} \frac{o(t)}{t} = 0$
$\mathrm{Proj}_\Omega(x)$	the Euclidean projection of x onto the set Ω
$\mathrm{dist}_\Omega(x) := \inf_{y \in \Omega} \|x - y\|$	distance function from vector x to set Ω
$\delta_A(\cdot) := \sup_{a \in A} \langle \cdot, a \rangle$	the support function of a set A
$\mathrm{dom}\, f := \{x \mid f(x) < \infty\}$	the effective domain of a convex function f
π_1 (π_2)	canonical projection of $\mathbb{R}^n \times \mathbb{R}^m$ onto $\mathbb{R}^n$ $(\mathbb{R}^m)$
$\triangle$	Laplace operator; $\triangle(u) := \frac{\partial^2 u^1}{\partial (x^1)^2} + \ldots + \frac{\partial^2 u^n}{\partial (x^n)^2}$

Multifunctions

$\operatorname{Dom}\Gamma$	the domain of a multifunction Γ
$\operatorname{Gph}\Gamma$	the graph of a multifunction Γ
$\operatorname{Im}\Gamma := \{y \in \Gamma(x) \mid x \in \operatorname{Dom}\Gamma\}$	the image of a multifunction Γ
Γ^{-1}	the inverse of a multifunction Γ

Sets

$\operatorname{conv} S$	convex hull of a set S
$\operatorname{cl} S$	the (topological) closure of a set S
$\operatorname{int} S$	the interior of a set S
$\operatorname{ri} S$	the relative interior of a set S
$\operatorname{lin} S$	the linear hull of a set S
$\times_{i=1}^{n} \Omega_i$	Cartesian product of sets Ω_i, $i = 1, 2, \ldots, n$
$S_1 \setminus S_2$	the difference of sets S_1 and S_2
$\|S\|$	the cardinality of a finite set S
$\mathbb{B}$	the unit ball
$\partial\mathbb{B}$	the unit sphere
$\arg\min_{x \in \Omega} f(x)$	the set of x attaining the minimum of the real-valued function f on the set Ω
$\arg\max_{x \in \Omega} f(x)$	the set of x attaining the maximum of the real-valued function f on the set Ω
$\operatorname{lev}_\alpha f = \{x \mid f(x) \le \alpha\}$	the level set of a real-valued function f associated with $\alpha \in \mathbb{R}$
$(S)^\perp$	the orthogonal complement of set S
$\{x\}^\perp$	the orthogonal complement of vector x
$[a, b]$	a closed interval in $\mathbb{R}$
(a, b)	an open interval in $\mathbb{R}$
$[x, y]$	a line segment in $\mathbb{R}^n$ $(x,\ y \in \mathbb{R}^n)$
(x, y)	an open line segment in $\mathbb{R}^n$ $(x,\ y \in \mathbb{R}^n)$
$\mathcal{P}(I)$	the set of all subsets of a finite set I
$\operatorname{meas}\Omega$	the Lebesgue measure of a set Ω
I	the index set of active inequality constraints
I^+	the index set of strongly active inequality constraints
I^0	the index set of weakly active inequality constraints
L	the index set of inactive inequality constraints
a, a^+, a^0	cardinality of I, I^+, I^0, respectively

Spaces of functions

Ω	domain in $\mathbb{R}^n$
$\partial\Omega$	boundary of Ω
$C_0^\infty(\Omega)$	the space of infinitely differentiable functions with a compact support in Ω
$C^{0,1}([a,b])$	the space of Lipschitz functions on $[a,b] \subset \mathbb{R}$
$L_2(\Omega)$	the space of measurable functions on Ω such that $\int_\Omega \lvert u(\xi)\rvert^2 \, d\xi < \infty$
$(\cdot,\cdot)_{L_2(\Omega)}$	inner product in $L_2(\Omega)$, $(u,v)_{L_2(\Omega)} := \int_\Omega u(\xi)v(\xi)\, d\xi$
$D^\alpha u$	$D^\alpha u := \dfrac{\partial^{\lvert\alpha\rvert} u}{\partial\xi_1^{\alpha_1} \ldots \partial\xi_n^{\alpha_n}}$, $u : \mathbb{R}^n \to \mathbb{R}$, $\alpha = (\alpha_1, \ldots, \alpha_n)$, $\lvert\alpha\rvert = \alpha_1 + \ldots + \alpha_n$
$H^m(\Omega)$	$H^m(\Omega) := \{v \in L_2(\Omega) \mid D^\alpha v \in L_2(\Omega)\ \forall \lvert\alpha\rvert \le m\}$
$(\cdot,\cdot)_{H^m(\Omega)}$	inner product in $H^m(\Omega)$, $(u,v)_{H^m(\Omega)} := \sum_{\lvert\alpha\rvert \le m} \int_\Omega D^\alpha u(\xi) D^\alpha v(\xi)\, d\xi$
$H_0^m(\Omega)$	the closure of $C_0^\infty(\Omega)$ in $H^m(\Omega)$
$\gamma_0 v$	trace of a function $v \in H^1(\Omega)$; $\gamma_0 : H^1(\Omega) \mapsto L_2(\partial\Omega)$

Optimum design problems

ξ	space coordinate
α	control variable, shape of the boundary
U_{ad}	set of admissible control variables
$\Omega(\alpha), \Omega_\alpha$	domain to be optimized (membrane, elastic body)
$\mathcal{J}$	cost (objective) functional in optimum design problems
φ	piecewise linear (Courant) basis function
ψ	piecewise constant basis function
M	number of finite elements
m	number of nodes in the discretization
n	size of the control vector $\boldsymbol{\alpha}$
h	discretization parameter
K_k	kth finite element
$\boldsymbol{\alpha}$	discrete control variable, $\boldsymbol{\alpha} \in \mathbb{R}^n$
$\mathbf{U_{ad}}$	set of discrete admissible control variables
$\boldsymbol{J}$	discrete cost functional in optimum design problems

Elasticity problems

$\mathbb{R}^{2\times 2}_s$	set of symmetric 2×2 matrices
$S(\Omega)$	set of matrix valued functions
n	outer normal
F	body forces
T	surface tractions
E	Young's modulus
ν	Poisson's ratio
λ, μ	Lamé coefficients
$\mathcal{H}$	generalized Hooke's law
$\mathcal{A}$	inverse Hooke's law
Φ	potential energy
Ψ	complementary energy
u	displacement vector
$\mathcal{U}$	set of admissible displacements
e	small strain tensor
σ	stress tensor
$\mathcal{E}(\Omega)$	set of stresses in weak equilibrium
$\mathcal{P}(\Omega)$	set of plastically admissible stresses
$\mathcal{M}(\Omega)$	set of admissible stresses for masonry materials
σ_E	limit stress
Υ	yield function
V	given volume for plasticity problem
λ	KKT vector for inequality constraints in masonry problem
$\boldsymbol{f}$	vector of discrete forces
$\boldsymbol{A}$	equilibrium matrix
$\boldsymbol{F}$	flexibility matrix
D	matrix in the discrete Hooke's law
$\boldsymbol{\Psi}$	discrete functional of complementary energy
$\boldsymbol{\sigma}$	vector of discrete stresses
$\mathbb{E}$	discrete set of stresses in weak equilibrium
$\mathbb{P}$	discrete set of plastically admissible stresses
$\mathbb{M}$	discrete set of admissible stresses for masonry material
$\boldsymbol{\lambda}$	discrete KKT vector for inequality constraints in masonry problem

List of Acronyms

ELICQ	Extended Linear Independence Constraint Qualification
ESCQ	Extended Slater Constraint Qualification
GE	Generalized Equation
GNE	Generalized Nash Equilibrium
ICP	Implicit Complementarity Problem
IMP	Implicit Programming
LICQ	Linear Independence Constraint Qualification
LCP	Linear Complementarity Problem
MF1	Mangasarian-Fromowitz constraint qualification for variational inequalities
MF2	Mangasarian-Fromowitz constraint qualification for complementarity problems
MPEC	Mathematical Program with Equilibrium Constraints
NCP	Nonlinear Complementarity Problem
NSE	Nonsmooth Equation
QP	Quadratic Program
QVI	Quasi–Variational Inequality
SCQ	Slater Constraint Qualification
SQP	Sequential Quadratic Programming
SRC	Strong Regularity Condition
SSOSC	Strong Second–Order Sufficient Optimality Condition
VI	Variational Inequality

I Theory

1 INTRODUCTION

The central subject of this book is an optimization problem with a special side constraint. Its definition and some illustrative examples are displayed in Section 1.1. Section 1.2 briefly describes various optimality conditions and numerical methods. Section 1.3 states two simple existence results.

1.1 THE MAIN PROBLEM

Let $C[\mathbb{R}^n \times \mathbb{R}^k \to \mathbb{R}^k]$ be a continuous map and Q a nonempty closed convex subset of $\mathbb{R}^k$. Further, let $N_Q(z)$ denote the standard normal cone to Q at $z \in \mathbb{R}^k$ in the sense of convex analysis; cf. Definition 2.6. These data define a what we call *perturbed generalized equation* (GE)

$$0 \in C(x, z) + N_Q(z) \tag{1.1}$$

in the variable $z \in \mathbb{R}^k$, the perturbation parameter being $x \in \mathbb{R}^n$. We shall see later that many equilibrium problems from natural sciences, engineering and economics can be written in the form (1.1) (with the perturbation parameter x possibly fixed). In particular, equilibria given as solutions to parameter-dependent optimization problems, variational inequalities or complementarity problems will, under some assumptions, fit the form (1.1) as shown in Chapter 5.

Now let U_{ad} be a nonempty closed subset of $\mathbb{R}^n$ and consider the map $S[U_{\text{ad}} \rightsquigarrow \mathbb{R}^k]$ which assigns each $x \in U_{\text{ad}}$ the solution set of (1.1). This so-called *solution map* or *solution multifunction* plays a crucial role in this book. Further, we are given an objective $f[\mathbb{R}^n \times \mathbb{R}^k \to \mathbb{R}]$ and a nonempty closed set $Z \subset \mathbb{R}^k$, constraining the choice of z. Our interest will primarily be with the optimization problem

$$\begin{array}{ll} \text{minimize} & f(x,z) \\ \text{subject to} & \\ & z \in S(x) \\ & x \in U_{\text{ad}},\ z \in Z. \end{array} \tag{1.2}$$

Problem (1.2) includes a subclass of so-called bilevel programs, where S assigns each $x \in U_{\text{ad}}$ the solution of a "lower-level" optimization problem, typically considered by an agent different from the one who considers (1.2). S can further be generated by generalized equations corresponding to variational inequalities and complementarity problems.

In all cases GEs represent certain equilibrium conditions and $S(x)$ restrains the set of equilibria. Hence the relation

$$z \in S(x)$$

is also called "equilibrium constraint". This led to the commonly used terminology *mathematical program with equilibrium constraints* (in short MPEC) (Luo et al., 1997).

We emphasize that the variables x and z in (1.1),(1.2) play different roles and will not be treated in like manner. Motivated by applications, we call x *control* or *design variable (parameter)* and z *state variable.* Accordingly, U_{ad} is the *set of feasible controls* or *design variables (parameters).*

Next come some examples:

Two-person Stackelberg games. Consider a game of two players, each of which tries to minimize his objective over his feasible strategy set. The first player is the Leader. This means that he acts first, knowing that the second player, called Follower, subsequently minimizes his payoff. Assume that the Follower selects a strategy from his solution set (if not a singleton) that is optimal for the Leader. Then we get the bilevel program

$$\begin{array}{ll} \text{minimize} & f_L(x,z) \\ \text{subject to} & \\ & z \in \operatorname{argmin}_{v \in \Omega_F} f_F(x,v) \\ & x \in \Omega_L, \end{array} \tag{1.3}$$

where f_L, f_F are the objectives and Ω_L, Ω_F are the feasible strategy sets of the Leader and the Follower, respectively. Note that the problem

$$\begin{array}{ll} \text{minimize} & f_F(x,v) \\ \text{subject to} & \\ & v \in \Omega_F \end{array}$$

is part of (1.3). If $f_F(x,\cdot)$ is convex and differentiable for all x, and Ω_F is convex and closed, then the Follower's problem is equivalent to the GE

$$0 \in \nabla_z f_F(x,z) + N_{\Omega_F}(z).$$

We arrive at an MPEC, where S assigns the solution set of the Follower's problem to the Leader's strategy x. S may have complicated structure even if the objective function is smooth and the strategy set is simple.

The Cournot oligopoly. Consider a market where n firms supply a homogeneous product in a noncooperative fashion. In such a situation, the well-known Nash concept characterizes market equilibrium: no individual deviation from equilibrium increases the profit of the firm in question. Under certain assumptions, this equilibrium can be described by a generalized equation. One firm, e.g. the largest producer, may now think of increasing its profit by changing its production; it assumes that the other firms will share the rest of the market again by the Nash equilibrium concept. This firm takes over the role of the Leader in the above Stackelberg game. The computation of its strategy leads to an MPEC, where f is the profit of the Leader, U_{ad} specifies its feasible production and the respective GE (1.1) characterizes the Nash equilibrium for the remaining firms. Chapter 12 will deal with this MPEC in detail.

Optimum design problems in mechanics. One wants to specify certain variables (e.g. forces, thicknesses, shape parameters) so that the resulting equilibrium is optimal with respect to a performance criterion (e.g. weight or compliance). If this equilibrium state is described by equations, we face a standard optimization (or optimal control) problem. However, as soon as we deal with elastic bodies in contact (giving raise to unilateral constraints) or with nonlinear (nonsmooth) constitutive laws (describing, e.g., plastic deformations), then equilibrium is described by a more complex model and we again face an MPEC. These MPECs are defined on function spaces so that their numerical solution first requires a suitable discretization.

Other MPECs are encountered in the design of transportation networks, where we work with so-called user's equilibrium. Then variational inequalities govern equilibrium flows along the links of the network (Harker and Pang, 1990; Qiu and Magnanti, 1992). Other source problems can be found in Luo et al., 1997.

1.2 SOME EXISTING APPROACHES TO OPTIMALITY CONDITIONS AND NUMERICAL METHODS

This section gives a brief overview of several existing approaches to MPECs. We regard both optimality conditions and numerical methods.

The equilibrium constraint (1.1) can be reformulated in many ways. In fact, some of the approaches described below are closely connected with the chosen reformulation. Assume for now that the GE (1.1) comes from a parameter-dependent nonlinear complementarity problem (NCP) (cf. Chapter 5):

Find $z \in \mathbb{R}^k$ such that

$$z \geq 0, \qquad F(x,z) \geq 0, \qquad \langle F(x,z), z \rangle = 0, \tag{1.4}$$

where $x \in \mathbb{R}^n$ and F maps $\mathbb{R}^n \times \mathbb{R}^k$ into $\mathbb{R}^k$ (here $\geq$ is meant componentwise). In this case MPEC amounts to a standard mathematical program with equality and inequality constraints and "abstract" constraints $x \in U_{\text{ad}}$, $z \in Z$. Unfortunately, the complementarity condition

$$\langle F(x,z), z \rangle = 0$$

makes this program rather difficult. As shown in Scheel and Scholtes, 1997, Mangasarian–Fromowitz constraint qualification is violated for the system

$$\begin{aligned} z &\geq 0 \\ F(x,z) &\geq 0 \\ \langle F(x,z), z\rangle &= 0 \end{aligned}$$

at all feasible points, and thus necessary optimality conditions cannot be derived in a standard way. Moreover, this also restricts the choice of possible numerical methods. The same situation arises if the GE (1.1) comes from a parameter-dependent variational inequality with the feasible set given by inequalities; cf. Chapter 5. In this case (1.1) can be replaced by the corresponding Karush–Kuhn–Tucker (KKT) conditions (cf. (4.11)) and we again arrive at a standard mathematical program with equality and inequality constraints. Unfortunately, the associated complementarity condition leads to the same difficulties. These difficulties were overcome first in case of special bilevel programs (Dempe, 1987; Vicente et al., 1994) and in Edmunds and Bard, 1991, where the complementarity condition is used in a special branch-and-bound algorithm. The results of Luo et al., 1996; Ye et al., 1997, however, indicate that the KKT reformulation can be very useful also in connection with some penalty approaches.

The difficulty with the complementarity constraint can be eliminated by the so called disjunctive or piecewise programming approach developed in Luo et al., 1997. Let

$$\mathcal{E} := \left\{(x,z) \in \mathbb{R}^n \times \mathbb{R}^k \mid 0 \in C(x,z) + N_Q(z)\right\}$$

be the set of feasible pairs (x, z) with respect to (1.1), and assume again that (1.1) comes from the NCP (1.4). With each partitioning of $\{1, 2, \ldots, k\}$ into two index sets I_i, J_i we can associate the set

$$\varepsilon_i := \left\{ (x,z) \in \mathbb{R}^n \times \mathbb{R}^k \;\middle|\; \begin{array}{ll} z^j \geq 0 \text{ and} & F^j(x,z) = 0 \text{ for } j \in I_i \\ z^j = 0 \text{ and} & F^j(x,z) \geq 0 \text{ for } j \in J_i \end{array} \right\}.$$

It is clear that ε is the union of the sets ε_i over all decompositions of $\{1, 2, \ldots, k\}$. Instead of (1.2), we can now consider the finite family of optimization problems

$$\begin{aligned} &\text{minimize} \quad f(x,z) \\ &\text{subject to} \\ &\qquad\qquad (x,z) \in \mathcal{E}_i \cap (U_{\text{ad}} \times Z)\,, \end{aligned} \tag{1.5}$$

$i = 1, 2, \ldots$, from which the difficult complementarity condition has been removed. Such a problem splitting can also be performed in the case of other equilibrium constraints and leads to useful optimality conditions. Moreover, on the basis of this decomposition of $\mathcal{E}$ a suitable modification of the powerful sequential quadratic programming (SQP) seems to be applicable to MPECs.

A different way to deal with the equilibrium constraint is to replace (1.1) by the equation $G(x, z) = 0$, where G is a suitable "gap" function. In the case of the NCP (1.4) we may use, for instance, the so-called NCP functions

$$G_1(x,z) := \min\nolimits_c\{F(x,z), z\} \tag{1.6}$$

or

$$G_2(x,z) := \begin{bmatrix} \Phi(F^1(x,z),z^1) \\ \Phi(F^2(x,z),z^2) \\ \vdots \\ \Phi(F^k(x,z),z^k) \end{bmatrix} \quad \text{with } \Phi(a,b) := \sqrt{a^2+b^2} - (a+b);$$

cf. Facchinei et al., 1995.

In bilevel programming such a gap function can be constructed by using the value function (marginal function) of the lower-level problem. This idea has been used in Outrata, 1990 for numerical purposes and in Ye and Zhu, 1995 for the derivation of optimality conditions. In this approach again the Mangasarian–Fromowitz constraint qualification does not hold in general. Instead, one works with a different condition, termed partial calmness, which holds true in some interesting classes of problems. In our nonsmooth approach we also make use of an equation equivalent to (1.1) and gained by means of a special gap function.

The major number of papers devoted to MPEC use a penalty technique. For instance, in MPECs corresponding to discretized shape optimization problems with variational inequalities, one converts the variational inequality to a smooth equation by help of an exterior penalty (Haslinger and Neittaanmäki, 1996). In the case of the NCP (1.4) this smooth equation attains the form

$$F(x,z) + r(z^-)^2 = 0, \tag{1.7}$$

where $r > 0$ is the penalty parameter. In Haslinger and Neittaanmäki, 1996 it is proved that for $r \to \infty$ the solutions of (1.7) (for fixed x) approach the solution set to (1.4).

Similarly, for a bilevel program Shimizu and Aiyoshi, 1981 propose to add the lower-level constraints to the lower-level objective by interior penalties. By setting the gradient of the augmented objective to zero, the lower-level problem becomes a (smooth) equation.

Another idea is used in Harker and Choi, 1987 and Marcotte and Zhu, 1996, where a gap function representing the equilibrium constraint is added to the objective in form of a penalty. Some gap functions lead, under certain conditions, even to exact penalties. This holds in particular for the gap function (1.6) if we have to do with equilibria described by the NCP (1.4). In this case we can add to the objective a norm of $G_1(x,z)$ multiplied by a suitable penalty parameter. In the recent papers Luo et al., 1996 and Ye et al., 1997 such penalization approaches are related to the well-developed theory of *error bounds* in mathematical programming. This enables to classify the used gap functions and to associate suitable penalties with particular classes of problems.

Let $(\hat{x},\hat{z})$ be a local solution of an MPEC. Further assume that to a gap function G there exist a neighbourhood $\mathcal{O}$ of $(\hat{x},\hat{z})$ and a modulus $\gamma > 0$ such that

$$\mathrm{dist}_{\mathcal{E}}(x,z) \le \gamma \|G(x,z)\| \quad \text{for all } (x,z) \in \mathcal{O}.$$

One says that G is a *(local) Lipschitz error bound* for the GE (1.1) near $(\hat{x},\hat{z})$ (Pang, 1997). Such a gap function can be used as an exact penalty for numerical purposes as well as in deriving optimality conditions. However, it might not always be easy to express the resulting conditions in terms of the original problem data.

The exact penalization in MPECs is also studied in detail in Scheel and Scholtes, 1997 and Scholtes and Stöhr, 1997, where the authors use recent results from the analysis of piecewise differentiable functions. In this approach the equilibrium constraint is written in so-called normal-equation form (cf. Robinson, 1991) and one works with a special "nonsmooth" Mangasarian–Fromowitz constraint qualification (Kuntz and Scholtes, 1994). This approach again leads to optimality conditions and a numerical method.

In connection with penalty methods, one should also mention the interior-point method, entitled Penalty Interior Point Algorithm (PIPA), that can be found in Luo et al., 1997. The convergence results require, however, that strict complementarity holds at the limit point, which can cause difficulties in some classes of MPECs.

In a large number of problems coming from applications, the map S is single-valued on U_{ad} and the state variable is not subject to any constraints (i.e., $Z = \mathbb{R}^k$). Then the MPEC is reduced to the optimization problem

$$\begin{array}{ll} \text{minimize} & \Theta(x) := f(x, S(x)) \\ \text{subject to} & \\ & x \in U_{\text{ad}} \end{array} \tag{1.8}$$

in the variable x only. The solution map S is now an implicit function defined by the GE (1.1). If S is Lipschitz and directionally differentiable, then the formulation (1.8) can be used with advantage for deriving optimality conditions and as a basis for numerical methods. In accordance with Luo et al., 1997, we will call techniques, growing out from (1.8), *implicit programming* approach to MPEC and use the abbreviation IMP. The assumptions for some IMP techniques need not be as severe as above (S is single-valued over U_{ad} and $Z = \mathbb{R}^k$). For establishing optimality conditions it suffices to assume that

(A) at the local solution $(\hat{x}, \hat{z})$ there exist neighbourhoods $\mathcal{U}$ of $\hat{x}$ and $\mathcal{V}$ of $\hat{z}$ and a directionally differentiable Lipschitz selection $\sigma[\mathcal{U} \to \mathbb{R}^k]$ of S such that $\sigma(\hat{x}) = \hat{z}$, and

$$S(x) \cap \mathcal{V} = \sigma(x) \quad \text{for all } x \in \mathcal{U}.$$

Note that (A) does not exclude the GE (1.1) to have multiple solutions; (A) only ensures that for x close to $\hat{x}$ the GE has unique solutions z in a neighbourhood of $\hat{z}$. Similarly, for numerical purposes it suffices to assume a certain globalized variant of (A).

To ensure (A), one can use stability and sensitivity results of parametric optimization and of parameter-dependent variational inequalities. In particular, a crucial role in this analysis plays the concept of strong regularity for GEs due to S.M. Robinson (Robinson, 1980), together with some recent results on nonunique KKT vectors (Ralph and Dempe, 1995; Pang and Ralph, 1996).

Pang et al., 1991 seems to be the first source of a numerical method for problems of type (1.8). This method computes a descent direction on the basis of directional derivatives of S and performs a line search along this direction.

In Luo et al., 1997 the authors examine the relation between (A) and the constraint qualifications needed in their optimality conditions. Further, they propose a conceptual algorithm to the solution of (1.8). In this method, however, one has to solve a sequence of optimization problems with a quadratic objective and an affine GE. This is a difficult and time-consuming task and an effective implementation may be rather complicated.

In the method proposed in Facchinei et al., 1997 a sequence of regular smooth programs is generated using an NCP function. These can be solved by standard software providing a sequence that converges to a solution of (1.8).

We believe that rather efficient IMP techniques can be developed using tools of nonsmooth analysis to (1.8). One speaks of the *nonsmooth approach* to MPEC. This is the actual subject of the book; we postpone a brief description to the end of this section.

A survey on MPEC has to mention the contributions of J. Morgan and coauthors (e.g. Loridan and Morgan, 1989b; Loridan and Morgan, 1989a) on the existence of solutions and approximation schemes to bilevel programs. These papers focus on infinite-dimensional problems and apply the theory of variational convergence. They are a valuable asset to the design of numerical methods.

Besides the above rigorous approaches to MPEC there is a number of numerical heuristics, that have been successfully set to work on various real-life problems; cf. Marcotte, 1986; Friesz et al., 1990. For instance, simulated annealing has been applied to large MPECs in Friesz et al., 1992.

In spite of all these efforts, the area of MPECs remains a rich source of interesting open theoretical questions and is still full of numerical challenges. Since most practical MPECs have complicated structure and large dimension, there is a strong need to improve the available theoretical and numerical tools and to bring into play fresh ideas.

This book grew out of a series of papers on MPECs, published in the years 1993–1997 (Outrata, 1993; Outrata, 1994; Kočvara and Outrata, 1994b; Kočvara and Outrata, 1994a; Kočvara and Outrata, 1994d; Outrata and Zowe, 1995b; Kočvara and Outrata, 1995b; Outrata and Zowe, 1995a; Kočvara and Outrata, 1997). These contributions systematically develop the nonsmooth approach to the analysis and the solution of (1.8). Outrata, 1993; Outrata, 1994 and Kočvara and Outrata, 1997 derive optimality conditions for various types of equilibria. In Outrata and Zowe, 1995b a numerical method is proposed based on bundle algorithms for nonsmooth minimization (Schramm and Zowe, 1992; Hiriart-Urruty and Lemaréchal, 1993). This method was applied in Kočvara and Outrata, 1994a; Kočvara and Outrata, 1994d; Kočvara and Outrata, 1995b to various nonacademic MPECs coming from continuum mechanics. Our book tries to unify the theoretical and applied results of these papers. To ensure (A) we use Robinson's concept of strong regularity. The relevant results are taken from several papers (Robinson, 1976; Robinson, 1980; Robinson, 1981; Robinson, 1991).

Optimality conditions work with the full generalized gradient (or an outer approximation), whereas numerical methods are more modest and manage with an arbitrary element of it. For this purpose we analyze the generalized Jacobians of the solution map S. Matrices from these generalized Jacobians are also helpful in the computation of $z = S(x)$ for $x \in U_{\text{ad}}$ by a suitable nonsmooth variant of Newton's method (Qi, 1993; Qi and Sun, 1993). Attention is also paid to the semismoothness of S, which is an property needed in both bundle and nonsmooth Newton's method. We successively study MPECs with equilibria described by variational inequalities, nonlinear complementarity problems and so-called implicit complementarity problems (ICP), which is a subclass of quasi-variational inequalities. For some of the results we also give a special version for bilevel programs.

For variational inequalities the made assumptions can be relaxed using recent results due to Pang and Ralph, 1996. In particular, the linear independence constraint qualification (LICQ) can be replaced by a weaker condition. These results rely on the use of the degree theory and would lead to a blow-up in the present text. Since we did not have any difficulties with (LICQ) in our mechanical applications, we avoid this generalization.

Another generalization of this approach is considered in Outrata, 1997 with the intention to extend the application area to network design problems (Marcotte, 1986; Harker and Pang, 1990; Qiu and Magnanti, 1992). In these MPECs either (A) is fulfilled, but the set Q has a complicated structure, or Q is standard, but (A) is violated. Unfortunately, the numerical evidence is not yet satisfactory and therefore we have not included this generalization either.

In the book we study in detail to the numerical solution of various nonacademic MPECs by the proposed method. These examples should demonstrate the efficiency of our method, but most of them are of interest in their own. In some of them one has to do with the state constraints $z \in Z$, which prevent a direct application of an IMP technique. To be able to apply our nonsmooth approach, we have added these constraints by means of exact penalties to the objective function.

To conclude, let us summarize that, due to our experience, the nonsmooth calculus is the proper toolbox for an IMP approach to a class of MPECs, which includes many important problems above all from mechanics and economic modelling. It may be, however, a nontrivial task to verify the assumptions needed by the optimality conditions and the numerical method. Besides the used bundle method, the choice of a suitable solver for the equilibrium problem (computation of $z = S(x)$ for a fixed $x \in U_{\mathrm{ad}}$) is also of vital importance for the efficiency of our numerical technique. All these issues are studied in this book in detail.

1.3 EXISTENCE OF SOLUTIONS

In general, the MPEC is a nonconvex optimization problem even if the objective f and the sets U_{ad} and Z are convex. Hence, the question of existence of solutions to MPEC is usually not easy to answer. Here we mention two elementary existence results which are easy to prove. For a more profound existence theory, e.g. in the framework of bilevel programming, we refer to Zhang, 1994; Loridan and Morgan, 1989b.

Proposition 1.1 *Let f be continuous and the graph GphS be closed. Assume further that there exists a pair $(x_0, z_0) \in \mathbb{R}^n \times \mathbb{R}^k$ for which the intersection*

$$\kappa := lev_f(x_0, z_0) \cap GphS \cap (U_{\mathrm{ad}} \times Z) \quad \textit{is nonempty and bounded.} \tag{1.9}$$

Then the MPEC possesses a solution pair $(\hat{x}, \hat{z})$.

Proof. Since f is continuous and the sets U_{ad}, Z are closed, we conclude that κ is compact. Since κ is nonempty, (1.2) amounts to the minimization of f over κ. Such a problem possesses a solution due to the Bolzano–Weierstraß Theorem. ∎

Assumption (1.9) is evidently satisfied if the intersection

$$\mathrm{Gph}S \cap (U_{\mathrm{ad}} \times Z)$$

is nonempty and bounded. The verification of the boundedness can be, however, a difficult task.

In most applications considered in this book, $Z = \mathbb{R}^k$ and S is single-valued on U_{ad}. Then another existence statement can be applied.

Proposition 1.2 *Let f be continuous, U_{ad} be compact and S be single-valued and continuous on an open set containing U_{ad}. Then the MPEC possesses a solution pair $(\hat{x}, \hat{z})$.*

Proof. It suffices to rewrite (1.2) in the form

$$\begin{aligned} &\text{minimize} \quad \Theta(x) := f(x, S(x)) \\ &\text{subject to} \\ &\qquad\qquad x \in U_{\mathrm{ad}}. \end{aligned} \tag{1.10}$$

The composite objective Θ is continuous over an open set containing U_{ad} and so again the Bolzano–Weierstraß Theorem applies. ∎

Since MPECs fall into the class of difficult nonconvex problems, global solutions are mostly out of reach of present methods. In fact, most numerical methods converge only to various stationary points which are local minima under additional assumptions. Nevertheless, even such a local optimization may be of great importance, if already the starting point for the iterative procedure reflects a considerable knowledge of the problem. This is definitely the case if we deal, e.g., with a shape optimization problem and the initial shape has been already suggested by a qualified designer or if we solve an economic problem and start from a verified strategy, successfully proven in such situations.

Vice versa, numerical experiments may sometimes help specialists to analyze the problem and get some hints concerning its solution.

2 AUXILIARY RESULTS

This chapter collects some nontrivial results which will be needed in the rest of the book. We try to keep the presentation as self-contained as possible.

Section 2.1 deals with multifunctions (set-valued maps). We give some basic definitions and then consider the properties which play an important role in our approach to mathematical programs with equilibrium constraints (MPECs). In particular, we shall concentrate on continuity properties of polyhedral multifunctions.

The major part of the chapter (Sections 2.2,2.3) comes from nonsmooth analysis. Some of the results are needed throughout the book. In Section 2.2 we introduce the basic notions and study some of their properties; the aim of Section 2.3 is to provide essential tools for establishing optimality conditions. As basic source we have used the monograph Clarke, 1983, but results of other works have also been included.

Section 2.4 is devoted to a local analysis of the projection onto polyhedral convex sets. This somewhat special topic is of a particular importance in the sensitivity and stability analysis of generalized equations, since they may be rewritten as equations involving projection operators. In this section, again, polyhedrality plays an important role.

2.1 SELECTED TOPICS FROM SET-VALUED ANALYSIS

Since our MPECs are formulated in finite-dimensional spaces, we limit ourselves to multifunctions (set-valued functions) which map points from $\mathbb{R}^n$ to subsets of $\mathbb{R}^m$. Such a multifunction $\Gamma[\mathbb{R}^n \leadsto \mathbb{R}^m]$ is characterized by its *graph*

$$\mathrm{Gph}\Gamma := \{(x,y) \in \mathbb{R}^n \times \mathbb{R}^m \mid y \in \Gamma(x)\}.$$

Its *domain* $\mathrm{Dom}\Gamma$ and *inverse* $\Gamma^{-1}[\mathbb{R}^m \leadsto \mathbb{R}^n]$ are defined by

$$\mathrm{Dom}\Gamma := \{x \in \mathbb{R}^n \mid \Gamma(x) \neq \emptyset\}$$

and

$$x \in \Gamma^{-1}(y) \Longleftrightarrow (x, y) \in \text{Gph}\Gamma.$$

Most multifunctions encountered in this book possess one or more of the properties defined next:

Definition 2.1 *(i) A multifunction* $\Gamma[\mathbb{R}^n \rightsquigarrow \mathbb{R}^m]$ *is said to be* closed at x *if for all sequences* $\{x_i\}$ *and* $\{y_i\}$ *with* $x_i \to x$ *and* $y_i \in \Gamma(x_i)$ *for all* i*, each accumulation point* y *of* $\{y_i\}$ *belongs to* $\Gamma(x)$.

(ii) We say that Γ *is* closed *if it is closed at each* $x \in \mathbb{R}^n$.

It is easy to see that $\Gamma[\mathbb{R}^n \rightsquigarrow \mathbb{R}^m]$ is closed if and only if $\text{Gph}\Gamma$ is a closed subset of $\mathbb{R}^n \times \mathbb{R}^m$.

There are various continuity concepts for multifunctions. The following will be useful to us:

Definition 2.2 *Let* $\Gamma[\mathbb{R}^n \rightsquigarrow \mathbb{R}^m]$ *be a multifunction.*

(i) Γ *is called* upper semicontinuous *at* $x \in \text{Dom}\Gamma$ *if for each neighbourhood* V *of* $\Gamma(x)$ *there exists a positive real* η *such that*

$$\Gamma(x') \subset V \quad \textit{for all } x' \in x + \eta \mathbb{B}.$$

(ii) Γ *is said to be* upper semicontinuous *if it is upper semicontinuous at each point of* $\text{Dom}\Gamma$.

(iii) Γ *is called* locally upper Lipschitz *at* $x \in \text{Dom}\Gamma$ *(with modulus* λ*), if there is a neighbourhood* U *of* x *and a real* $\lambda \geq 0$ *such that*

$$\Gamma(x') \subset \Gamma(x) + \lambda \|x' - x\| \mathbb{B} \quad \textit{for all } x' \in U.$$

(iv) Γ *is called* Lipschitz *on a set* $V \subset \text{Dom}\Gamma$ *(with modulus* λ*), if*

$$\Gamma(x_1) \subset \Gamma(x_2) + \lambda \|x_1 - x_2\| \mathbb{B} \quad \textit{for all } x_1, x_2 \in V.$$

Clearly, a multifunction Γ, which is Lipschitz on an open set $V \in \text{Dom}\Gamma$, will be locally upper Lipschitz at each $x \in V$.

Definition 2.3 *A map* $\Gamma[\mathbb{R}^n \rightsquigarrow \mathbb{R}^m]$ *is declared* uniformly compact *near* $x \in \text{Dom}\Gamma$ *if there is a neighbourhood* U *of* x *such that the set* $\text{cl} \bigcup_{z \in U} \Gamma(z)$ *is compact.*

Under uniform compactness we get a simple relation between upper semicontinuity and closedness at a point, cf. Hogan, 1973.

Theorem 2.1 *Let* $\Gamma[\mathbb{R}^n \rightsquigarrow \mathbb{R}^m]$ *be uniformly compact near* $x \in \text{Dom}\Gamma$ *and let the set* $\Gamma(x)$ *be closed. Then the following statements are equivalent:*

(i) Γ *is upper semicontinuous at* x.

(ii) Γ *is closed at* x.

Proof. (i) $\Rightarrow$ (ii) Assume that $x_i \to x$, $y_i \in \Gamma(x_i)$ and a subsequence $\{y_{i'}\}$ converges to a point y which does not belong to $\Gamma(x)$. Since $\Gamma(x)$ is closed, there is a neighbourhood V of $\Gamma(x)$ such that $y_{i'} \notin V$ for all sufficiently large i'. This contradicts the upper semicontinuity.

(ii) $\Rightarrow$ (i) Suppose there exists an open neighbourhood V of $\Gamma(x)$ such that each ball $x + \frac{1}{i}\mathbb{B}$, $i = 1, 2, \ldots$, contains a point x_i for which $\Gamma(x_i) \not\subset V$. Choose a sequence $y_i \in \Gamma(x_i) \setminus V$. By the uniform compactness of Γ near x we can select a subsequence $\{y_{i'}\}$ of $\{y_i\}$ converging to a point y. This point does not belong to V whence not to $\Gamma(x)$, which contradicts the closedness of Γ at x. ■

We are especially interested in *polyhedral* multifunctions. Recall that a *polyhedral convex set* is the intersection of finitely many (closed) half-spaces. Since we will exclusively work with polyhedral convex sets, we will omit the adjective convex.

Definition 2.4 *A multifunction $\Gamma[\mathbb{R}^n \rightsquigarrow \mathbb{R}^m]$ is called* polyhedral, *if its graph is the union of finitely many polyhedral sets, called* components *of* Γ.

Lemma 2.2 *Let $P[\mathbb{R}^n \rightsquigarrow \mathbb{R}^m]$ be a polyhedral multifunction with components G_i, $i = 1, 2, \ldots, k$. Suppose that $x \in \mathrm{Dom}P$ and define the index set*

$$J(x) := \{i \in \{1, 2, \ldots, k\} \mid x \in \pi_1(G_i)\},$$

where π_1 denotes the canonical projection of $\mathbb{R}^n \times \mathbb{R}^m$ onto $\mathbb{R}^n$. Then there is a neighbourhood U of x such that

$$(U \times \mathbb{R}^m) \cap \mathrm{Gph}P \subset \bigcup_{i \in J(x)} G_i. \tag{2.1}$$

Proof. The affine subspace $\{x\} \times \mathbb{R}^m$ and the components G_i, $i = 1, 2, \ldots, k$, are nonempty polyhedral subsets of $\mathbb{R}^n \times \mathbb{R}^m$. If $i \notin J(x)$, the intersection of $\{x\} \times \mathbb{R}^m$ and G_i is empty and these two sets can be strongly separated (see, e.g., Rockafellar, 1970, Cor. 19.3.3). Hence, there are neighbourhoods U_i of x such that

$$(U_i \times \mathbb{R}^m) \cap G_i = \emptyset \quad \text{for } i \notin J(x).$$

Thus $U := \bigcap_{i \notin J(x)} U_i$ is also a neighbourhood of x and

$$(U \times \mathbb{R}^m) \cap \mathrm{Gph}P \subset \left(\bigcup_{i=1}^{k} G_i\right) \setminus \left(\bigcup_{i \notin J(x)} G_i\right) \subset \bigcup_{i \in J(x)} G_i,$$

as required. ■

According to the above lemma, each point (x', y') with $x' \in U$ and $y' \in P(x')$ belongs to a component G_i which also contains (x, y) for some $y \in P(x)$. We further need a generalization of Hoffman's Theorem about linear inequalities, where again π_1, π_2 denote the canonical projections of $\mathbb{R}^n \times \mathbb{R}^m$ onto $\mathbb{R}^n$ and $\mathbb{R}^m$, respectively.

Lemma 2.3 *Let G be a nonempty polyhedral set in $\mathbb{R}^n \times \mathbb{R}^m$. For $z = (x, y) \in \pi_1(G) \times \pi_2(G)$ define*

$$d_x(z, G) := \min\{\|x' - x\| \mid (x', y) \in G\}$$

and

$$d_y(z, G) := \min\{\|y' - y\| \mid (x, y') \in G\},$$

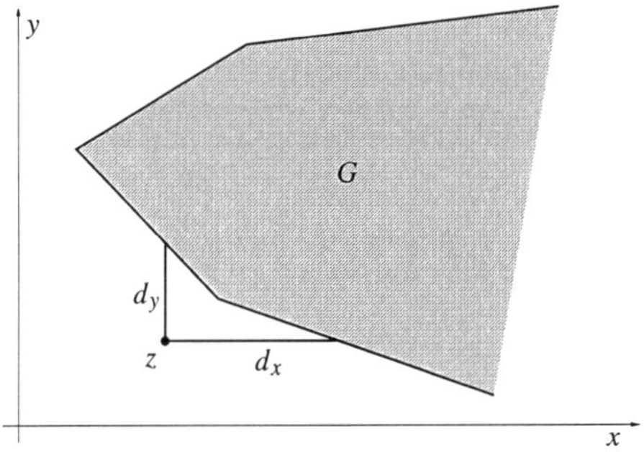

Figure 2.1.

the "horizontal" and the "vertical" distance of z *to* G*, respectively (cf. Fig. 2.1). Then there exist nonnegative reals* ξ, η *such that*

$$d_x(z,G) \leq \eta d_y(z,G) \quad \text{and} \quad d_y(z,G) \leq \xi d_x(z,G) \quad \text{for all } z \in \pi_1(G) \times \pi_2(G). \tag{2.2}$$

Proof. The convex polyhedral G can be represented in the form

$$G = \{(x,y) \in \mathbb{R}^n \times \mathbb{R}^m \,|\, Ax + By \leq c\}$$

with an $\ell \times n$ matrix A, an $\ell \times m$ matrix B, and $c \in \mathbb{R}^\ell$ for some ℓ. By the standard form of Hoffman's Theorem (Hoffmann, 1952) there are reals α and β such that for each $a \in \mathcal{R}(A) + \mathbb{R}^\ell_+$, $b \in \mathcal{R}(B) + \mathbb{R}^\ell_+$, $x_0 \in \mathbb{R}^n$ and $y_0 \in \mathbb{R}^m$ one has

$$\text{dist}_{\{x' | Ax' \leq a\}}(x_0) \leq \alpha \|(Ax_0 - a)^+\|$$

and

$$\text{dist}_{\{y' | By' \leq b\}}(y_0) \leq \beta \|(By_0 - b)^+\|. \tag{2.3}$$

Put $\xi := \beta\|A\|$, $\eta := \alpha\|B\|$ and choose any $z := (x,y) \in \pi_1(G) \times \pi_2(G)$. Then we get from (2.3) that

$$d_y(z,G) = \text{dist}_{\{y' | By' \leq c - Ax\}}(y) \leq \beta \|(Ax + By - c)^+\|. \tag{2.4}$$

For $\tilde{x}$ closest to x in the set $\{x' \mid Ax' \leq c - By\}$ one has

$$\|(Ax + By - c)^+\| \leq \|(Ax + By - c) - (A\tilde{x} + By - c)\|, \tag{2.5}$$

which yields

$$\|(Ax + By - c)^+\| \leq \|A\| \|x - \tilde{x}\|. \tag{2.6}$$

But $\|x - \tilde{x}\| = d_x(z,G)$ by construction and thus, combining (2.4), (2.5) and (2.6), we get

$$d_y(z,G) \leq \beta \|(Ax + By - c)^+\| \leq \beta \|A\| \|x - \tilde{x}\| = \xi d_x(z,G).$$

The first inequality in (2.2) is proven in the same way. ∎

Lemmas 2.2, 2.3 help to state and prove the most important result of this section.

Theorem 2.4 *Let $P[\mathbb{R}^n \rightsquigarrow \mathbb{R}^m]$ be a polyhedral multifunction. Then there is a constant λ such that P is locally upper Lipschitz with modulus λ at each $x \in \mathrm{Dom}P$.*

Proof. Let G_i, $i = 1, 2, \ldots, k$, be the components of P. With the constants ξ_i associated with G_i according to Lemma 2.3 we put $\lambda := \max\{\xi_1, \xi_2, \ldots, \xi_k\}$. Now consider some arbitrary but fixed $x \in \mathrm{Dom}P$ and the index set

$$J(x) := \{i \in \{1, 2, \ldots, k\} \mid x \in \pi_1(G_i)\}.$$

By Lemma 2.2 there is a neighbourhood U of x such that

$$(U \times \mathbb{R}^m) \cap \mathrm{Gph}P \subset \bigcup_{i \in J(x)} G_i.$$

For $x' \in U$ with $x' \notin \mathrm{Dom}P$ nothing has to be shown. Hence let $x' \in \mathrm{Dom}P$ and $y' \in P(x')$. Then we have

$$(x', y') \in [(U \times \mathbb{R}^m) \cap \mathrm{Gph}P] \subset \bigcup_{i \in J(x)} G_i,$$

which implies $(x', y') \in G_i$ for some $i \in J(x)$. For this i we get

$$\begin{aligned}
\mathrm{dist}_{P(x)}(y') &= \mathrm{dist}_{\{v \mid (x,v) \in GphP\}}(y') \\
&\leq \mathrm{dist}_{\{v \mid (x,v) \in G_i\}}(y') \\
&= d_y((x, y'), G_i) \\
&\leq \xi_i d_x((x, y'), G_i) \\
&= \xi_i \mathrm{dist}_{\{u \mid (u,y') \in G_i\}}(x) \\
&\leq \xi_i \|x' - x\| \\
&\leq \lambda \|x' - x\|.
\end{aligned}$$

Since $P(x)$ is closed and y' was arbitrary in $P(x')$, it follows that

$$P(x') \subset P(x) + \lambda \|x' - x\| \mathbb{B}$$

and we are done. ∎

Note that the size of the neighbourhood U in the above proof depends on x, whereas the Lipschitz modulus λ depends only on P and not on x. Hence, λ is of global nature. This will be used in the proof of the following corollary, where we specialize the above result to single-valued P.

Recall that a function $F[\mathbb{R}^n \to \mathbb{R}^m]$ is said to be *Lipschitz* on a set $A \subset \mathbb{R}^n$ if there exists $\lambda \geq 0$ such that

$$\|F(x_1) - F(x_2)\| \leq \lambda \|x_1 - x_2\| \quad \text{for all } x_1, x_2 \in A.$$

Corollary 2.5 *Let the polyhedral multifunction* $P[\mathbb{R}^n \rightsquigarrow \mathbb{R}^m]$ *be single-valued on a convex subset* A *of* $\mathrm{Dom}P$. *Then* P *is Lipschitz on* A.

Proof. By Theorem 2.4 the map P is locally upper Lipschitz with modulus λ at each $x \in \mathrm{Dom}P$. Consider two arbitrary points $x_1, x_2 \in A$ and assign to each $x \in [x_1, x_2]$ an open neighbourhood U_x such that

$$\|P(x') - P(x)\| \le \lambda \|x' - x\| \quad \text{for all } x' \in U_x. \tag{2.7}$$

By compactness, a finite sub-family of these neighbourhoods covers $[x_1, x_2]$ and an appropriate addition of inequalities (2.7) proves the claim. ∎

Definition 2.5 *Let* Ω *be a convex subset of* $\mathbb{R}^n$ *and* $x \in \mathrm{cl}\Omega$. *Then*

$$T_\Omega(x) := \mathrm{cl} \bigcup_{\lambda > 0} \lambda(\Omega - x). \tag{2.8}$$

is called the tangent cone *to* Ω *at* x.

Associated with $T_\Omega(x)$ is a "dual" object which we define next:

Definition 2.6 *Let* Ω *be a convex subset of* $\mathbb{R}^n$ *and* $x \in \mathrm{cl}\Omega$. *Then*

$$N_\Omega(x) := \{\xi \in \mathbb{R}^n \mid \langle \xi, y - x\rangle \le 0 \textit{ for all } y \in \Omega\}, \tag{2.9}$$

is said to be the normal cone *to* Ω *at* x.

The duality between the tangent and the normal cone becomes apparent in the following equality, which is easily verified: for $x \in \mathrm{cl}\Omega$ one has

$$N_\Omega(x) = -(T_\Omega(x))^*.$$

Here, for a set $K \subset \mathbb{R}^n$, K^* denotes the the polar cone to K, i.e.,

$$K^* := \{\xi \in \mathbb{R}^n \mid \langle \xi, k\rangle \ge 0 \text{ for all } k \in K\}.$$

We will often work with the multifunction

$$y \rightsquigarrow N_\Omega(y),$$

defined on all of $\mathbb{R}^n$ with the convention that $N_\Omega(y) = \emptyset$ whenever $y \notin \mathrm{cl}\Omega$. We will prove that for polyhedral Ω this multifunction is also polyhedral and thus enjoys the useful local upper Lipschitz continuity. The proof requires certain auxiliary results of Rockafellar, 1970 and the following lemma.

Lemma 2.6 *Assume that* Ω *is a nonempty closed convex subset of* $\mathbb{R}^n$ *and* $y, z \in \Omega$. *Then for each* $x \in (y, z)$ *one has* $N_\Omega(x) \subset N_\Omega(y)$. *Moreover,* $N_\Omega(\cdot)$ *is constant on the relative interior* $\mathrm{ri}C$ *of each convex subset* C *of* Ω.

Proof. We can express $x \in (y, z)$ as $(1-\lambda)y + \lambda z$ with $\lambda \in (0, 1)$. Let $x^* \in N_\Omega(x)$ and c be an arbitrary point of Ω. By convexity, $(1-\lambda)c + \lambda z \in \Omega$ and therefore

$$\langle x^*, [(1-\lambda)c + \lambda z] - x\rangle \le 0.$$

Now $[(1-\lambda)c+\lambda z]-x=(1-\lambda)[c-(1-\lambda)^{-1}(x-\lambda z)]=(1-\lambda)(c-y)$ and thus

$$\langle x^*, c-y\rangle \leq 0.$$

This shows $x^* \in N_\Omega(y)$ and consequently $N_\Omega(x) \subset N_\Omega(y)$.

To prove the second assertion let $x, y \in \mathrm{ri}C$ with $x \neq y$. By the definition of the relative interior there exist points $\hat{x}, \hat{y} \in C$ such that $x \in \mathrm{ri}(\hat{x}, y)$ and $y \in \mathrm{ri}(\hat{y}, x)$. Thus $N_\Omega(x) \subset N_\Omega(y)$ and $N_\Omega(y) \subset N_\Omega(x)$, which implies that $N_\Omega(x) = N_\Omega(y)$ and so $N_\Omega(\cdot)$ is constant on $\mathrm{ri}C$. ■

We are now prepared for the last result of this section.

Theorem 2.7 *Let Ω be a nonempty polyhedral convex set in $\mathbb{R}^n$. Then N_Ω is a polyhedral multifunction.*

In the proof we will make use of the following definition.

Definition 2.7 *A subset C' of a convex set $C \subset \mathbb{R}^n$ is called a* face *of C, if it is convex and if for each line segment $[x,y] \subset C$ with $(x,y) \cap C' \neq \emptyset$ one has $x, y \in C'$.*

Proof. (of Theorem 2.7) By Rockafellar, 1970, Thms. 19.1,18.2, there are finitely many faces $F_1, \ldots, F_m$ of Ω, such that $\{\mathrm{ri}F_i\}_{i=1}^m$ form a partition of Ω, i.e.,

$$\Omega = \bigcup_{i=1}^{m} \mathrm{ri}F_i, \quad \mathrm{ri}F_i \cap \mathrm{ri}F_j = \emptyset \quad \text{for all } i,j \in \{1,2\ldots,m\} \text{ with } i \neq j. \tag{2.10}$$

Lemma 2.6 implies that, on each set $\mathrm{ri}F_i$, $N_\Omega(\cdot)$ has a constant value denoted P_i. By Rockafellar, 1970, Thm. 23.10 these P_i are nonempty polyhedral convex cones. Summarizing, we get

$$\mathrm{Gph}N_\Omega = \bigcup_{i=1}^{m} [\mathrm{ri}F_i \times P_i] \tag{2.11}$$

and thus

$$\mathrm{Gph}N_\Omega \subset \bigcup_{i=1}^{m} (F_i \times P_i). \tag{2.12}$$

To prove equality for (2.12), suppose that

$$(x, x^*) \in \bigcup_{i=1}^{m} (F_i \times P_i).$$

Then for some $i \in \{1, 2, \ldots, m\}$ one has $x \in F_i$ and $x^* \in P_i$. If $x \in \mathrm{ri}F_i$ then, by (2.11), $(x, x^*) \in \mathrm{Gph}N_\Omega$.

Hence suppose $x \notin \mathrm{ri}F_i$ and choose $z \in \mathrm{ri}F_i$ and $y \in F_i$ such that $z \in (x, y)$. Then by Lemma 2.6

$$P_i = N_\Omega(z) \subset N_\Omega(x). \tag{2.13}$$

On the other hand, we know that $x \in \mathrm{ri}F_j$ for a suitable j and thus $N_\Omega(x) = P_j$, which together with (2.13) implies

$$P_i \subset P_j.$$

It follows $x^* \in P_i \subset P_j$ and thus

$$(x, x^*) \in \mathrm{ri} F_j \times P_j \subset \mathrm{Gph} N_\Omega,$$

which proves that (2.12) is an equality.

It remains to observe that each face of the polyhedral set Ω itself is a polyhedral set and so $F_i \times P_i$ is polyhedral. Consequently, $N_\Omega(\cdot)$ is a polyhedral multifunction. ■

The multifunction $N_\Omega(\cdot)$ arises in generalized equation (1.1) and so Theorems 2.4 and 2.7 play a key role in the stability analysis of equilibrium problems (Chapter 5).

2.2 LIPSCHITZ ANALYSIS

It is always desirable to have quantitative estimates of continuity. Therefore we begin with the basic concepts of Lipschitz continuity, used throughout the book.

Definition 2.8 *Let Ω be a subset of $\mathbb{R}^n$, F an operator from Ω into $\mathbb{R}^m$ and λ a nonnegative real number.*

(i) We say that F is Lipschitz *(with modulus λ) on Ω, if*

$$\|F(x_1) - F(x_2)\| \le \lambda \|x_1 - x_2\| \quad \textit{for all } x_1, x_2 \in \Omega.$$

(ii) F is called Lipschitz near x *(with modulus λ) if, for some $\varepsilon > 0$, F is Lipschitz with modulus λ on $x + \varepsilon \mathbb{B}$.*

(iii) If F is Lipschitz near each $x \in \Omega$, we say that F is locally Lipschitz *on Ω.*

Note that an operator F which is Lipschitz near x need not be differentiable at x; even the classic directional derivative of F at x in a direction $h \in \mathbb{R}^n$

$$F'(x; h) := \lim_{t \downarrow 0} \frac{F(x + th) - F(x)}{t}$$

may fail to exist.

For real-valued Lipschitz functions $f[\mathbb{R}^n \to \mathbb{R}]$, a generalized derivative proposed by Clarke proves to be useful.

Definition 2.9 *We call*

$$f^0(x; h) = \limsup_{\substack{x' \to x \\ t \downarrow 0}} \frac{f(x' + th) - f(x')}{t}.$$

the generalized directional derivative *of f at x in the direction h.*

The above limit enjoys many good properties summarized in the theorem below. Note that f, being Lipschitz near x, is in fact Lipschitz near all y close to x. Hence $f^0(\cdot;\cdot)$ in the statement below is defined on all of $U \times \mathbb{R}^n$ for some suitable neighbourhood of x.

Theorem 2.8 *Let $f[\mathbb{R}^n \to \mathbb{R}]$ be Lipschitz near x with modulus λ. Then:*

(i) The function $h \mapsto f^0(x; h)$ is finite, positively homogeneous, subadditive on $\mathbb{R}^n$ and

$$|f^0(x; h)| \le \lambda \|h\|.$$

(ii) The function $f^0(\cdot;\cdot)$ is upper semicontinuous at (x,h) for each $h \in \mathbb{R}^n$ and the function $h \mapsto f^0(x;h)$ is Lipschitz with modulus λ on $\mathbb{R}^n$.

Proof. (i) From the Lipschitz continuity we get

$$\frac{|f(x'+th)-f(x')|}{t} \le \lambda\|h\|,$$

whenever x' is sufficiently close to x and $t > 0$ is sufficiently small. Hence the upper limit of the expression on the left-hand side exists as a finite value and $f^0(x;h) \le \lambda\|h\|$.

For $\gamma > 0$ one has

$$f^0(x;\gamma h) = \gamma \limsup_{\substack{x' \to x \\ t \downarrow 0}} \frac{f(x'+t\gamma h)-f(x')}{t\gamma} = \gamma f^0(x;h),$$

so $f^0(x;\cdot)$ is positively homogeneous. To prove the subadditivity, simply observe that for $v, w \in \mathbb{R}^n$

$$\begin{aligned} f^0(x;v+w) &= \limsup_{\substack{x' \to x \\ t \downarrow 0}} \frac{f(x'+tv+tw)-f(x')}{t} \\ &\le \limsup_{\substack{x' \to x \\ t \downarrow 0}} \frac{f(x'+tv+tw)-f(x'+tw)}{t} + \limsup_{\substack{x' \to x \\ t \downarrow 0}} \frac{f(x'+tw)-f(x')}{t}. \end{aligned}$$

The first term in the last expression is $f^0(x;v)$, because each y tending to x may be expressed as $x'+tw$ with $x' \to x$ and $t \downarrow 0$. Thus

$$f^0(x;v+w) \le f^0(x;v) + f^0(x;w),$$

which proves the subadditivity.

(ii) Let $\{x_i\}$ and $\{h_i\}$ be arbitrary sequences converging to x and h. By definition of the generalized directional derivative we can find for each $i = 1, 2, \ldots$ some $y_i \in \mathbb{R}^n$ and a positive real t_i such that

$$\|y_i - x_i\| + t_i < \frac{1}{i}$$

and

$$\begin{aligned} f^0(x_i;h_i) &\le \frac{f(y_i+t_ih_i)-f(y_i)}{t_i} + \frac{1}{i} \\ &= \frac{f(y_i+t_ih)-f(y_i)}{t_i} + \frac{f(y_i+t_ih_i)-f(y_i+t_ih)}{t_i} + \frac{1}{i}. \end{aligned}$$

The Lipschitz continuity implies that

$$\left\| \frac{f(y_i+t_ih_i)-f(y_i+t_ih)}{t_i} \right\| \le \lambda\|h_i - h\|$$

for all i sufficiently large and thus

$$\limsup_{i \to \infty} f^0(x_i;h_i) \le f^0(x;h),$$

which establishes the upper semicontinuity.

To prove the second part of (ii), let v and w in $\mathbb{R}^n$ be given. Again by Lipschitz continuity

$$f(x' + tv) - f(x') \leq f(x' + tw) - f(x') + \lambda \|v - w\| t,$$

provided x' is sufficiently close to x and $t > 0$ is sufficiently small. Dividing by t and taking upper limits for $x' \to x$ and $t \downarrow 0$, we get

$$f^0(x; v) \leq f^0(x; w) + \lambda \|v - w\|.$$

The same inequality also holds for interchanged v and w and the proof is complete. ■

With the help of $f^0(x; \cdot)$ we now introduce a tool which serves as a substitute for the gradient of a smooth function.

Definition 2.10 *Let $f[\mathbb{R}^n \to \mathbb{R}]$ be Lipschitz near x. The* generalized gradient *of f at x, denoted by $\partial f(x)$, is the set*

$$\partial f(x) = \left\{ \xi \in \mathbb{R}^n \mid \langle \xi, h \rangle \leq f^0(x; h) \text{ for all } h \in \mathbb{R}^n \right\}.$$

The elements of $\partial f(x)$ are called subgradients *of f at x.*

Remark. For f continuously differentiable at x one has $f^0(x; h) = \langle \nabla f(x), h \rangle$ and thus $\partial f(x) = \nabla f(x)$. This justifies the name generalized gradient.

Theorem 2.9 *Let $f[\mathbb{R}^n \to \mathbb{R}]$ be Lipschitz near x with modulus λ. Then:*

(i) $\partial f(x)$ is a nonempty convex compact set and $\partial f(x) \subset \lambda B$;

(ii) For each $h \in \mathbb{R}^n$, one has

$$f^0(x; h) = \max \{ \langle \xi, h \rangle \mid \xi \in \partial f(x) \}.$$

(iii) The multifunction $y \mapsto \partial f(y)$ upper semicontinuous (and closed) at x.

Proof. (i) By Theorem 2.8, the functional $f^0(x; \cdot)$ is sublinear, i.e, positively homogeneous and subadditive. The analytic form of the Hahn–Banach Theorem tells us that the sublinear functional $f^0(x; \cdot)$ majorizes some linear functional on $\mathbb{R}^n$. Therefore there exists $\xi \in \mathbb{R}^n$ with

$$f^0(x; h) \geq \langle \xi, h \rangle \quad \text{for all } h \in \mathbb{R}^n$$

and, consequently, $\xi \in \partial f(x)$, i.e, $\partial f(x) \neq \emptyset$.

The convexity and closedness of $\partial f(x)$ immediately follow by rewriting $\partial f(x)$ in the form

$$\partial f(x) = \bigcap_{h \in \mathbb{R}^n} \{ \xi \in \mathbb{R}^n \mid \langle \xi, h \rangle \leq f^0(x; h) \}.$$

From Theorem 2.8(i) we get

$$\|\xi\|^2 \leq \lambda \|\xi\| \quad \text{for all } \xi \in \partial f(x)$$

and thus $\|\xi\| \leq \lambda$. Hence $\partial f(x)$ is compact and further $\partial f(x) \subset \lambda B$.

(ii) By definition of $\partial f(x)$ we have

$$f^0(x; h) \geq \max \{ \langle \xi, h \rangle \mid \xi \in \partial f(x) \} \quad \text{for all } h \in \mathbb{R}^n. \tag{2.14}$$

To prove equality, fix some arbitrary $\bar{h}$ and consider the linear functional $\gamma\bar{h} \mapsto \gamma f^0(x;\bar{h})$ defined on the one-dimensional subspace $\{\gamma\bar{h} | \gamma \in \mathbb{R}\}$. By the analytic form of the Hahn–Banach Theorem, this functional can be extended to a linear functional defined on the whole $\mathbb{R}^n$ and majorized by $f^0(x;\cdot)$. Thus there exists some $\bar{\xi} \in \mathbb{R}^n$ such that

$$\langle\bar{\xi},\gamma\bar{h}\rangle = \gamma f^0(x;\bar{h}) \quad \text{for all } \gamma \in \mathbb{R}$$

and

$$\langle\bar{\xi},h\rangle \leq f^0(x;h) \text{ for all } h \in \mathbb{R}^n.$$

This yields $\bar{\xi} \in \partial f(x)$ and equality in (2.14) for $\bar{\xi}$ and $\bar{h}$.

(iii) We first prove that $\partial f(\cdot)$ is closed at x. Let $\{x_i\}$ and $\{\xi_i\}$ be arbitrary sequences such that $x_i \to x$ and $\xi_i \in \partial f(x_i)$ for $i = 1, 2, \ldots$. Assume that ξ is an accumulation point of $\{\xi_i\}$ and that h is an arbitrary vector from $\mathbb{R}^n$. From the sequence $\{\langle\xi_i, h\rangle\}$ we can select a subsequence $\{\langle\xi_{i'}, h\rangle\}$ converging to $\langle\xi, h\rangle$. As $f^0(x_{i'}; h) \geq \langle\xi_{i'}, h\rangle$ and f^0 is upper semicontinuous, it follows that $f^0(x;h) \geq \langle\xi,h\rangle$. Hence $\xi \in \partial f(x)$ and the map $\partial f(\cdot)$ is closed at x.

By property (i) there is a neighbourhood U of x such that

$$\text{cl} \bigcup_{y\in U} \partial f(y) \subset \lambda I\!B.$$

Hence $\partial f(\cdot)$ is uniformly compact near x and claim (iii) follows from Theorem 2.1. ■

The next statement generalizes the classic concept of stationarity.

Proposition 2.10 *Suppose that $f[\mathbb{R}^n \to \mathbb{R}]$ is Lipschitz near x and x is a local minimizer or a local maximizer of f. Then*

$$0 \in \partial f(x). \tag{2.15}$$

Proof. In the case of a local minimizer we get for all $h \in \mathbb{R}^n$

$$0 \leq \liminf_{t\downarrow 0} \frac{f(x+th) - f(x)}{t} \leq f^0(x;h).$$

If x is a local maximizer, then one has for all $h \in \mathbb{R}^n$

$$0 \leq \limsup_{t\downarrow 0} \frac{-f(x+th)+f(x)}{t} \leq \limsup_{\substack{x'\to x\\ t\downarrow 0}} \frac{f(x'-th) - f(x')}{t} = f^0(x;-h).$$

In both cases the claim follows immediately from the definition of the generalized gradient. ■

Points satisfying (2.15) are called (Clarke's) *stationary points* in the optimization problem

$$\begin{aligned} &\text{minimize} \quad f(x) \\ &\text{subject to} \\ &\qquad\qquad\quad x \in \mathbb{R}^n. \end{aligned}$$

Definition 2.11 *For a set $A \subset \mathbb{R}^n$ the function $\delta_A[\mathbb{R}^n \to \mathbb{R} \cup +\infty]$ defined by*

$$\delta_A(h) := \sup\{\langle a, h\rangle | a \in A\} \qquad \textit{for } h \in \mathbb{R}^n$$

is called the support function *of A.*

Theorem 2.9(ii) says that the *generalized directional derivative* of f at x is the support function of its *generalized gradient* at x, i.e.,

$$f^0(x; h) = \delta_{\partial f(x)}(h) \quad \text{for all } h \in \mathbb{R}^n. \tag{2.16}$$

A simple technical result helps to characterize inclusions of sets by means of their support functions.

Proposition 2.11 *Let $A, B \subset \mathbb{R}^n$ be nonempty closed convex sets. Then*

$$A \subset B \quad \textit{if and only if} \quad \delta_A(h) \le \delta_B(h) \quad \textit{for all } h \in \mathbb{R}^n. \tag{2.17}$$

Proof. The direction "$\Rightarrow$" comes from the definition of a support function. To prove the opposite direction suppose that $\delta_A(\cdot) \le \delta_B(\cdot)$ but there exists $\bar{a} \in A$ with $\bar{a} \notin B$. By a separation argument (e.g. Rockafellar, 1970, Cor. 11.4.2), the closed convex set B and $\{\bar{a}\}$ can be strongly separated, i.e., there exists a vector $\bar{h} \in \mathbb{R}^n$ and a real α such that

$$\langle \bar{a}, \bar{h}\rangle > \alpha \ge \langle b, \bar{h}\rangle \quad \text{for all } b \in B.$$

We end up with the contradiction

$$\delta_A(\bar{h}) > \delta_B(\bar{h}),$$

which proves the claim. ■

As a trivial consequence, let us add for later use:

Corollary 2.12 *Let $A, B \subset \mathbb{R}^n$ be closed convex sets. Then*

$$A = B \quad \textit{if and only if} \quad \delta_A(h) = \delta_B(h) \quad \textit{for all } h \in \mathbb{R}^n.$$

The next result gives an upper estimate of the generalized gradient of the sum of two functions.

Proposition 2.13 *Consider two functions $f_1, f_2[\mathbb{R}^n \to \mathbb{R}]$.*

(i) If f_1 and f_2 are Lipschitz near x, then

$$\partial(f_1 + f_2)(x) \subset \partial f_1(x) + \partial f_2(x). \tag{2.18}$$

(ii) If f_1 is Lipschitz near x and f_2 is continuously differentiable on a neighbourhood of x, then

$$\partial(f_1 + f_2)(x) = \partial f_1(x) + \nabla f_2(x). \tag{2.19}$$

Proof. (i) Obviously,

$$(f_1 + f_2)^0(x;h) \le f_1^0(x;h) + f_2^0(x;h) \tag{2.20}$$

and thus by (2.16)

$$\delta_{\partial(f_1+f_2)(x)}(h) \le \delta_{\partial f_1(x)}(h) + \delta_{\partial f_2(x)}(h) \quad \text{for all } h \in \mathbb{R}^n.$$

Since the sum of support functions is equal to the support function of the sum of the defining sets, we can continue

$$\delta_{\partial(f_1+f_2)(x)}(h) \le \delta_{\partial f_1(x)+\partial f_2(x)}(h) \quad \text{for all } h \in \mathbb{R}^n$$

which, because of (2.17), proves (2.18).

(ii) Returning to Definition 2.9 and using the continuous differentiability of f_2, we can sharpen (2.20) to

$$(f_1 + f_2)^0(x;h) = f_1^0(x;h) + \nabla f_2(x)^T h. \tag{2.21}$$

Equation (2.19) results from Corollary 2.12. ∎

A version of the mean-value theorem will be useful when computing generalized gradients of composite functions. To simplify the notation, we will sometimes work with sets as arguments in analytic expressions. For example, we write

$$\langle \partial f(u), y - x\rangle \quad \text{for} \quad \{\langle \xi, y - x\rangle \mid \xi \in \partial f(u)\}.$$

Theorem 2.14 *Let $f[\mathbb{R}^n \to \mathbb{R}]$ be Lipschitz on an open set containing the line segment $[x, y]$. Then there is a point $u \in (x, y)$, such that*

$$f(y) - f(x) \in \langle \partial f(u), y - x\rangle.$$

Proof. First consider the function $\psi[[0,1] \to \mathbb{R}]$ defined by $\psi(\lambda) = f(x + \lambda(y - x))$. By assumption on f, this $\psi(\lambda)$ is Lipschitz on $(0,1)$. We claim that for $\lambda \in (0,1)$

$$\partial\psi(\lambda) \subset \langle \partial f(x + \lambda(y - x)), y - x\rangle. \tag{2.22}$$

By Theorem 2.9(i), the sets in (2.22) are closed intervals in $\mathbb{R}$. Hence, according to Proposition 2.11, it suffices to prove

$$\delta_{\partial\psi(\lambda)}(\beta) \le \delta_{\langle \partial f(x+\lambda(y-x)), y-x\rangle}(\beta) \quad \text{for } \beta = \pm 1$$

or, because of (2.16),

$$\psi^0(\lambda;\beta) \le \max\{\langle \xi, y - x\rangle\beta \mid \xi \in \partial f(x + \lambda(y - x))\} \quad \text{for } \beta = \pm 1. \tag{2.23}$$

By definition,

$$\begin{aligned}\psi^0(\lambda;\beta) &= \limsup_{\substack{\lambda' \to \lambda \\ t \downarrow 0}} \frac{\psi(\lambda' + t\beta) - \psi(\lambda')}{t} \\ &= \limsup_{\substack{\lambda' \to \lambda \\ t \downarrow 0}} \frac{f(x + [\lambda' + t\beta](y - x)) - f(x + \lambda'(y - x))}{t}.\end{aligned}$$

Therefore, replacing the real variable λ' by a vector variable y', we see that

$$\begin{aligned}\psi^0(\lambda;\beta) &\leq \limsup_{\substack{y'\to x+\lambda(y-x)\\ t\downarrow 0}} \frac{f(y'+t\beta(y-x))-f(y')}{t}\\ &= f^0(x+\lambda(y-x);(y-x)\beta)\\ &= \max\{\langle\xi,y-x\rangle\beta \mid \xi\in\partial f(x+\lambda(y-x))\},\end{aligned}$$

which proves (2.23) and thus (2.22).

Now consider the function $\theta[[0,1]\to\mathbb{R}]$ defined by

$$\theta(\lambda)=f(x+\lambda(y-x))+\lambda[f(x)-f(y)].$$

As θ is continuous and $\theta(0)=\theta(1)=f(x)$, there is a point $\lambda^*\in(0,1)$ at which θ attains a local minimum or maximum. By (2.15) we have $0\in\partial\theta(\lambda^*)$ and due to (2.22) and (2.19) we get

$$0\in\langle\partial f(x+\lambda^*(y-x)),y-x\rangle+f(x)-f(y).$$

Thus the claim holds with $u=x+\lambda^*(y-x)$. ■

Rademacher's Theorem (Rademacher, 1919) states that each operator $F[\mathbb{R}^n\to\mathbb{R}^m]$, which is Lipschitz on a set $\Omega\subset\mathbb{R}^n$, is differentiable almost everywhere on Ω (in the Lebesgue sense). This property, which furnishes another characterization of the generalized gradient, is very helpful when we want to compute elements from $\partial f(x)$. Letting Ω_f denote the subset of $\mathbb{R}^n$ where f fails to be differentiable, i.e.,

$$\Omega_f:=\{x\in\mathbb{R}^n\,|\,f \text{ is not differentiable at } x\},$$

we have:

Lemma 2.15 *Let $f[\mathbb{R}^n\to\mathbb{R}]$ be Lipschitz near x. Then for all $h\in\mathbb{R}^n$*

$$f^0(x;h)\leq\limsup\left\{\langle\nabla f(x'),h\rangle \mid x'\to x, x'\notin\Omega_f\right\}. \tag{2.24}$$

Proof. Fix $\varepsilon>0$ and $h\in\mathbb{R}^n$ and write α for the right-hand side of (2.24). Then there is a $\delta>0$ such that

$$\langle\nabla f(x'),h\rangle\leq\alpha+\varepsilon \quad \text{for all} \quad x'\in x+\delta\mathbb{B}, \quad x'\notin\Omega_f.$$

Making $\delta>0$ smaller, if necessary, we can assume that f is Lipschitz on $x+\delta\mathbb{B}$ and that, by Rademacher's Theorem, the set $\Omega_f\cap\{x+\delta\mathbb{B}\}$ has Lebesgue measure zero. Now consider the line segments

$$\mathbb{L}_{x'}:=\{x'+th \mid 0<t<\delta/(2\|h\|)\}$$

which all lie in the ball $x+\delta\mathbb{B}$ for $x'\in x+(\delta/2)\mathbb{B}$. Let χ be the characteristic function of Ω_f, i.e.,

$$\chi(y):=\begin{cases}1 & \text{if } f \text{ is not differentiable at } y\\ 0 & \text{elsewhere.}\end{cases}$$

It comes from Fubini's Theorem (e.g. Rudin, 1974) that for almost all $x' \in x + (\delta/2)I\!B$ one gets for the one-dimensional integrals

$$\int_{I\!L_{x'}} \chi(s)\mathrm{d}s = 0,$$

i.e., $I\!L_{x'} \cap \Omega_f$ is of measure zero in $\{x' + th | t \in \mathbb{R}\}$. Let x' be such a point and $t \in (0, \delta/(2\|h\|))$. Then the integral

$$f(x' + th) - f(x') = \int_0^t \langle \nabla f(x' + \theta h), h\rangle \mathrm{d}\theta$$

is well-defined, since ∇f exists almost everywhere on $I\!L_{x'}$. It follows that

$$f(x' + th) - f(x') \leq t(\alpha + \varepsilon). \tag{2.25}$$

This inequality holds for almost all $x' \in x + (\delta/2)I\!B$ and for all $t \in (0, \delta/(2\|h\|))$. However, since f is continuous, it must hold for all $x' \in x + (\delta/2)I\!B$ and (2.25) shows

$$f^0(x; h) \leq \alpha + \varepsilon.$$

This proves (2.24), since $\varepsilon > 0$ was arbitrary. ■

The announced characterization of the generalized gradient follows.

Theorem 2.16 *Let $f[\mathbb{R}^n \to \mathbb{R}]$ be Lipschitz near x. Then*

$$\partial f(x) = \text{conv}\partial_B f(x),$$

where

$$\partial_B f(x) := \left\{ \lim_{i\to\infty} \nabla f(x_i) \mid x_i \to x, x_i \notin \Omega_f \right\}.$$

Proof. Note first that by Rademacher's Theorem we can approach x by sequences $\{x_i\}$ avoiding Ω_f. Since $\nabla f(x_i) \in \partial f(x_i)$ (for all $h \in \mathbb{R}^n$ one has $\langle \nabla f(x_i), h\rangle = f'(x_i; h) \leq f^0(x_i; h)$), we get from the uniform compactness of $\partial f(\cdot)$ near x that the sequences $\{\nabla f(x_i)\}$ possess convergent subsequences. By Theorem 2.9(iii) the limit of any such subsequence must belong to $\partial f(x)$ and thus $\partial_B f(x) \subset \partial f(x)$. Since $\partial f(x)$ is convex (Theorem 2.9(i)), we have

$$\text{conv}\partial_B f(x) \subset \partial f(x).$$

Conversely, note that $\partial_B f(x)$ is compact since it is obviously closed and bounded as a subset of $\partial f(x)$. The convex hull of a compact set is compact in finite dimension and thus closed. Hence, by Proposition 2.11, it remains to prove that $f^0(x; \cdot)$ does not exceed the support function of $\text{conv}\partial_B f(x)$. This is precisely the message of Lemma 2.15. ■

The last theorem suggests extensions of the concept of generalized gradients to vector-valued functions $F[\mathbb{R}^n \to \mathbb{R}^m]$, Lipschitz on a set $\Omega \subset \mathbb{R}^n$. As in the case of real-valued functions we introduce the set

$$\Omega_F := \{x \in \Omega | \mathcal{J}F(x) \text{ does not exist}\}.$$

By Rademacher's Theorem, Ω_F has Lebesgue measure zero.

Definition 2.12 *Let $F[\mathbb{R}^n \to \mathbb{R}^m]$ be Lipschitz near x. The* generalized Jacobian *of $F[\mathbb{R}^n \to \mathbb{R}^m]$ at x, denoted $\partial F(x)$, is the subset of $\mathbb{R}^{m\times n}$ (space of $[m \times n]$ matrices) given by*

$$\partial F(x) = \text{conv}\partial_B F(x),$$

where

$$\partial_B F(x) := \left\{ \lim_{i\to\infty} \mathcal{J}F(x_i) \mid x_i \to x, x_i \notin \Omega_f \right\}.$$

Remark. For $m = 1$ the generalized Jacobian coincides with the *transpose* of the generalized gradient. Hence there is a certain inconsistency in the $\partial-$notation for $m = 1$ and $m > 1$. We will, however, stick to this notation since it is common in literature.

The next result parallels Theorem 2.9. Note again that F, which is Lipschitz near x with modulus λ, is in fact Lipschitz with modulus λ near all y from some neighbourhood U of x. Hence the following statements (i)–(iii) hold true for all y from this neighbourhood U of x. Here $\mathbb{R}^{m\times n}$ is endowed with the Frobenius norm

$$\|A\|_{m\times n} := (\sum_{i=1}^{m} \|A^i\|^2)^{1/2}.$$

Theorem 2.17 *Let $F[\mathbb{R}^n \to \mathbb{R}^m]$ be Lipschitz near x with modulus λ. Then:*

(i) $\partial F(x)$ is a nonempty convex compact set and $\partial F(x) \subset \lambda \mathbb{B}$.

(ii) The multifunction $y \mapsto \partial F(y)$ is upper semicontinuous at x.

(iii) $\partial F(x) \subset (\partial F^1(x), \partial F^2(x), \cdots, \partial F^m(x))^T$, where $\partial F^i(x)$ is the generalized gradient of the real-valued function F^i at x for $i = 1, 2, \ldots, m$.

Proof. (i) The proof becomes straightforward using similar arguments as in the proof of Theorem 2.9(i).

(ii) Suppose that, for some $\varepsilon > 0$, we can choose $y_i \in x + \frac{1}{2i}\mathbb{B}$ for $i = 1, 2, \ldots$ with

$$\partial F(y_i) \not\subset \partial F(x) + \varepsilon \mathbb{B}. \tag{2.26}$$

Since the right-hand side of (2.26) is convex, we deduce from Definition 2.12 the existence of a matrix $M_i \in \partial_B F(y_i)$ such that

$$M_i \notin \partial F(x) + \varepsilon \mathbb{B}.$$

Hence, there must be $\tilde{y}_i \in x + \frac{1}{i}\mathbb{B}$ at which F is differentiable and

$$\mathcal{J}F(\tilde{y}_i) \notin \partial F(x) + \frac{\varepsilon}{2}\mathbb{B}. \tag{2.27}$$

From (i) we know that $\partial F(\cdot)$ is uniformly compact near x. Hence by passing to a subsequence, if necessary, we can assume that $\mathcal{J}F(\tilde{y}_i)$ converges to a matrix Z. By construction, $Z \in \partial F(x)$, which contradicts (2.27).

Assertion (iii) comes immediately from Theorem 2.16. ∎

Our next aim is to derive a chain rule for the generalized gradients/Jacobians of composite functions. We start with the following mean-value theorem in which we make use of the self-explanatory notation:

$$\partial F([x,y]) = \bigcup_{z\in[x,y]} \partial F(z),$$

$$\partial F([x,y])h = \bigcup_{Z\in\partial F([x,y])} Zh.$$

Theorem 2.18 *Let $F[\mathbb{R}^n \to \mathbb{R}^m]$ be Lipschitz on an open convex set U in $\mathbb{R}^n$ and x, y be two points in U. Then*

$$F(y) - F(x) \in \operatorname{conv}\left(\partial F([x,y])(y-x)\right). \tag{2.28}$$

Proof. Let x be fixed. From the same reason as in the proof of Lemma 2.15 for almost all y the line segment $[x, y]$ meets Ω_F in a set of one-dimensional measure zero. For such y, one has

$$F(y) - F(x) = \int_0^1 \mathcal{J}F(x + t(y-x))(y-x)\mathrm{d}t, \tag{2.29}$$

which expresses $F(y) - F(x)$ as a "continuous" convex combination of points from $\partial F([x,y])(y-x)$. To show that $F(y) - F(x)$ can also be expressed as a usual (finite) convex combination of points from $\partial F([x,y])(y-x)$, we use Theorem 2 from Yosida, 1968, Chap. 0.3, according to which to each $\varepsilon > 0$ there exists a continuous function $C_\varepsilon(t)$ in $(0,1)$ such that

$$\int_0^1 |\mathcal{J}F(x + t(y-x))(y-x) - C_\varepsilon(t)|\mathrm{d}t < \varepsilon.$$

Further, $\int_0^1 C_\varepsilon(t)\mathrm{d}t$ can be approximated by Riemann sums (cf. Dieudonné, 1960) in form of the required finite convex combination. The Carathéodory Theorem (see, e.g., Rockafellar, 1970) together with a limiting argument leads to the required expression.

Consider now a point $\bar{y}$ such that the line segment $[x, \bar{y}]$ does not meet Ω_F in a set of one-dimensional measure zero. Then, by the above argument and by the continuity of F and the upper semicontinuity of the map $x \rightsquigarrow \partial F(x)$, the difference $F(\bar{y}) - F(x)$ can also be expressed in the form (2.28). This completes the proof. ■

For composite functions we now get:

Theorem 2.19 *Let $F[\mathbb{R}^n \to \mathbb{R}^m]$ be Lipschitz near x and $g[\mathbb{R}^m \to \mathbb{R}]$ Lipschitz near $F(x)$. Then also the composite function $\varphi = g \circ F$ is Lipschitz near x and*

$$\partial\varphi(x) \subset \operatorname{conv}\left\{Z^{\mathrm{T}}\xi \mid Z \in \partial F(x), \xi \in \partial g(F(x))\right\}. \tag{2.30}$$

Proof. It is easy to verify that φ is Lipschitz near x. To prove (2.30) we show that that for each $h \in \mathbb{R}^n$

$$\varphi^0(x; h) \le \langle \xi_0, Z_0 h\rangle \quad \text{with suitable } Z_0 \in \partial F(x) \text{ and } \xi_0 \in \partial g(F(x)). \tag{2.31}$$

Indeed, if (2.31) holds, then the support function $\varphi^0(x;h)$ is less than or equal to the support function of the (closed) set on the right-hand side of (2.30), which proves inclusion (2.30) in virtue of Proposition 2.11.

We use Theorem 2.14 to see that for each x' close to x and positive t sufficiently small we can choose $u \in [F(x'), F(x'+th)]$ and $\xi \in \partial g(u)$ such that

$$\frac{\varphi(x'+th)-\varphi(x')}{t} = \langle \xi, \frac{F(x'+th)-F(x')}{t} \rangle.$$

By Theorem 2.18,

$$\frac{F(x'+th)-F(x')}{t} = w$$

for some $w \in \text{conv}\,(\partial F([x', x'+th])h)$ and thus with some $Z \in \partial F([x', x'+th])$ one has

$$\langle \xi, w \rangle \leq \langle \xi, Zh \rangle.$$

Combining these relations, we get

$$\frac{\varphi(x'+th)-\varphi(x')}{t} \leq \langle \xi, Zh \rangle \tag{2.32}$$

with $\xi \in \partial g(u)$ and $Z \in \partial F([x', x'+th])$.

Now choose sequences $x_i \to x$, $t_i \downarrow 0$, for which the difference quotient on the left-hand side of (2.32) converges to $\varphi^0(x;h)$. Let u_i, ξ_i, Z_i be the quantities for which (2.32) holds with $x' := x_i$ and $t := t_i$. Evidently, u_i must converge to $F(x)$, and the line segments $[x_i, x_i + t_i h]$ shrink to x. Suppose that $\{x_{i'}\}$ and $\{t_{i'}\}$ are subsequences of $\{x_i\}$, $\{t_i\}$ for which the vectors $\xi_{i'}$ converge to a vector ξ_0 and the matrices $Z_{i'}$ to a matrix Z_0. By the upper semicontinuity of ∂g and ∂F one has $\xi_0 \in \partial g(F(x))$ and $Z_0 \in \partial F(x)$. Inequality (2.32) thus implies (2.31) and the assertion follows. ∎

The case when the outer function g is continuously differentiable is of particular interest. Then we can strengthen the inclusion (2.30) to become an equality.

Theorem 2.20 *Let the assumptions of Theorem 2.19 hold and suppose that g is continuously differentiable on a neighbourhood of $F(x)$. Then*

$$\partial\varphi(x) = \left\{ Z^T \nabla g(F(x)) \mid Z \in \partial F(x) \right\}. \tag{2.33}$$

Proof. Put $\xi := \nabla g(F(x))$. By the continuous differentiability of g, for each $\varepsilon > 0$ there exists $\delta > 0$ such that F is Lipschitz on $x + \delta I\!\!B$ and, for all $y \in x + \delta I\!\!B$, $y \notin \Omega_F$, one has

$$\nabla\varphi(y) \in (\mathcal{J}F(y))^T \xi + \varepsilon I\!\!B. \tag{2.34}$$

Consider now the maximum

$$q := \max\left\{ \langle \xi, Zh \rangle \mid Z \in \partial F(x) \right\},$$

where $h \in I\!\!R^n$ is arbitrary but fixed. Obviously, q is the value of the support function of the set on the right-hand side of (2.33) at h.On the other hand, we get from the definition of $\partial F(x)$

$$\begin{aligned} q &= \text{limsup}\left\{ \langle \xi, \mathcal{J}F(y)h \rangle \mid y \to x, y \notin \Omega_F \right\} \\ &\leq \text{limsup}\left\{ \langle \nabla\varphi(y)h \rangle \mid y \to x, y \notin \Omega_{g\circ F} \right\} \end{aligned}$$

using (2.34) and the inclusion $\Omega_{g\circ F} \subset \Omega_F$. The last quantity is $\varphi^0(x;h)$ as a simple consequence of Theorem 2.16. Thus, in view of Theorem 2.19, the support functions of the sets on both sides of (2.33) are equal and the claim follows from Corollary 2.12. ■

Many Lipschitz mappings possess an important property, called semismoothness, which relates directional derivatives to generalized Jacobians.

Definition 2.13 *We say that $F[\mathbb{R}^n \to \mathbb{R}^m]$ is* semismooth *at x or* weakly semismooth *at x, respectively, if F is Lipschitz near x and the limit*

$$\lim_{\substack{V\in\partial F(x+th')\\ h'\to h,t\downarrow 0}} \{Vh'\} \tag{2.35}$$

or

$$\lim_{\substack{V\in\partial F(x+th)\\ t\downarrow 0}} \{Vh\} \tag{2.36}$$

exists for all $h \in \mathbb{R}^n$.

Evidently, semismoothness implies weak semismoothness. The relation of the limit in (2.36) to the directional derivative of F is clarified next.

Proposition 2.21 *Let $F[\mathbb{R}^n \to \mathbb{R}^m]$ be weakly semismooth at x. Then the directional derivative*

$$F'(x;h) := \lim_{t\downarrow 0} \frac{F(x+th)-F(x)}{t}$$

exists for all $h \in \mathbb{R}^n$ and

$$F'(x;h) = \lim_{\substack{V\in\partial F(x+th)\\ t\downarrow 0}} \{Vh\}.$$

Proof. The difference quotient $(F(x+th)-F(x))/t$ is bounded. So, there is a sequence $t_i \downarrow 0$ and some $\ell \in \mathbb{R}^m$ such that

$$\frac{F(x+t_ih)-F(x)}{t_i} \to \ell.$$

It suffices to show that ℓ equals the limit in (2.36). By Proposition 2.18

$$\frac{F(x+t_ih)-F(x)}{t_i} \in \operatorname{conv}\left(\partial F([x,x+t_ih])h\right).$$

Now use Carathéodory's Theorem to see that there exist numbers $t_i^{(k)} \in [0,t_i]$, coefficients of a convex combination $\lambda_i^{(k)}$ and matrices $V_i^{(k)} \in \partial F(x_i+t_i^{(k)}h)$ for $k=1,2,\ldots,m+1$, such that

$$\frac{F(x+t_ih)-F(x)}{t_i} = \sum_{k=1}^{m+1} \lambda_i^{(k)} V_i^{(k)} h, \quad \sum_{k=1}^{m+1} \lambda_i^{(k)} = 1.$$

By passing to a subsequence, if necessary, we can assume that $\lambda_i^{(k)} \to \lambda^{(k)}$ as $i \to \infty$. Clearly, $\lambda^{(k)} \in [0,1]$ and $\sum_{k=1}^{m+1} \lambda^{(k)} = 1$. Then

$$\begin{aligned} \ell &= \lim_{i\to\infty} \left\{ \sum_{k=1}^{m+1} \lambda_i^{(k)} V_i^{(k)} h \right\} = \sum_{k=1}^{m+1} \lim_{i\to\infty} \lambda_i^{(k)} \lim_{i\to\infty} \{V_i^{(k)} h\} \\ &= \sum_{k=1}^{m+1} \lambda^{(k)} \lim_{\substack{V\in\partial F(x+th)\\ t\downarrow 0}} \{Vh\} = \lim_{\substack{V\in\partial F(x+th)\\ t\downarrow 0}} \{Vh\} \end{aligned}$$

as required. ■

As in Theorem 2.8 we can prove that $F'(x;\cdot)$ is Lipschitz on $\mathbb{R}^n$. This property will be used in the following characterization of semismoothness.

Theorem 2.22 *The following statements are equivalent for $F[\mathbb{R}^n \to \mathbb{R}^m]$:*

(i) F is semismooth at x.

(ii) The convergence in (2.35) is uniform for all h with unit norm.

(iii) The convergence in (2.36) is uniform for all h with unit norm.

(iv) For any $V \in \partial F(x+h)$, $h \to 0$,

$$Vh - F'(x;h) = o(\|h\|). \tag{2.37}$$

Proof. (i) $\to$ (ii). Suppose (ii) does not hold. Then there exist $\varepsilon > 0$ and sequences $\{h_i\}$, $\{\bar{h}_i\}$, $t_i \downarrow 0$ and $V_i \in \partial F(x + t_i\bar{h}_i)$ such that

$$\|h_i\| = 1 \quad \text{and} \quad \|\bar{h}_i - h_i\| \to 0$$

and

$$\|V_i\bar{h}_i - F'(x;h_i)\| \geq 2\varepsilon \quad \text{for } i = 1,2,\ldots \tag{2.38}$$

By passing to a subsequence, if necessary, we get $h_i \to h$ and thus $\bar{h}_i \to h$ as well. Since $F'(x;\cdot)$ is Lipschitz on $\mathbb{R}^n$, inequalities (2.38) say that

$$\|V_i\bar{h}_i - F'(x;h)\| \geq \varepsilon$$

for sufficiently large i. By Proposition 2.21 this contradicts (i).

The implications (ii) $\Rightarrow$ (iii) $\Rightarrow$ (iv) are obvious.

(iv) $\Rightarrow$ (i). Suppose that F is not semismooth at x. Then there exist $\varepsilon > 0$, $h \in \mathbb{R}^n$ and sequences $h_i \to h$, $t_i \downarrow 0$ and $V_i \in \partial F(x + t_ih_i)$ such that

$$\|V_ih_i - F'(x;h)\| \geq 2\varepsilon \quad \text{for } i = 1,2,\ldots$$

By the Lipschitz continuity of $F'(x;\cdot)$, these inequalities say that

$$\|V_ih_i - F'(x;h_i)\| \geq \varepsilon \tag{2.39}$$

for sufficiently large i. Let d_i denote t_ih_i for $i = 1,2,\ldots$ Then we derive from (2.39) that

$$\|V_id_i - F'(x;d_i)\| \geq t_i\varepsilon$$

for sufficiently large i. Since $V_i \in \partial F(x + d_i)$, this contradicts (2.37). ■

The equivalence of (i) and (iii) will be used in Section 6.3 to prove the semismoothness of a rather complicated Lipschitz map.

2.3 CONICAL APPROXIMATIONS AND OPTIMALITY CONDITIONS

Recall definition (2.8) of the *tangent cone* of a convex set $\Omega \subset \mathbb{R}^n$ at $x \in \text{cl}\Omega$:

$$T_\Omega(x) := \text{cl} \bigcup_{\lambda>0} \lambda(\Omega - x).$$

The following result (cf. Aubin and Frankowska, 1990) gives another characterization of T_Ω.

Proposition 2.23 *Let $\Omega \subset \mathbb{R}^n$ be convex and $x \in \Omega$. Then*

$$T_\Omega(x) = \left\{ h \in \mathbb{R}^n \;\middle|\; \begin{array}{l} \text{there exist sequences } h_i \to h, t_i \downarrow 0 \\ \text{such that } x + t_i h_i \in \Omega \text{ for } i = 1, 2, \dots \end{array} \right\}.$$

Proof. Denote by $\mathcal{B}$ the cone on the right-hand side of the above equality. The inclusion $T_\Omega \supset \mathcal{B}$ is an immediate consequence of the definition of T_Ω. To prove the converse inclusion, let

$$h := \frac{1}{t}(y - x) \quad \text{with } y \in \Omega \text{ and } t > 0$$

be given. Choose sequences $y_i \to y$ and $t_i \downarrow 0$ with $y_i \in \Omega$ and put $h_i := \frac{1}{t}(y_i - x)$. Then $h_i \to h$ and, by convexity of Ω,

$$x + t_i h_i = (1 - \frac{t_i}{t})x + \frac{t_i}{t} y_i \in \Omega \quad \text{for all sufficiently large } i,$$

which proves

$$\bigcup_{\lambda>o} \lambda(\Omega - x) \subset \mathcal{B}.$$

It is easily seen that $\mathcal{B}$ is closed and thus also $T_\Omega(x) \subset \mathcal{B}$. ∎

In convex programming, Ω is often the "feasible" set

$$\Omega = \left\{ y \in \mathbb{R}^n \mid h^i(y) = 0,\ i = 1, 2, \dots, \ell,\ g^j(y) \le 0,\ j = 1, 2, \dots, s \right\}, \tag{2.40}$$

where the functions $h^i[\mathbb{R}^n \to \mathbb{R}]$, $i = 1, 2, \dots, \ell$, are affine and the functions $g^j[\mathbb{R}^n \to \mathbb{R}]$, $j = 1, 2, \dots, s$, are convex and continuously differentiable. For $x \in \Omega$ let

$$I(x) := \{i \in \{1, 2, \dots, s\} \mid g^i(x) = 0\}$$

be the index set of *active inequalities* and put

$$H(\cdot) := \left[h^1(\cdot), h^2(\cdot), \dots, h^\ell(\cdot)\right]^T, \qquad G(\cdot) := \left[g^1(\cdot), g^2(\cdot), \dots, g^s(\cdot)\right]^T.$$

Let $J \subset \{1, 2, \dots, s\}$ and recall that, according to our general notation, $G_J(x)$ is the subvector of $G(x)$ with components $g^i(x), i \in J$.

Theorem 2.24 *Let Ω, H, G be as above and assume that there exists $\overline{x} \in \Omega$ with $G(\overline{x}) < 0$. Then for each $x \in \Omega$*

$$T_\Omega(x) = \text{Ker}(\mathcal{J}H(x)) \cap \left\{ d \in \mathbb{R}^n \mid \mathcal{J}G_{I(x)}(x)d \le 0 \right\}. \tag{2.41}$$

Proof. Let $x \in \Omega$ be arbitrary but fixed. Put $I := I(x)$ and let us introduce the abbreviations

$$\begin{aligned} K &:= \{d \in \mathbb{R}^n \,|\, \mathcal{J}G_I(x)d \leq 0\} \\ L &:= \{d \in \mathbb{R}^n \,|\, \mathcal{J}G_I(x)d < 0\}. \end{aligned}$$

We start by showing that

$$T_\Omega(x) \subset \mathrm{Ker}(\mathcal{J}H(x)) \cap K. \tag{2.42}$$

To this purpose, let $d \in T_\Omega(x)$. By Proposition 2.23 there exist sequences $d_i \to d$ and $t_i \downarrow 0$ such that

$$x + t_i d_i \in \Omega \quad \text{for } i = 1, 2, \ldots.$$

Thus, since H is affine,

$$0 = H(x + t_i d_i) = H(x) + t_i \mathcal{J}H(x)d_i = t_i \mathcal{J}H(x)d_i$$

which implies $d \in \mathrm{Ker}(\mathcal{J}H(x))$. Further, by convexity,

$$\begin{aligned} 0 &\geq G_I(x + t_i d_i) \\ &\geq G_I(x) + t_i \mathcal{J}G_I(x)d_i \\ &= t_i \mathcal{J}G_I(x)d_i, \end{aligned}$$

which implies $d \in K$. This proves inclusion (2.42), i.e., one "half" of (2.41).

For the converse inclusion we first show that

$$\mathrm{Ker}(\mathcal{J}H(x)) \cap L \subset T_\Omega(x). \tag{2.43}$$

Let d be such that $\mathcal{J}H(x)d = 0$ and $\mathcal{J}G_I(x)d < 0$. Then for all sufficiently small $t > 0$

$$\begin{aligned} G_I(x + td) &= G_I(x) + t\mathcal{J}G_I(x)d + o(t) < 0 \\ H(x + td) &= H(x) + t\mathcal{J}H(x)d = 0 \end{aligned}$$

and thus $x + td \in \Omega$, i.e., $d \in \frac{1}{t}(\Omega - x)$ for small $t > 0$ which proves (2.43). Since T_Ω is closed by definition, we can continue

$$\mathrm{cl}(\mathrm{Ker}(\mathcal{J}H(x) \cap L)) \subset T_\Omega(x). \tag{2.44}$$

To finish the proof we show that under the regularity condition "$G(\overline{x}) < 0$ for some $\overline{x} \in \Omega$" one has

$$\mathrm{Ker}(\mathcal{J}H(x)) \cap K \subset \mathrm{cl}(\mathrm{Ker}(\mathcal{J}H(x)) \cap L).$$

Consider a vector $d \in \mathrm{Ker}(\mathcal{J}H(x)) \cap K$, i.e., $\mathcal{J}H(x)d = 0$ and $\mathcal{J}G_I(x)d \leq 0$. From the subgradient inequality for convex functions (cf. (3.5)) we get for $\overline{d} := \overline{x} - x$

$$\mathcal{J}(x)\overline{d} \leq G_I(\overline{x}) - G_I(x)$$

and, since $G_I(x) = 0$ and $G_I(\overline{x}) < 0$,

$$\mathcal{J}(x)\overline{d} < 0.$$

Furthermore, since H is affine, we have

$$\mathcal{J}H(x)\overline{d} = H(\overline{x}) - H(x) = 0.$$

This implies for the convex combination of d and $\overline{d}$

$$\mathcal{J}G_I(x)(\lambda\overline{d} + (1-\lambda)d) < 0, \quad \mathcal{J}H(x)(\lambda\overline{d} + (1-\lambda)d) = 0, \quad \text{for } 0 < \lambda \leq 1,$$

i.e.,

$$\lambda\overline{d} + (1-\lambda)d \in \operatorname{Ker}(\mathcal{J}H(x)) \cap L \quad \text{for } 0 < \lambda \leq 1.$$

It follows that $d \in \operatorname{cl}(\operatorname{Ker}(\mathcal{J}H(x)) \cap L)$ and thus, together with (2.44),

$$\operatorname{Ker}(\mathcal{J}H(x)) \cap K \subset T_\Omega(x).$$

■

Corollary 2.25 *Under the assumptions of Theorem 2.24 one has for each $x \in \Omega$*

$$N_\Omega(y) = \left\{ \sum_{i=1}^{\ell} \mu^i \nabla h^i(y) + \sum_{i=1}^{s} \lambda^i \nabla g^i(y) \mid \mu \in \mathbb{R}^\ell, \lambda \in \mathbb{R}^s_+, \right. \left. \lambda^i g^i(y) = 0,\ i = 1, 2, \ldots, s \right\}. \tag{2.45}$$

Proof. By Theorem 2.24 and by Rockafellar, 1970, Cor. 23.8.1

$$N_\Omega(x) = N_1 + N_2, \tag{2.46}$$

where $N_1 = (\operatorname{Ker}\mathcal{J}H(x))^\perp = \mathcal{R}((\mathcal{J}H(x))^T)$ and N_2 is the negative polar cone to the cone $T := \{d \in \mathbb{R}^n \mid \mathcal{J}G_{I(x)}(x)d \leq 0\}$. Consider the cone

$$\begin{aligned}\widetilde{N}_2 &:= \{\xi \in \mathbb{R}^n \mid \xi = (\mathcal{J}G_{I(x)}(x))^T\kappa, \kappa \geq 0\} \\ &= \left\{ \sum_{i=1}^{s} \lambda^i \nabla g^i(x) \mid \lambda \in \mathbb{R}^s_+, \lambda^i = 0 \text{ for } i \notin I(x) \right\}\end{aligned}$$

which is obviously closed. Clearly, $(\widetilde{N}_2)^* = -T$. Since $\widetilde{N}_2$ is closed, we get by the Bipolar Theorem (see, e.g., Aubin and Frankowska, 1990)

$$-T^* = (\widetilde{N}_2)^{**} = \widetilde{N}_2$$

and so $N_2 = \widetilde{N}_2$. Thus, in view of (2.46), formula (2.45) has been established. ■

Later we will need another approximating cone whose definition is based on the Lipschitz continuity of the distance function for a subset Ω of $\mathbb{R}^n$:

$$\operatorname{dist}_\Omega(x) := \inf\{\|y - x\| \,|y \in \Omega\} \quad \text{for } x \in \mathbb{R}^n.$$

To see that $\operatorname{dist}_\Omega(\cdot)$ is Lipschitz, let x and y be given. For each $\varepsilon > 0$ we can choose $c \in \Omega$ such that $\|y - c\| \leq \operatorname{dist}_\Omega(y) + \varepsilon$ and thus

$$\operatorname{dist}_\Omega(x) \leq \|x - c\| \leq \|x - y\| + \|y - c\| \leq \|x - y\| + \operatorname{dist}_\Omega(y) + \varepsilon.$$

Since ε is arbitrary and the arguments x, y may be interchanged, $\mathrm{dist}_\Omega(\cdot)$ is Lipschitz on $\mathbb{R}^n$ with modulus 1. Hence, the directional derivative $\mathrm{dist}^0_\Omega(x; h)$ is well-defined for all x and h and the following definition makes sense.

Definition 2.14 *Let Ω be a subset of $\mathbb{R}^n$ and $x \in \Omega$. Then the* Clarke's tangent cone *to Ω at x, denoted $C_\Omega(x)$, is defined as*

$$C_\Omega(x) := \{h \in \mathbb{R}^m \mid \mathrm{dist}^0_\Omega(x; h) = 0\}.$$

It is an immediate consequence of Theorem 2.8(i) that $C_\Omega(x)$ is a closed convex cone. For convex Ω this cone coincides with the tangent cone introduced in Definition 2.5 (see, e.g., Mäkelä and Neittaanmäki, 1992). With the help of $C_\Omega(x)$ we define another cone which for convex Ω gives back Definition 2.6.

Definition 2.15 *Let Ω be a subset of $\mathbb{R}^n$ and $x \in \Omega$. Then the* Clarke's normal cone *to Ω at x, denoted $K_\Omega(x)$, is defined as*

$$K_\Omega(x) := -(C_\Omega(x))^* = \{\xi \in \mathbb{R}^m \mid \langle \xi, h\rangle \le 0 \text{ for all } h \in C_\Omega(x)\}.$$

$K_\Omega(x)$ admits the following characterization in terms of the distance function of Ω.

Proposition 2.26

$$K_\Omega(x) = \mathrm{cl} \bigcup_{\lambda \ge 0} \lambda \partial \mathrm{dist}_\Omega(x). \tag{2.47}$$

Proof. By definition of $C_\Omega(x)$ and by Theorem 2.9

$$h \in C_\Omega(x) \quad \text{if and only if} \quad \langle h, \xi\rangle \le 0 \quad \text{for all } \xi \in \partial\mathrm{dist}_\Omega(x)$$

and thus

$$C_\Omega(x) = (-\partial \mathrm{dist}_\Omega(x))^*.$$

It follows

$$K_\Omega(x) = -(C_\Omega(x))^* = -(-(\partial\mathrm{dist}_\Omega(x))^*)^* = (\partial\mathrm{dist}_\Omega(x))^{**}.$$

Relation (2.47) is now a direct consequence of the Bipolar Theorem. ∎

We close this section with an optimality condition for a constrained optimization problem.

Theorem 2.27 *Let Ω be a closed subset of $\mathbb{R}^n$ and $x \in \Omega$. Assume that $f[\mathbb{R}^n \to \mathbb{R}]$ is Lipschitz near x with modulus λ and x is a (constrained) local minimizer of f on Ω. Then for each $r \ge \lambda$, the function $f(\cdot) + r\mathrm{dist}_\Omega(\cdot)$ attains an (unconstrained) local minimum at x. In particular, one has*

$$0 \in \partial f(x) + K_\Omega(x). \tag{2.48}$$

Proof. Suppose there exists a sequence $x_i \to x$ such that

$$f(x_i) + r\mathrm{dist}_\Omega(x_i) < f(x) \qquad \text{for all } i.$$

Then $\text{dist}_\Omega(x_i) > 0$ for sufficiently large i since otherwise $x_i \in \Omega$ and the above inequality contradicts the optimality of x. Choose y_i in the (closed) Ω such that

$$\|x_i - y_i\| = \text{dist}_\Omega(x_i).$$

Since $\text{dist}_\Omega(x_i) \downarrow 0$, for i sufficiently large the Lipschitz modulus λ applies in a ball with center x_i and a radius greater than $\text{dist}_\Omega(x_i)$. Therefore,

$$f(y_i) \leq f(x_i) + \lambda\|x_i - y_i\| \leq f(x_i) + r\text{dist}_\Omega(x_i) < f(x),$$

which contradicts the optimality of x because $y_i \to x$ and $y_i \in \Omega$. Propositions 2.10, 2.13 and 2.26 imply immediately the optimality condition (2.48). ■

Using 2.48, first-order necessary optimality conditions can be derived for various optimization problems with Lipschitz functions in the objective and in the constraints. In particular, the results of Section 7.1, where we derive optimality conditions for MPEC, are based on (2.48). Points satisfying (2.48) are called (Clarke's) *stationary points* in the optimization problem

$$\begin{array}{ll} \text{minimize} & f(x) \\ \text{subject to} & x \in \Omega. \end{array}$$

2.4 PROJECTION ONTO POLYHEDRAL SETS

In this section we study the local behaviour of the projection onto polyhedral sets. Let $\Omega \subset \mathbb{R}^n$ be such a set and let Proj_Ω denote the projection map

$$\text{Proj}_\Omega(x) := \arg\min_{y \in \Omega} \|y - x\|.$$

We start with two well-known and easy-to-verify results (Zarantonello, 1971) (which hold true for arbitrary closed convex Ω):

- Proj_Ω is Lipschitz on $\mathbb{R}^n$ with modulus 1;
- for each $x \in \mathbb{R}^n$ one has $y = \text{Proj}_\Omega(x)$ if and only if $x \in y + N_\Omega(y)$, see Figure 2.2.

Our aim is to study the behaviour of $\text{Proj}_\Omega(x + k)$ for a given x and for k close to 0. Let $y = \text{Proj}_\Omega(x)$. Since $x - y \in N_\Omega(y)$, the set

$$F = \{c \in \Omega \mid \langle x - y, c - y \rangle = 0\} \tag{2.49}$$

is a face of the polyhedral Ω containing y; cf. Definition 2.7. (F is the intersection of Ω and the supporting hyperplane to Ω at y given by $\{c \in \mathbb{R}^n \mid \langle x - y, c \rangle = \langle x - y, y \rangle\}$). We call F the *critical face* of Ω corresponding to $x - y$. From (2.49) it follows that

$$F = \{c \in \Omega \mid \langle x - y, c \rangle = \delta_\Omega(x - y)\}.$$

Further, we introduce as *critical cone* of Ω corresponding to y and $x - y$ the set

$$K := T_\Omega(y) \cap \{x - y\}^\perp. \tag{2.50}$$

It is easy to see that $K = T_F(y)$.

The proof of the main result of this section will make use of the following two lemmas.

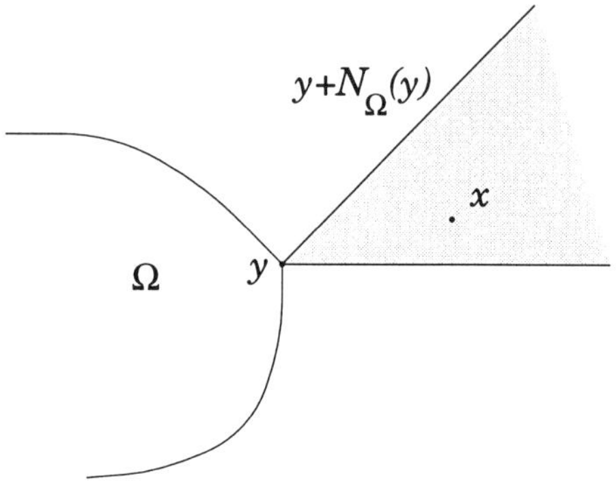

Figure 2.2. Normal cone vs. Projection

Lemma 2.28 *Let $P[\mathbb{R}^n \leadsto \mathbb{R}^m]$ be a polyhedral multifunction. If C is a bounded subset of*

$$\mathrm{Im}P := \{y \in \mathbb{R}^m \mid y \in P(x) \textit{ for some } x \in \mathbb{R}^n\},$$

then there is a bounded set $M \subset \mathbb{R}^n$ such that $C \subset P(M)$.

Proof. $\mathrm{Im}P$ is closed so that $\mathrm{cl}C$ is also a bounded subset of $\mathrm{Im}P$. Hence, without loss of generality, assume that C is compact. As in Section 2.1 we denote by G_i, $i = 1, 2, \ldots, k$, the components of $\mathrm{Gph}P$ (see Figure 2.3).

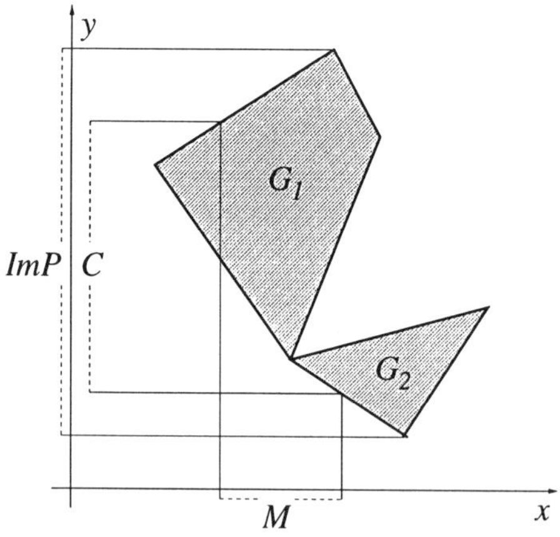

Figure 2.3.

Now fix some arbitrary $y \in C$ and put

$$J(y) := \{i \in \{1, 2, \ldots, k\} \mid y \in \pi_2(G_i)\},$$

where π_2 denotes the canonical projection of $\mathbb{R}^n \times \mathbb{R}^m$ onto $\mathbb{R}^m$. By Lemma 2.2 there is a neighbourhood V of y such that

$$(V \times \mathbb{R}^n) \cap \mathrm{Gph}P \subset \bigcup_{i \in J(y)} G_i,$$

which means

$$V \cap \mathrm{Im}P = \bigcup_{i \in J(y)} (V \cap \pi_2(G_i)). \tag{2.51}$$

Now assume that $W \subset V$ is a bounded polyhedral convex neighbourhood of y and define the sets

$$H_i := W \cap \pi_2(G_i) \quad \text{for } i \in J(y).$$

These sets are compact and polyhedral and for the (convex) functions

$$f_i(v) := \inf\{\|x\| \mid (x, v) \in G_i\}, \quad i \in J(y)$$

one has $H_i \subset \mathrm{dom} f_i$. Therefore, for each $i \in J(y)$, the function f_i attains a maximum on H_i, say μ_i (see, e.g., Rockafellar, 1970, Cor. 32.3.3). Let $\mu = \mu(y) = \max_{i \in J(y)} \mu_i$. If $v \in W \cap \mathrm{Im}P$ then, because of (2.51), $v \in H_i$ for some $i \in J(y)$ and thus there exists x with $(x, v) \in G_i$ such that

$$\|x\| = f_i(v) \leq \mu_i \leq \mu.$$

This implies

$$W \cap \mathrm{Im}P \subset P(\mu \mathbb{B}).$$

Such a relative neighbourhood of the form $W \cap \mathrm{Im}P$ can be constructed for each $y \in C$. By compactness already a finite number of these neighbourhoods covers C. Since each of them is contained in $P(\lambda \mathbb{B})$ for some positive real λ, the proof is complete. ■

The next lemma is fundamental since it reveals an essential property of the critical face F. For $x' \in \mathbb{R}^n$ and $z' := x' - \mathrm{Proj}_\Omega(x')$ we introduce the sets

$$\mathcal{F}_\Omega(z') := \{y' \in \Omega \mid \langle z', y' \rangle = \delta_\Omega(z')\} \tag{2.52}$$

and

$$\mathcal{F}_F(z') := \{y' \in F \mid \langle z', y' \rangle = \delta_F(z')\}, \tag{2.53}$$

where F is the critical face of Ω corresponding to $z := x - y$ with $y = \mathrm{Proj}_\Omega(x)$. Clearly, $\mathcal{F}_\Omega(z')$ and $\mathcal{F}_F(z')$ are the critical faces of Ω and F, corresponding to z'. Note that $\mathcal{F}_\Omega(z) = F$.

Lemma 2.29 *Let $\Omega \subset \mathbb{R}^n$ be a polyhedral set, $x \in \mathbb{R}^n$ and $y = \mathrm{Proj}_\Omega(x)$. Then there is a neighbourhood U of x such that for all $x' \in U$ one has*

$$\mathcal{F}_\Omega(z') = \mathcal{F}_F(z'), \quad \textit{where} \quad z' = x' - \mathrm{Proj}_\Omega(x').$$

Proof. Let x' be close to x so that z' is close to z.

(i) We show that $\mathcal{F}_\Omega(z') \subset \mathcal{F}_F(z')$. By (2.52), if $y' \in \mathcal{F}_\Omega(z')$, then for each $c \in \Omega$ one has $\langle z', c - y' \rangle \leq 0$. This holds in particular if $c \in F$ and so, if additionally $y' \in F$, we get $y' \in \mathcal{F}_F(z')$. Thus it suffices to prove that there is a neighbourhood of z such that for all z' in that neighbourhood $\mathcal{F}_\Omega(z') \subset F$. Suppose by contradiction the existence of a sequence $\{z_i\}$ converging to z such that for each i there exists $y'_i \in \mathcal{F}_\Omega(z_i) \setminus F$. Each set $\mathcal{F}_\Omega(z_i)$ is a face of Ω, but there are only finitely many such faces, since Ω is polyhedral (Rockafellar, 1970, Thm. 19.1). It follows that a face, say G, of Ω appears infinitely often among the faces $\mathcal{F}_\Omega(z_i)$. Let now $g \in G$; then for infinitely many i's the pair $(z_i, g) \in \mathrm{Gph}\mathcal{F}_\Omega$ (which is a closed set). Then, however, (z, g) belongs to this graph as well, i.e. $g \in \mathcal{F}_\Omega(z)$, and consequently $G \subset F$. This contradicts the existence of points y'_i, and thus $\mathcal{F}_\Omega(z') \subset \mathcal{F}_F(z')$ for all z' in some neighbourhood, say U_1, of z.

(ii) To prove the converse inclusion, we observe first that $(\mathcal{F}_F)^{-1}(\cdot) = N_F(\cdot)$, where $N_F(y') = \emptyset$ for $y' \notin F$ (cf. Section 2.1). Indeed, the inverse map $(N_F)^{-1}$ assigns to each direction z those points y of F, where $z \in N_F(y)$; this is exactly $\mathcal{F}_F(z)$. If $\xi \in F$ then, near 0, the sets $\Omega - \xi$ and $F - \xi$ coincide with $T_\Omega(\xi)$ and $T_F(\xi)$, respectively. Let L denote the half-line $\{-\lambda z \mid \lambda \geq 0\}$; then by definition

$$F - \xi = (\Omega - \xi) \cap (-L^*).$$

This implies that for a neighbourhood V of 0 one has

$$V \cap T_F(\xi) = V \cap (F - \xi) = V \cap (\Omega - \xi) \cap (-L^*) = V \cap T_\Omega(\xi) \cap (-L^*).$$

Since $T_F(\xi)$ and $T_\Omega(\xi) \cap (-L^*)$ are cones, then in fact $T_F(\xi) = T_\Omega(\xi) \cap (-L^*)$ and hence, by polyhedrality, $N_F(\xi) = N_\Omega(\xi) + L$. Thus for each $\xi \in \Omega$ we have

$$N_F(\xi) = \begin{cases} N_\Omega(\xi) + L & \text{if } \xi \in F \\ \emptyset & \text{otherwise.} \end{cases} \tag{2.54}$$

In the next step we show that for small elements of $N_F(\xi)$ only small elements of L need be used in the representation (2.54). By Lemma 2.6 N_Ω is constant on the relative interior of all faces of Ω. Since these relative interiors create a partition of Ω (Rockafellar, 1970, Thm. 18.2), N_Ω takes only finitely many distinct values. Denote these values (which are cones) by $K_1, K_2, \ldots, K_N$. We now introduce for each $i = 1, 2, \ldots, N$ a convex polyhedral multifunction $G_i[\mathbb{R}^1 \rightsquigarrow \mathbb{R}^n]$ such that

$$G_i(t) = K_i - tz \qquad \text{for } t \in \mathbb{R}_+.$$

Then, evidently, $K_i + L = G_i(\mathbb{R}_+)$ and thus by Lemma 2.28 there is some $\alpha_i > 0$ such that

$$\mathbb{B} \cap (K_i + L) = \mathbb{B} \cap G_i(\mathbb{R}_+) \subset G_i([0, \alpha_i]).$$

It follows that

$$(\alpha_i^{-1}\mathbb{B}) \cap (K_i + L) \subset G_i([0, 1])$$

and consequently, with $\alpha := \max_{i=1,2,\ldots,N} \alpha_i$, we get

$$(\alpha^{-1}\mathbb{B}) \cap (K_i + L) \subset G_i([0, 1]) \quad \text{for all } i = 1, 2, \ldots, N. \tag{2.55}$$

We conclude that in order to represent elements of $N_F(\xi)$ with norm not greater than α^{-1}, we do not need to use elements of L with $\lambda > 1$.

We now define a neighbourhood U of z with U_1 from (i) as follows:

$$U := U_1 \cap (z + \alpha^{-1}\mathbb{B}).$$

If $z' \in U$, then the critical face $\mathcal{F}_F(z')$ is either empty or nonempty. In the former case the inclusion $\mathcal{F}_F(z') \subset \mathcal{F}_\Omega(z')$ holds trivially and so assume that $y' \in \mathcal{F}_F(z')$, i.e. $z' \in N_F(y')$. In virtue of (2.53), this implies that

$$\langle z', w - y' \rangle \leq 0 \quad \text{for all } w \in F.$$

But then also

$$\langle z' - z, w - y' \rangle \leq 0 \quad \text{for all } w \in F$$

since $\langle z, w - y' \rangle = 0$ (both w and y' belong to F). Thus $z' - z \in N_F(y')$ so that, due to (2.54) and (2.55),

$$z' - z \in G_i([0,1]) \quad \text{for some } i \in \{1, 2, \ldots, N\}.$$

Therefore there is a $t \in [0,1]$ such that $z' - z \in N_\Omega(y') - tz$. However, we have also $(1-t)z \in N_\Omega(y')$ since $y' \in F(= \mathcal{F}_\Omega(z))$. By combining these two relations we get $z' \in N_\Omega(y')$, i.e. $y' \in \mathcal{F}_\Omega(z')$. It follows that for $z' \in U$

$$\mathcal{F}_F(z') \subset \mathcal{F}_\Omega(z')$$

and the proof is complete. ■

The preceding lemma is essential for the proof of the next statement.

Proposition 2.30 *Let $\Omega \subset \mathbb{R}^n$ be a polyhedral set, $y \in \Omega$ and $z \in N_\Omega(y)$. Then there is a neighbourhood V of the origin in $\mathbb{R}^n \times \mathbb{R}^n$ such that for $(h,k) \in V$ the following statements are equivalent:*

(i) $(y+h, z+k) \in \mathrm{Gph} N_\Omega$;

(ii) $(h,k) \in \mathrm{Gph} N_K$.

Proof. By Lemma 2.29 there exists a neighbourhood U of $0 \in \mathbb{R}^n$ such that for $k \in U$ we have

$$\mathcal{F}_\Omega(z+k) = \mathcal{F}_F(z+k). \tag{2.56}$$

Since F is polyhedral and K is the tangent cone to F at y, there is an open neighbourhood W of $0 \in \mathbb{R}^n$ such that

$$(y+W) \cap F = y + (W \cap K).$$

This implies that for $h \in W$ one has $N_F(y+h) = N_K(h)$. Now let $V = W \times U$ and $(h,k) \in V$. Since $(\mathcal{F}_\Omega)^{-1}(\cdot) = N_\Omega(\cdot)$ and $(\mathcal{F}_F)^{-1}(\cdot) = N_F(\cdot)$, we infer from (2.56) that

$$z + k \in N_\Omega(y+h) \tag{2.57}$$

is equivalent to

$$z + k \in N_F(y+h). \tag{2.58}$$

Since $h \in W$, relation (2.58) is equivalent to $z + k \in N_K(h)$. By definition, $K \subset \{z\}^\perp$ and hence z is orthogonal to $\mathrm{lin} K$. Consequently,

$$z \in N_K(h) \cap -N_K(h)$$

(i.e. z belongs to the lineality space of $N_K(h)$). Then, however, the relation $z+k \in N_K(h)$ is equivalent to $k \in N_K(h)$ and we are done. ■

We are now in a position to describe the behaviour of $\mathrm{Proj}_\Omega(x+k)$ for small perturbations k.

Theorem 2.31 *Let $\Omega \subset \mathbb{R}^n$ be a polyhedral set, $x \in \mathbb{R}^n$ and $y = \mathrm{Proj}_\Omega(x)$. Assume that K is the critical cone of Ω corresponding to y and $x - y$. Then there is a neighbourhood U of $0 \in \mathbb{R}^n$ such that for all $k \in U$*

$$\mathrm{Proj}_\Omega(x+k) = y + \mathrm{Proj}_K(k). \tag{2.59}$$

Proof. Put again $z := x - y$ and for an arbitrary $k \in \mathbb{R}^n$ define h by

$$y + h = \mathrm{Proj}_\Omega(x+k). \tag{2.60}$$

We have to show that $h = \mathrm{Proj}_K(k)$ for small k.

Clearly, equality (2.60) is equivalent to

$$(z + k - h) \in N_\Omega(y+h). \tag{2.61}$$

Further, since the map $\mathrm{Proj}_\Omega(\cdot)$ is nonexpansive (Zarantonello, 1971), there is a neighbourhood U of $0 \in \mathbb{R}^n$ such that for each $k \in U$ we have $(h, k-h) \in V$, where V is the neighbourhood introduced in Proposition 2.30. Due to Proposition 2.30, statement (2.61) is equivalent to $(h, k-h) \in \mathrm{Gph} N_K$, i.e. $k - h \in N_K(h)$. This yields, however, that $h = \mathrm{Proj}_K(k)$ as required. ■

The projection operator is generally not directionally differentiable as shown by a counterexample in Kruskal, 1969. For polyhedral Ω, however, we get from Theorem 2.31 a simple and convenient formula

$$\mathrm{Proj}'_\Omega(x;h) = \lim_{t \downarrow 0} \frac{\mathrm{Proj}_K(th)}{t}$$

and, since the projection onto a cone is positively homogeneous,

$$\mathrm{Proj}'_\Omega(x;h) = \mathrm{Proj}_K(h).$$

This formula is widely used in Chapters 5 and 6.

Bibliographical notes

The results on polyhedral multifunctions, collected in Section 2.1, have been taken from Robinson, 1976; Robinson, 1981; Hogan, 1973. As already mentioned, the main source for Sections 2.2 and 2.3 was the monograph Clarke, 1983. Another interesting treatment of this subject can be found in Rockafellar, 1981 and Aubin and Frankowska, 1990. The concept of semismoothness was introduced for real-valued functions in Mifflin, 1977 and proved to be useful in the study of numerical methods for nonsmooth optimization (Schramm and Zowe, 1992). In Qi and Sun, 1993 this concept has been generalized to the form presented here. Proposition 2.26 and Theorem 2.27 stem from Qi and Sun, 1993. The role of semismoothness in nonsmooth Newton techniques will become apparent in the next chapter. The differentiability of the projection map, has been studied in many papers starting with Haraux, 1977 up to recent contributions of Shapiro dealing with the projection onto nonconvex sets. It is well-known that this map need not be directionally differentiable, not even for convex sets in finite dimension (Kruskal, 1969). The presented results concerning polyhedral sets were taken from Robinson, 1984; Pang, 1990a and Robinson, 1991.

3 ALGORITHMS OF NONSMOOTH OPTIMIZATION

In our applications we have to deal with two types of nonsmooth problems: the minimization of a nonsmooth functional f and the solution of nonsmooth equations. This chapter presents in short two basic algorithms which can cope with the difficulty caused by the nondifferentiability. These two codes are working horses in the numerical part of this book.

After a short discussion of two "historical" nonsmooth approaches for minimization of nonsmooth f (the subgradient concept and the cutting plane idea), we will concentrate on the bundle strategy, which was developed in the seventies by Lemaréchal, 1975 and Wolfe, 1975. One of the first implementations of this idea was the code M1FC1 by Lemaréchal and Imbert, 1985. This code proved to be very successful when dealing with nonsmooth objective functions and is still widely used in practice. Its weak point is the line search, which usually is responsible for a break-down in delicate situations. A way to (partly) overcome this difficulty lies in a marriage of the bundle idea with the trust region technique. Since the motivating ideas behind the algorithm are of geometric nature and all stem from convexity, we will assume throughout Sections 3.1 and 3.2 that f is convex. The necessary technical modifications for nonconvex f will shortly be touched in Section 3.3.

BT being one leg in our numerical part, we need as second leg a nonsmooth variant of the Newton's method to deal with nonsmooth equations. Section 3.4 presents this algorithm without going into details or discussing the relations to similar proposals. In the application part, this method proved to be an efficient tool.

3.1 CONCEPTUAL IDEA

We study the minimization of a nonsmooth functional

$$\text{minimize } f(x) \quad \text{where } f : \mathbb{R}^n \to \mathbb{R}, \tag{3.1}$$

under the general assumption (which is standard in this context) that

$$f \text{ is locally Lipschitz on } \mathbb{R}^n. \tag{3.2}$$

For such f, the generalized gradient $\partial f(x)$ at x is a well-defined non-empty subset of $\mathbb{R}^n$ with the analytic representation (see Theorems 2.9(i) and 2.16)

$$\partial f(x) = \operatorname{conv}\left\{\lim_{i\to\infty} \nabla f(x_i) \mid x_i \to x, x_i \notin \Omega_f\right\}. \tag{3.3}$$

Recall that the elements g from $\partial f(x)$ are also called *subgradients* of f at x. Quite naturally, these subgradients serve as substitute of the gradients in the C^1-case. Hence, parallel to what is standard for smooth optimization codes, we require throughout that we dispose of a subroutine which

$$\text{computes } f(x) \text{ and one (arbitrary) } g \in \partial f(x) \text{ for a given } x. \tag{3.4}$$

Although this seems to be a modest (and minimal) requirement, the reader should be aware that in the real-life problems of Chapters 9–12 already the computation of one such $g \in \partial f(x)$ can be all but easy.

Since the motivating algorithmic ideas all stem from the convex situation, we will assume in Sections 3.1 and 3.2 a convex f. We start with some standard facts from convex folklore which can be found in most textbooks on convex analysis; see, e.g., Rockafellar, 1970; Hiriart-Urruty and Lemaréchal, 1993. A convex $f[\mathbb{R}^n \to \mathbb{R}]$ is locally Lipschitz on all of $\mathbb{R}^n$ and thus (3.2) holds automatically. For convex f, the generalized directional derivative $f^0(x;h)$ from Definition 2.9 coincides with the classic directional derivative $f'(x;h)$, i.e.,

$$f^0(x;h) = f'(x;h).$$

Since the difference quotient $(f(x+\lambda h) - f(x))/\lambda$ of convex f is monotone in λ, we can continue

$$f^0(x;h) = f'(x;h) = \inf\left\{\frac{f(x+\lambda h) - f(x)}{\lambda} \mid \lambda > 0\right\}.$$

It follows from Definition 2.10 that $g \in \partial f(x)$ can be characterized by an inequality:

$$g \in \partial f(x) \quad \text{if and only if} \quad \langle g, y - x\rangle \le f(y) - f(x) \quad \text{for all } y \in \mathbb{R}^n. \tag{3.5}$$

This so-called *subgradient inequality* for convex f plays a crucial role in what is going to come.

3.1.1 Subgradient methods

Among the first methods that could deal with (3.1) for convex f and under assumption (3.4) were the *subgradient methods* (also called *Kiev methods*); see e.g. Ermoliev, 1976; Poljak, 1978 and Shor, 1985. At iterate x_k one makes a step along a negative subgradient with some off-line chosen steplength

$$x_{k+1} := x_k + \lambda_k d_k \text{ where } d_k := -g_k/\|g_k\| \text{ with } g_k \in \partial f(x_k). \tag{3.6}$$

It can be shown that, under rather suggestive assumptions on the solution set and for $\lambda_k \downarrow 0$ and $\sum_{k=1}^{\infty} \lambda_k = \infty$, the x_k from (3.6) converge to an optimal point. The simple structure of these subgradient methods still makes them widely used, although they suffer from some serious drawbacks: the methods do not guarantee a descent at each step, they lack an implementable stopping criterion and the convergence speed is extremely poor (less than linear). The last disadvantage can be partly overcome by premultiplying d_k in (3.6) with some variable metric matrix H_k, which is updated in a simple way at each iteration. Linear convergence in the function values can be established for a member of this class (see e.g. Shor, 1985); the additional H_k makes, however, the method very cumbersome for large n.

3.1.2 Bundle concept

A huge step forward in nonsmooth optimization was the *bundle concept* introduced in Wolfe, 1975 and Lemaréchal, 1989 which can handle convex and nonconvex f. The convexity assumption only serves for motivation purposes.

All variants of the bundle concept carry two distinctive features:

(i) At the current iterate x_k they make use of the *bundle* of information $(f(x_k), g_k)$, $(f(x_{k-1}), g_{k-1}), \ldots$ collected so far to build up a model of f close to x_k.

(ii) If, due to the kinky structure of f, this model is not yet an adequate one, then they mobilize more subgradient information close to x_k.

Recipe (i) leads in a natural way to the *cutting plane* (CP) *model* of f at x_k:

$$\max_{1 \le i \le k} \{g_i^T(x - x_i) + f(x_i)\}. \tag{3.7}$$

(3.7) is a piecewise linear approximation of the convex f from below, which coincides with f at all x_i. For short we put $d := x - x_k$ in (3.7) and use the notation

$$f_{\mathrm{CP}}(x_k; d) := \max_{1 \le i \le k} \{g_i^T d + g_i^T(x_k - x_i) + f(x_i)\} \text{ for } d \in \mathbb{R}^n. \tag{3.8}$$

Obviously there is no reason to trust this substitute for f far away from x_k. Therefore a stabilizing term $\frac{1}{2t_k} d^T d$ with positive t_k is added in (3.8), when minimizing this CP-model of f. If f_{CP} models f close to x_k good enough, then the minimizer d_k of

$$f_{\mathrm{CP}}(x_k; d) + \frac{1}{2t_k} d^T d$$

is a descent direction for f and a line search along $x_k + \lambda d_k$ for $\lambda \ge 0$ provides some x_{k+1} with $f(x_{k+1}) < f(x_k)$. For a nonsmooth f it may happen, however, that f_{CP} is such a poor approximation of f that d_k is not a descent direction for f or that the line search only leads to a marginal decrease in f; think e.g. of $f(x) = |x|$, $x_i < 0$ for $i = 1, \ldots, k$ and x_k close to the kink 0. Here strategy (ii) comes up: obviously f_{CP} does not copy f on the halfline $x_k + \lambda d_k$, $\lambda \ge 0$; to master this lack of information one stays at x_k and enriches the model by including one more subgradient from $\partial f(x_k + \lambda d_k)$ for small $\lambda > 0$. Omitting all details we obtain the

Iteration step $x_k \to x_{k+1}$**:** (3.9)

(1) *Compute* $d_k := d(t_k) := \arg\min\{f_{\text{CP}}(x_k; d) + \frac{1}{2t_k} d^T d \mid d \in \mathbb{R}^n\}$.

(2) *Perform a line search for* f *along* $x_k + \lambda d_k$, $\lambda \geq 0$.

   (a) *If the line search leads to a "sufficient decrease" in* f*, then make a* **Serious Step**: *Put* $x_{k+1} := x_k + \lambda_k d_k$ *with* $\lambda_k \in \arg\min_{\lambda > 0} f(x_k + \lambda d_k)$ *and compute* $g_{k+1} \in \partial f(x_{k+1})$.

   (b) *If the line search yields an "insufficient decrease" only, then make a* **Null Step**: *Put* $x_{k+1} := x_k$ *and compute* $g_{k+1} \in \partial f(x_k + \lambda d_k)$ *for suitable small* $\lambda > 0$.

Different from the subgradient approach, the above iteration guarantees a decrease for each Serious Step. Further one disposes of an implementable stopping criterion: x_k is "optimal" as soon as d_k in (1) is "close" to 0. And, since the line search adjusts the steplength λ_k to the chosen d_k, one has a considerably faster convergence speed. All this can be made precise and a detailed convergence analysis exists for convex and also for nonconvex f; see Lemaréchal et al., 1981; Mifflin, 1982 or the monograph Kiwiel, 1985.

The above concept has been implemented by a number of authors. We mention in particular the Fortran code M1FC1 by Lemaréchal and Imbert, 1985, which is widely used in nonsmooth optimization and which proved its efficiency in a huge number of real-life problems. Two weak points remain in the above iteration step (3.9). First, the success of this step depends in a delicate way on the parameter t_k in step (1) of (3.9) (actually some "dual" parameter ε_k is used in M1FC1); a bad guess for t_k (or ε_k, respectively) may lead to a "bad" search direction d_k and step (2) breaks down with line search difficulties. Second, for $f \in C^1$ and $t_k \to 0$, (3.9) is reduced to one step of the *Steepest Descent* method, which is only linearly convergent. Numerical experiments confirm this first-order behaviour of M1FC1 and related bundle implementations. We will discuss how one can bypass the first shortcoming in practice; further, it will become obvious how to deal in principle with the second problem and reach faster convergence.

3.1.3 Bundle trust region concept

We start with a simple observation: with d_k from step (1) of iteration (3.9) and $\rho_k := \frac{1}{2} d_k^T d_k$ the minimization in (3.9)(1) becomes "equivalent" to

$$\text{compute } d(\rho_k) := \arg\min\{f_{\text{CP}}(x_k; d) \mid \frac{1}{2} d^T d \leq \rho_k\}. \tag{3.10}$$

This follows by a comparison of the Karush-Kuhn-Tucker (KKT) conditions for the two problems. A closer inspection shows that there is even a monotone correspondence between t_k and ρ_k. Replacing (3.9)(1) by (3.10) gives an idea how to bypass the above discussed first difficulty. Instead of working with some a priori and more or less randomly chosen ρ_k (respectively t_k) one should follow the *trust region* philosophy: we decrease and/or increase ρ_k in a systematic way (*trust region part*) and improve f_{CP} by Null Steps (*bundle part*), until we reach some f_{CP} together with a ρ_k-ball, on which we can *trust* this model, i.e., the optimal d_k from (3.10) leads to a substantial decrease in f. The advantage of this procedure is twofold: first, it suggests a way how to choose ρ_k, and, second, it releases us at the same time from the need for a line search. Obviously we can apply just the same strategy to (3.9) and tune the t_k in (3.9)(1) instead of ρ_k in (3.10). The reason to work with (3.9) is of purely numerical nature. We will see that (3.9)(1) leads to a quadratic program for which there exists a lot of reliable software (e.g. Kiwiel, 1986). This is not so for (3.10) because of the quadratic constraint. In schematic terms we obtain

Iteration step $x_k \to x_{k+1}$: (3.11)

(1) *Compute* $d_k := d(t_k) := \arg\min\{f_{\mathrm{CP}}(x_k; d) + \frac{1}{2t_k} d^T d \mid d \in \mathbb{R}^n\}$.

(2) *If* $f(x_k + d_k)$ *is "sufficiently smaller" than* $f(x_k)$, *then either*

(a) *enlarge* t_k *and go back to* (1), *or*

(b) *make a* **Serious Step**: *Put* $x_{k+1} := x_k + d_k$, *compute* $g_{k+1} \in \partial f(x_{k+1})$.

If $f(x_k + d_k)$ *is "not sufficiently smaller" than* $f(x_k)$, *then either*

(c) *reduce* t_k *and go back to* (1), *or*

(d) *make a* **Null Step**: *Put* $x_{k+1} := x_k$ *and compute some* $g_{k+1} \in \partial f(x_k + d_k)$.

How to "realize" the alternatives (a)–(b) and (c)–(d) will be seen in the precise statement of the algorithm.

The above variant was implemented in Schramm and Zowe, 1992 under the name BT (Bundle Trust region) algorithm; see, e.g., Schramm, 1989; Schramm and Zowe, 1992; Schramm and Zowe, 1991. Extensive testing (in particular on some real life problems, which are known as tough nuts) proved the code to be robust and efficient. This experience was confirmed when working with BT in Chapters 9–12.

Let us briefly return to the second drawback of existing bundle implementations, namely the linear (hence slow) convergence. There is hope that the trust region approach can also help with this difficulty by tuning the bilinear form $d^T d$ (=$d^T I d$) in step (1) of (3.11) to account for the compiled knowledge about the level sets of f. A whole series of recent papers encircles this challenging item and tries to gain control on such curvature (hence 2nd order information) in the bundle trust region framework or in the related proximal point concept; see, e.g., Lemaréchal and Sagastizábal, 1997; Lukšan and Vlček, 1996. However, no essential numerical break-through has been reported till now. We close by mentioning that our BT concept and its implementation has benefited a lot from a long cooperation with Lemaréchal and from the many innovative contributions by Kiwiel in this area.

3.2 BT-ALGORITHM: THE CONVEX CASE

As already said, the motivation of the BT model and the key arguments in the convergence proof are all based on convexity. Hence we will treat in this section in detail the convex case: iteration (3.11) together with the overall algorithm will be specified and and we will present the convergence analysis for convex f. Section 3.3 sketches the necessary technical modifications for nonconvex f; for more details we refer to Schramm, 1989.

In this section the subscript k always refers to the sequence of iterates $x_1, x_2, \ldots$, whereas the superscript j will be used in the inner iteration, which leads from x_k to x_{k+1}. If J is a set of indices, then $|J|$ denotes its cardinality. Further we put

$$\Lambda(n) := \{\lambda \in \mathbb{R}^n \mid \lambda^i \geq 0, 1 \leq i \leq n, \text{ and } \sum_{i=1}^{n} \lambda^i = 1\}.$$

At the iterate x_k we have at our disposal the sequence $x_1, x_2, \ldots, x_k$ and a collection of auxiliary points y_i together with subgradients $g_i \in \partial f(y_i)$ for $i \in J_k$; here J_k is some nonempty set of indices. On first reading the reader may think of J_k as a subset of $\{1, \ldots, k\}$ and assume $y_i = x_i$. This bundle of information leads to the *cutting plane model* of f at x_k:

$$\max_{i \in J_k} \{g_i^T(x - y_i) + f(y_i)\} \quad \text{for } x \in \mathbb{R}^n.$$

With the *linearization errors* (the gap at x_k between f and its linear approximation $f(y_i) + g_i^T(\cdot - y_i)$)

$$\alpha_{k,i} := \alpha(x_k, y_i) := f(x_k) - (f(y_i) + g_i^T(x_k - y_i)) \tag{3.12}$$

and the new variable $d := x - x_k$ we can write this model in a condensed form

$$\max_{i \in J_k} \{g_i^T d - \alpha_{k,i}\} + f(x_k) \text{ for } d \in \mathbb{R}^n.$$

For convenience let us skip the constant $f(x_k)$ and put

$$f_{\mathrm{CP}}(x_k; d) := \max_{i \in J_k} \{g_i^T d - \alpha_{k,i}\} \text{ for } d \in \mathbb{R}^n. \tag{3.13}$$

Step (1) from iteration (3.11) becomes for *suitable t* (which still has to be chosen appropriately!)

$$\text{compute } d := d(t) = \arg\min\{f_{\mathrm{CP}}(x_k; d) + \frac{1}{2t}\|d\|^2 \mid d \in \mathbb{R}^n\}. \tag{3.14}$$

This can equivalently be written as a quadratic program in $\mathbb{R}^1 \times \mathbb{R}^n$:

$$\begin{aligned} \text{compute } (v, d) &:= (v(t), d(t)) \\ &= \arg\min\{v + \tfrac{1}{2t}\|d\|^2 \mid v \geq g_i^T d - \alpha_{k,i} \text{ for } i \in J_k\}. \end{aligned} \tag{3.15}$$

(3.14) is a strictly convex problem with a unique minimizer $d(t)$; the same holds for the minimizing $(v(t), d(t))$ in (3.15). From the KKT conditions for (3.15) one easily obtains a representation of $d(t)$ and $v(t)$:

Lemma 3.1 *For the solution* $(v(t), d(t))$ *of* (3.15) *there exists* $\lambda(t) \in \Lambda(|J_k|)$ *such that*

$$\lambda^i(t)(-v(t) + g_i^T d(t) - \alpha_{k,i}) = 0 \text{ for } i \in J_k, \tag{3.16}$$

$$d(t) = -t \sum_{i \in J_k} \lambda^i(t) g_i, \tag{3.17}$$

$$v(t) = -t\|\sum_{i \in J_k} \lambda^i(t) g_i\|^2 - \sum_{i \in J_k} \lambda^i(t)\alpha_{k,i} \tag{3.18}$$

$$= -\frac{1}{t}\|d(t)\|^2 - \sum_{i \in J_k} \lambda^i(t)\alpha_{k,i}. \tag{3.19}$$

Proof. We skip the fixed $t > 0$ as argument in λ, d and v. Let (v, d) be a solution of the quadratic problem (3.15). The KKT conditions imply the existence of reals $\lambda^i \geq 0$ for $i \in J_k$ such that

$$\nabla_{v,d}(v + \frac{1}{2t}\|d\|^2 + \sum_{i\in J_k} \lambda^i \nabla_{v,d}(-v + g_i^T d - \alpha_{k,i})) = 0_{\mathbb{R}^{1+n}} \tag{3.20}$$

$$\lambda^i(-v + g_i^T d - \alpha_{k,i}) = 0 \quad \text{for } i \in J_k. \tag{3.21}$$

Taking the partial derivatives with respect to v and d in (3.20) we get

$$\begin{aligned} 1 - \sum_{i\in J_k} \lambda^i &= 0_{\mathbb{R}}, \\ \frac{1}{t}d + \sum_{i\in J_k} \lambda^i g_i &= 0_{\mathbb{R}^n}. \end{aligned}$$

This shows

$$\lambda := (\lambda^i)_{i\in J_k} \in \Lambda(|J_k|) \text{ and } d = -t \sum_{i\in J_k} \lambda^i g_k,$$

i.e., (3.17) holds. By adding (3.21) (= (3.16)) over $i \in J_k$ and using (3.17) one realizes that also (3.19) holds. ∎

Since (3.15) is a convex problem with linear constraints, the KKT conditions (and the equivalent relations (3.16)–(3.19)) are also sufficient for optimality of a feasible $(v(t), d(t))$.

Thanks to convexity, all $\alpha_{k,i}$ are nonnegative (consequence from (3.5)),

$$\alpha_{k,i} \geq 0 \quad \text{for} \quad i \in J_k. \tag{3.22}$$

Now add $\alpha_{k,i} - [f(x_k) - (f(y_i) + g_i^T(x_k - y_i))] = 0$ to the subgradient inequality

$$g_i^T(x - y_i) \leq f(x) - f(y_i)$$

to obtain after simple reordering

$$g_i^T(x - x_k) \leq f(x) - f(x_k) + \alpha_{k,i} \quad \text{for all } x \in \mathbb{R}^n \text{ and } i \in J_k. \tag{3.23}$$

Thus the linearization error $\alpha_{k,i}$ "measures" how good $g_i \in \partial f(y_i)$ satisfies the subgradient inequality at x_k and the $\alpha_{k,i}$'s take care that the influence of g_i in (3.15) and (3.29) below will be the less the greater the "penalty" $\alpha_{k,i}$ is.

Now fix some $\lambda \in \Lambda(|J_k|)$, multiply (3.23) by the components λ^i of λ and sum up over i. We obtain the useful formula, which holds with arbitrary $\lambda \in \Lambda(|J_k|)$:

$$(\sum_{i\in J_k} \lambda^i g_i)^T(x - x_k) \leq f(x) - f(x_k) + \sum_{i\in J_k} \lambda^i \alpha_{k,i} \quad \text{for all } x \in \mathbb{R}^n. \tag{3.24}$$

Further note as a direct consequence of (3.19) and (3.22):

$$v(t) \leq 0 \text{ for the optimal } v(t) \text{ from (3.15).} \tag{3.25}$$

As expected, $v(t) = 0$ characterizes optimality of x_k. This follows immediately if we put in the next result $\varepsilon = 0$ and use (3.19). The lemma itself is an immediate consequence of inequality (3.24).

Lemma 3.2 *Suppose there exists* $\lambda \in \Lambda(|J_k|)$ *with*

$$\|\sum_{i\in J_k} \lambda^i g_i\| \leq \varepsilon \text{ and } \sum_{i\in J_k} \lambda^i \alpha_{k,i} \leq \varepsilon. \tag{3.26}$$

Then x_k *is* ε*-optimal in the sense that*

$$f(x_k) \leq f(x) + \varepsilon\|x - x_k\| + \varepsilon \text{ for all } x \in \mathbb{R}^n. \tag{3.27}$$

■

For later use we add a continuity result on $(v(t), d(t))$ which follows from standard perturbation theory:

$$\text{The solution } (v(t), d(t)) \text{ of (3.15) depends continuously on } t \in (0, \infty). \tag{3.28}$$

Remark. Due to the simple structure of (3.15), the last statement can be strengthened substantially. We add this interesting result without proof since we will not make use of it; a detailed treatment can be found in Schramm, 1989.

(i) There exists a finite sequence $0 = t^0 < t^1 < \ldots < t^m = \infty$ and $a^i, b^i \in \mathbb{R}^n$, such that $d(t) = a^i + tb^i$ for $t \in (t^{i-1}, t^i]$ and i=1,2,. . .,m.

(ii) It holds $a^1 = 0$ and b^1 is the projection of the origin onto $\text{conv}\{g_i \mid i \in J_k, \alpha_{k,i} = 0\}$.

(iii) There exists a CP-solution d_{CP} (i.e. d_{CP} minimizes $f_{\text{CP}}(x_k; .)$) if and only if $a^m = d_{CP}$ and $b^m = 0$.

Remark. For an efficient solution of (3.15) two devices become important:
(a) The index set J_k (i.e. the number of subgradients carried along) should be kept at reasonable size as $k \to \infty$. Hence from time to time we clean up the bundle. The convergence analysis requires $|J_k| \geq 3$ together with a certain *Reset Strategy*.
(b) (3.15) is a quadratic program in $1 + n$ variables and $|J_k|$ linear constraints. Since, typically, $|J_k|$ will be much smaller than the dimension n of the $d-$space, we replace (3.15) by its *dual* in $|J_k|$ variables and $|J_k| + 1$ constraints:

$$\min\{\frac{1}{2}\|\sum_{i\in J_k} \lambda^i g_i\|^2 + \frac{1}{t}\sum_{i\in J_k} \lambda^i \alpha_{k,i} \mid \lambda \in \Lambda(|J_k|)\}. \tag{3.29}$$

Some standard duality arguments show that the solution λ of (3.29) and the $\lambda(t)$ from Lemma 3.1 are the same.

The next step discusses how to find an appropriate t for (3.15). Then we summarize the overall algorithm and Subsection 3.2.3 presents the convergence analysis.

3.2.1 Inner iteration $x_k \to x_{k+1}$

We fix an upper bound T for t, parameters $0 < m_1 < m_2 < 1$, $0 < m_3 < 1$, some small $\nu > 0$ and a stopping parameter $\varepsilon \geq 0$. Suppose we are at the iterate x_k and let J_k, y_i, $g_i \in \partial f(y_i)$ and $\alpha_{k,i}$ be as discussed above. Then we specialize (3.11) as follows. Here the superscript j is the running index, the subscript k is kept fixed. The stopping rule in step (1) is based on (3.17), (3.19) and Lemma 3.2. Finally, the decisive criteria **SS** and **NS** will be specified below.

Inner Iteration $x_k \to x_{k+1}$**:** (3.30)

(0) *Choose* $t^1 := t_{k-1}$. *Set* $l^1 := 0$, $u^1 := T$ and $j := 1$.

(1) *Compute the solution* $(v^j, d^j) = (v(t^j), d(t^j))$ *of* (3.15). *If* $\frac{1}{t^j}\|d^j\| \leq \varepsilon$ *and* $-\frac{1}{t^j}\|d^j\|^2 - v^j \leq \varepsilon$, *then stop:* x_k *is* ε*-optimal in the sense of Lemma 3.2. Otherwise put* $y^j := x_k + d^j$ *and compute* $g^j \in \partial f(y^j)$.

(2) (a) *If* **SS**(i) *and* **SS**(ii) *hold, then make a* **Serious Step:** *Put* $x_{k+1} := y_{k+1} := y^j$, $g_{k+1} := g^j$ *and stop.*

(b) *If* **SS**(i) *holds but not* **SS**(ii), *then put* $l^{j+1} := t^j$, $u^{j+1} := u^j$, $t^{j+1} := \frac{1}{2}(u^{j+1} + l^{j+1})$, $j := j + 1$ *and go back to* (1).

(c) *If* **NS**(i) *and* **NS**(ii) *hold, then make a* **Null Step:** *Put* $x_{k+1} := x_k$, $y_{k+1} := y^j$, $g_{k+1} := g^j$ *and stop.*

(d) *If* **NS**(i) *holds but not* **NS**(ii), *then put* $u^{j+1} := t^j$, $l^{j+1} := l^j$, $t^{j+1} := \frac{1}{2}(u^{j+1} + l^{j+1})$, $j := j + 1$ *and go back to* (1).

Let v_k, d_k and t_k be the values, with which we leave (3.30) in case of a Serious Step or a Null Step. Then $d_k = -t_k \sum_{i \in J_k} \lambda_k^i g_i$ for suitable $\lambda_k = (\lambda_k^i) \in \Lambda(|J_k|)$ (see (3.17)). With this λ_k we define for later use

$$z_k := \sum_{i \in J_k} \lambda_k^i g_i \quad \text{and} \quad \sigma_k := \sum_{i \in J_k} \lambda_k^i \alpha_{k,i}. \tag{3.31}$$

We now present the criteria, which determine whether a Serious Step or a Null Step is taken (for $k = 1$ put in **NS**(ii) $z_0 := g_1$, $\sigma_0 := 0$):

SS(i) $f(y^j) - f(x_k) < m_1 v^j$,

SS(ii) $(g^j)^T d^j \geq m_2 v^j$ or $t^j \geq T - \nu$

NS(i) $f(y^j) - f(x_k) \geq m_1 v^j$,

NS(ii) $\alpha(x_k, y^j) \leq m_3 \sigma_{k-1}$ or $|f(x_k) - f(y^j)| \leq \|z_{k-1}\| + \sigma_{k-1}$.

Discussion of SS and NS. *Ad* (2)(a) *and* (b): Condition **SS**(i) ensures for a Serious Step a decrease of at least m_1-times $v_k[= f'_{\text{CP}}(x_k; d_k)$ = decrease in the CP-model]. The first part of **SS**(ii) takes care of a substantial change in the CP-model; this follows from (we use $x_{k+1} = y_{k+1}$, $v_k < 0$ and $m_2 < 1$)

$$g_{k+1}^T d_k - \alpha_{k+1,k+1} = g_{k+1}^T d_k \geq m_2 v_k > v_k \geq g_i^T d_k - \alpha_{k,i} \quad \text{for } i \in J_k, \tag{3.32}$$

which implies that, after a Serious Step, the updated model (3.15) will provide some (v, d) in step $k+1 \to k+2$, which differs from the present (v_k, d_k). If the first part of **SS**(ii) does not hold (and thus such a change in the model cannot be guaranteed) and if t is still smaller than some upper bound T (this is taken care of by the second condition under **SS**(ii)), then we prefer to try some larger t, even if **SS**(i) holds. This motivates steps (2)(a) and (b).

Ad (2)(c) *and* (d): Now suppose **NS**(i) (the negation of **SS**(i)) holds. Then either f_{CP} is not yet an adequate model and/or we were too optimistic with respect to t. The obvious way out: try some smaller t in (1); this is step (2)(d). If, however, also the first condition under **NS**(ii) holds, then a Null Step makes sense as well and we prefer this option. The reason: after such a Null Step we get from (3.23) for $k+1$ and $i = k+1$ (use $x_{k+1} = x_k$)

$$g_{k+1}^T(x - x_k) \leq f(x) - f(x_k) + \alpha_{k+1,k+1},$$

where $\alpha_{k+1,k+1} = \alpha(x_k, y_{k+1}) \leq m_3\sigma_{k-1}$ and $m_3 < 1$. We conclude that g_{k+1} is "close" to $\partial f(x_k)$ and thus it makes sense to add g_{k+1} to the bundle at x_k. Condition **NS**(i) guarantees that this g_{k+1} contributes non-redundant information. This follows from the next inequality, which serves the same purpose as (3.32) in case of a Serious Step,

$$\begin{aligned} g_{k+1}^T d_k - \alpha_{k+1,k+1} &= f(y_{k+1}) - f(x_k) \geq m_1 v_k \\ &> v_k \geq g_i^T d_k - \alpha_{k,i} \text{ for } i \in J_k; \end{aligned} \tag{3.33}$$

consequently, in iteration $k+1 \to k+2$ the enriched model f_{CP} will yield some direction d which differs from the unsuccessful present d_k. This, taken together, explains one half of (2)(c); for technical reasons (which will become clear in Proposition 3.7 below) we also make a Null Step, if **NS**(i) holds together with the second condition under **NS**(ii).

We summarize (3.30) in a flow chart (see Figure 3.1).

In our implementation of (3.30) we replace the simple bisection rule for t by a more sophisticated heuristic strategy and work with a safeguarded variation of t, which takes into account the change of the function value. In step (0) we choose the initial $t^1 = t_{k-1}$ only in case of a Null Step; in case of a Serious Step we choose $t^1 \geq t_{k-1}$.

The t-variation in (3.30) replaces the line search in standard bundle implementations as in M1FC1. The crucial difference: in M1FC1, for instance, one makes an a priori decision on t_k (respectively on some dual quantity ε_k) which results in a *fixed direction* d_k. Then, in the line search, one "minimizes" $f(x_k + \cdot d_k)$. Contrary to this, t in (3.30) is variable and we thus try *different directions* $d(t)$ when "minimizing" $f(x_k + d(\cdot))$. Experience shows that this can be a decisive advantage.

The next result supplies the actual justification for what we are doing.

Theorem 3.3 *The inner iteration* (3.30) *ends after finitely many cycles either with a Serious Step or a Null Step or the information that x_k is ε-optimal.*

Proof. Three cases can happen:

(i) $l^j = 0$ for all j; (ii) $u^j = T$ for all j; (iii) neither (i) nor (ii) holds.

Ad (i): We are all the time on the right branch in Figure 3.1 and thus $t^{j+1} = \frac{1}{2}(0+t^j) \downarrow 0$ and $y^j \to x_k$ as $j \to \infty$. Hence **NS**(ii) will hold for large enough j; since **NS**(i) is satisfied by construction on the right branch, (3.30) stops with a Null Step.

Ad (ii): Now we are all the time on the left branch and thus $t^{j+1} = \frac{1}{2}(t^j + T) \uparrow T$ for $j \to \infty$, i.e. **SS**(ii) holds for large enough j. Since **SS**(i) is automatically satisfied on the left branch, we will stop with a Serious Step.

Ad (iii): In this case $0 < l^j < u^j < T$ for all sufficiently large j and a monotonicity argument implies $l^j \uparrow t^*$ and $u^j \downarrow t^*$ for some $t^* \in (0, T)$. A continuity argument (recall (3.28)) together with **SS**(i) and **NS**(i) yields for $d^* := d(t^*)$ and $v^* := v(t^*)$

$$f(x_k + d^*) - f(x_k) = m_1 v^*. \tag{3.34}$$

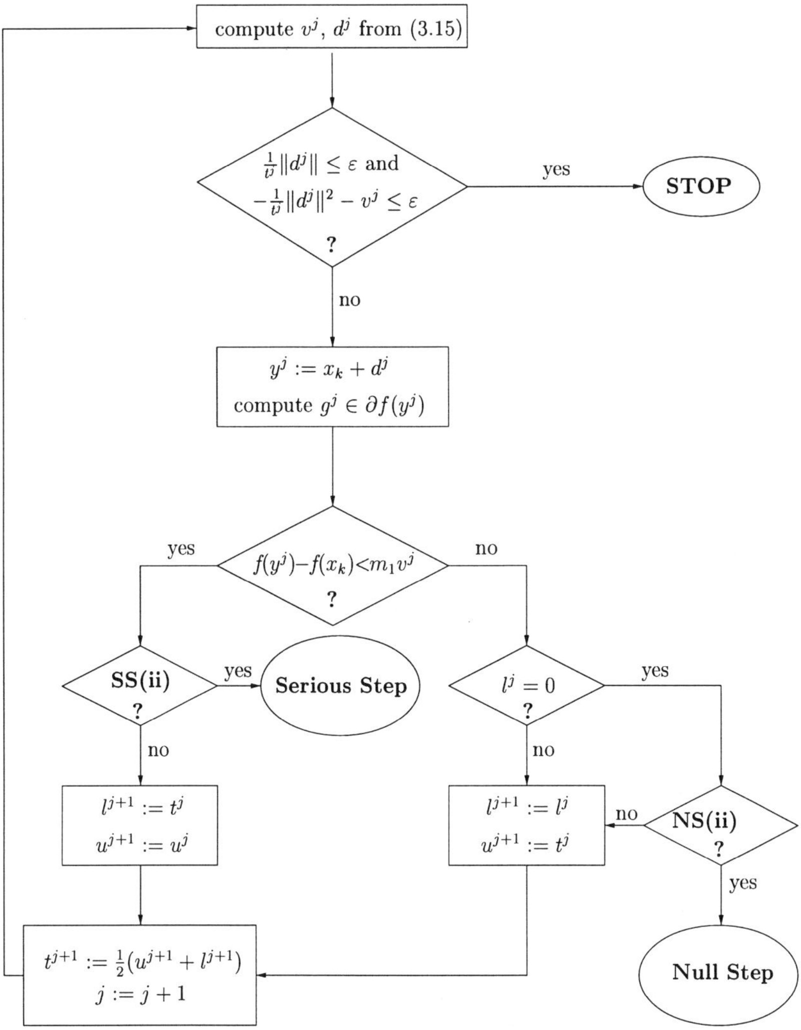

Figure 3.1. Flow chart for inner iteration (convex f)

Let $j(1)$, $j(2)$, ... be the subsequence of indices, for which **SS**(i) holds; this is an infinite sequence since otherwise $l^{j(m)} = t^*$ for some m and then (3.34) would contradict **SS**(i). Since $l^j \uparrow t^*$, the $g^{j(i)}$ have a cluster point g^* which belongs to $\partial f(x_k + d^*)$ (Theorem 2.9(iii)). Hence

$$(g^*)^T(x_k - (x_k + d^*)) \le f(x_k) - f(x_k + d^*)$$

and, because of (3.34),

$$(g^*)^T d^* \geq m_1 v^*.$$

Now $v^* < 0$ (otherwise (3.30) would have stopped because of ε-optimality) together with $0 < m_1 < m_2$ shows $g^* d^* > m_2 v^*$. A continuity argument implies for sufficiently large i

$$(g^{j(i)})^T d^{j(i)} \geq m_2 v^{j(i)},$$

and we will stop with a Serious Step. ■

As a byproduct of the proof of Theorem 3.3 we note:

$$\text{If } f(y^j) - f(x_k) < m_1 v^j \text{ for some } j, \quad \text{then one leaves the inner iteration (3.30) with a Serious Step.} \tag{3.35}$$

Hence it suffices to check **NS**(ii), as already depicted in Figure 3.1, only as long as $l^j = 0$.

3.2.2 Overall algorithm

We shortly summarize the overall algorithm for convex f with *Reset Strategy*.

BT-Algorithm: *Choose a starting point* $x_1 \in \mathbb{R}^n$ *and parameters* $T > 0$, $0 < m_1 < m_2 < 1$, $0 < m_3 < 1$, $\nu > 0$, $\varepsilon \geq 0$ *and an upper bound* $j_{max} \geq 3$ *for* $|J_k|$. (3.36)

(0) *Compute* $f(x_1)$, $g_1 \in \partial f(x_1)$ *and put* $y_1 := x_1$, $J_1 := \{1\}$ *and* $k := 1$.

(1) *INNER ITERATION: Compute* x_{k+1} *and* g_{k+1} *as in* (3.30) *or realize that* x_k *is* ε*-optimal (in which case we stop).*

(2) *If* $|J_k| = j_{max}$ *then go to* (3)*; otherwise put* $J := J_k$ *and go to* (4).

(3) *RESET: Choose* $J \subset J_k$ *with* $|J| \leq j_{max} - 2$ *and* $\max\{i \mid i \in J_k, \alpha_{k,i} = 0\} \in J$. *Introduce some additional index* $\tilde{k}$ *and define with* z_k, σ_k *from* (3.31)

$$g_{\tilde{k}} := z_k, \alpha_{k,\tilde{k}} := \sigma_k, J := J \cup \{\tilde{k}\}.$$

(4) *UPDATE: If the outcome of* (3.30) *was a Serious Step, then put*

$$\alpha_{k+1,i} := \alpha_{k,i} + f(x_{k+1}) - f(x_k) - g_i^T d_k \text{ for } i \in J, \ \alpha_{k+1,k+1} := 0.$$

If the outcome of (3.30) *was a Null Step, then put*

$$\alpha_{k+1,i} := \alpha_{k,i} \text{ for } i \in J, \ \alpha_{k+1,k+1} := \alpha(x_k, y_{k+1}).$$

Put $J_{k+1} := J \cup \{k+1\}$ *and go to* (1).

Remark. We add a comment on the index $\tilde{k}$ in step (3) and the update formula in step (4). (a) The $g_{\tilde{k}}$ defined in the reset step corresponds to the *aggregate subgradient technique* introduced in Kiwiel, 1985. Usually there will be no $y_{\tilde{k}}$ with $g_{\tilde{k}} \in \partial f(y_{\tilde{k}})$ and thus

$\alpha_{k,\bar{k}}$ has no interpretation as linearization error. It follows, however, from (3.24) that the "synthetic" $\alpha_{k,\bar{k}}$ again satisfies

$$g_{\bar{k}}^T(x - x_k) \leq f(x) - f(x_k) + \alpha_{k,\bar{k}} \text{ for all } x,$$

which is actually what is needed from subgradients in our context.
(b) One easily checks that for the indices i which correspond to points y_i, the update formula in (4) above is in accordance with (3.13). The update strategy dispenses the need to carry along the x_i's and y_i's.

3.2.3 Convergence analysis

The proof technique below is largely based on ideas which go back to Kiwiel, 1985. Throughout we work with the stopping parameter $\varepsilon = 0$. Let x_k, $k = 1, 2, \ldots$, be the iterates generated by (3.36) and recall the abbreviations introduced in (3.31)

$$z_k = \sum_{i \in J_k} \lambda_k^i g_i \quad \text{and} \quad \sigma_k = \sum_{i \in J_k} \lambda_k^i \alpha_{k,i}.$$

In terms of z_k and σ_k the crucial relations (3.17) and (3.19) become

$$d_k = -t_k z_k \quad \text{and} \quad v_k = -t_k \|z_k\|^2 - \sigma_k. \tag{3.37}$$

Further let us denote the minimal value in (3.29) by w_k, i.e.

$$w_k = \frac{1}{2}\|z_k\|^2 + \frac{1}{t_k}\sigma_k, \tag{3.38}$$

and put

$$X^* := \{x^* \in \mathbb{R}^n \mid f(x^*) \leq f(x) \text{ for all } x \in \mathbb{R}^n\}.$$

Finally we mention a technical assumption, needed for our auxiliary results, and saying that the sequence $\{f(x_k)\}$ is bounded below:

$$\textit{There exists } \bar{x} \textit{ such that } f(\bar{x}) \leq f(x_k) \textit{ for all } k. \tag{3.39}$$

Note that (3.39) holds whenever $X^* \neq \emptyset$. As a foretaste of what we will prove, we summarize the final convergence result:

$f(x_k)$ converges to $\inf_x f(x)$ $(\geq -\infty)$ and,
if $X^* \neq \emptyset$, then x_k converges to some $x^* \in X^*$.

We start with an observation on the sequence $\{x_k\}$:

Lemma 3.4 *If (3.39) holds then for each $\delta > 0$ there exists $n_0(\delta) \in \mathbb{N}$ such that*

$$\|\bar{x} - x_{k+1}\|^2 \leq \|\bar{x} - x_m\|^2 + \delta \quad \text{for} \quad k \geq m \geq n_0(\delta). \tag{3.40}$$

Proof. The "subgradient inequality" (3.24) becomes in terms of z_k and σ_k

$$z_k^T(\bar{x} - x_k) \leq f(\bar{x}) - f(x_k) + \sigma_k$$

and, since by assumption $f(\bar{x}) \leq f(x_k)$, we conclude

$$z_k^T(\bar{x} - x_k) \leq \sigma_k.$$

If we put

$$\delta_k := \begin{cases} 1, & \text{if } k \to k+1 \text{ is a Serious Step,} \\ 0, & \text{if } k \to k+1 \text{ is a Null Step,} \end{cases}$$

then $x_{k+1} - x_k = \delta_k d_k = -\delta_k t_k z_k$ for all k and thus

$$-(\bar{x} - x_k)^T(x_{k+1} - x_k) = \delta_k t_k (\bar{x} - x_k)^T z_k \leq \delta_k t_k \sigma_k.$$

It follows

$$\begin{aligned} \|\bar{x} - x_{k+1}\|^2 &= \|\bar{x} - x_k\|^2 + \|x_k - x_{k+1}\|^2 - 2(\bar{x} - x_k)^T(x_{k+1} - x_k) \\ &\leq \|\bar{x} - x_k\|^2 + \|x_k - x_{k+1}\|^2 + 2\delta_k t_k \sigma_k. \end{aligned}$$

Hence for all $m \in \mathbb{N}$ and $k \geq m$

$$\|\bar{x} - x_{k+1}\|^2 \leq \|\bar{x} - x_m\|^2 + \sum_{i=m}^{k}(\|x_i - x_{i+1}\|^2 + 2\delta_i t_i \sigma_i). \tag{3.41}$$

To finish the poof, consider the sum in (3.41). From $f(x_{i+1}) - f(x_i) \leq \delta_i m_1 v_i$ we obtain for arbitrary $l > 1$

$$f(x_1) - f(x_l) = f(x_1) - f(x_2) + f(x_2) - \ldots + f(x_{l-1}) - f(x_l) \geq -m_1 \sum_{i=1}^{l} \delta_i v_i,$$

and thus for $l \to \infty$ (we use (3.19), (3.17) and (3.37))

$$\infty > f(x_1) - f(\bar{x}) \geq -m_1 \sum_{i=1}^{\infty} \delta_i v_i = m_1 \sum_{i=1}^{\infty} \delta_i (t_i \|z_i\|^2 + \sigma_i).$$

Since $\delta_i t_i^2 \|z_i\|^2 = \|x_{i+1} - x_i\|^2$ we can continue

$$\infty > m_1 \sum_{i=1}^{\infty} (\frac{1}{t_i}\|x_{i+1} - x_i\|^2 + \delta_i \sigma_i),$$

and thus (we use that by construction $t_i \leq T$)

$$\infty > m_1 \sum_{i=1}^{\infty} (\|x_{i+1} - x_i\|^2 + \delta_i t_i \sigma_i).$$

Consequently we can make the sum in (3.41) as small as we want by choosing m large enough. This proves the claim. ∎

The next lemma is an almost immediate consequence of (3.40).

Lemma 3.5 *If* (3.39) *holds then the x_k converge to some $\tilde{x}$, for which*

$$f(\tilde{x}) \leq f(x_k) \quad \text{for all } k.$$

Proof. By (3.40) the x_k-sequence is bounded and has a cluster point, say $\tilde{x}$. Since, by construction, $f(x_k)$ is monotonically decreasing we see

$$f(\tilde{x}) \le f(x_k) \quad \text{for all } k.$$

Hence Lemma 3.4 applies once more (now with $\bar{x}$ replaced by $\tilde{x}$) and for given $\varepsilon > 0$ we can choose $n_0(\frac{\varepsilon}{2})$ such that

$$\|\tilde{x} - x_{k+1}\|^2 \le \|\tilde{x} - x_m\|^2 + \frac{\varepsilon}{2} \quad \text{for } k \ge m \ge n_0(\frac{\varepsilon}{2}).$$

Since $\tilde{x}$ is a cluster point of the x_k-sequence, one can choose $\tilde{m} \ge n_0(\frac{\varepsilon}{2})$ such that $\|\tilde{x} - x_{\tilde{m}}\|^2 \le \frac{\varepsilon}{2}$ and we end up with

$$\|\tilde{x} - x_{k+1}\|^2 \le \frac{\varepsilon}{2} + \frac{\varepsilon}{2} = \varepsilon \quad \text{for all } k \ge \tilde{m}.$$

■

In the following we will show that the limit $\tilde{x}$ in Lemma 3.5 is indeed optimal. For this aim we prove for a suitable subsequence:

$$z_{k(i)} \to 0 \text{ and } \sigma_{k(i)} \to 0 \quad \text{for } i \to \infty. \tag{3.42}$$

The optimality of $\tilde{x}$ follows from inequality (3.24) (cf. Theorem 3.10).

We start with the crucial observation that besides $\{x_k\}$ also the auxiliary sequences $\{w_k\}$ etc. are bounded in situation (3.39).

Lemma 3.6 *If* (3.39) *holds then the sequences of* w_k, z_k, σ_k, d_k, y_k, g_k *and* $\alpha_{k,k}$, $(k = 1, 2, \ldots)$, *are bounded.*

Proof. We combine (3.37) and (3.38) to see

$$0 \le w_k = -\frac{1}{t_k}(v_k + \frac{1}{2t_k}\|d_k\|^2),$$

i.e.

$$0 \ge -w_k = \frac{1}{t_k}\left(\max_{i \in J_k}\{g_i^T d_k - \alpha_{k,i}\} + \frac{1}{2t_k}\|d_k\|^2\right).$$

Now choose $i(k) \in J_k$ such that $\alpha_{k,i(k)} = 0$ (such $i(k)$ exists because of our reset strategy) and continue the last inequality

$$0 \ge -w_k \ge \frac{1}{t_k}\min_{d \in \mathbb{R}^n}\{g_{i(k)}^T d + \frac{1}{2t_k}\|d\|^2\}.$$

With the minimizer $d := -t_k g_{i(k)}$ we obtain

$$0 \ge -w_k \ge -\frac{1}{2}\|g_{i(k)}\|^2. \tag{3.43}$$

The choice $\alpha_{k,i(k)} = 0$ guarantees $g_{i(k)} \in \partial f(x_k)$ (consequence from (3.23) and (3.5)). This together with the convergence of the x_k (Lemma 3.5) and the closedness of the map $x \to \partial f(x)$ (see Theorem 2.9(iii)) yields the boundedness of $\{g_{i(k)}\}_{k \in \mathrm{N}}$ and because of (3.43) the boundedness of $\{w_k\}_{k \in \mathrm{N}}$. A look at (3.38) and (3.37) convinces the reader that also the sequences of z_k, σ_k,d_k and $y_k = x_k + d_k$ are bounded (we use $t_k \leq T$). Consequently also the $g_k \in \partial f(y_k)$ are bounded, since $\partial f(\cdot)$ is a closed map. This, together with the convergence of the x_k and the continuity of f, finally proves the boundedness of the $\alpha_{k,k}$. ■

In our two key Propositions 3.8, 3.9 below we will prove $v_{k(i)} \to 0$ (and thus $w_{k(i)} \to 0$) for a suitable subsequence. A glance at (3.37) and (3.38) shows that this implies the crucial relation (3.42), provided $t_k \geq \underline{t} > 0$ with fixed $\underline{t}$. The situation $t_k \to 0$ has to be treated as a special case in the next proposition. Here the role of the second condition in **NS**(ii) becomes clear: it is needed to ensure (3.42) even if $t_k \to 0$.

Proposition 3.7 *Suppose* (3.39) *holds and* 0 *is a cluster point of the sequence* $\{t_k\}$. *Then for a suitable subsequence*

$$\lim_{i\to\infty} z_{k(i)} = 0 \quad \textit{and} \quad \lim_{i\to\infty} \sigma_{k(i)} = 0. \tag{3.44}$$

Proof. (By contradiction). Suppose there is $\delta > 0$ with

$$\|z_k\| + \sigma_k \geq \delta \quad \text{for all } k.$$

Now denote by $d_k(t)$ the solution in step (1) of (3.30) for variable t and variable k. The Lipschitz continuity of the convex f, the convergence of the x_k (Lemma 3.5) and (3.17) together with the boundedness of the g_i (Lemma 3.6) imply the existence of $L > 0$, $C > 0$ and $0 < \tilde{T} \leq T$ such that

$$|f(x_k + d_k(t)) - f(x_k)| \leq L\|d_k(t)\| \leq tLC \quad \text{forall } k \text{ and } t \leq \tilde{T}.$$

By making $\tilde{T}$ smaller, if necessary, we can guarantee that

$$|f(x_k + d_k(t)) - f(x_k)| \leq \delta \leq \|z_{k-1}\| + \sigma_{k-1} \quad \text{for } k \geq 2 \text{ and } t \leq \tilde{T}. \tag{3.45}$$

Hence, whenever we are on the right branch in Figure 3.1, then we will leave (3.30) with a Null Step as soon as for the first time $t^j \leq \tilde{T}$. Since t is increased on the left branch, we conclude from the bisection update rule for t that $t_k \geq \frac{1}{2}\tilde{T}$ for all k. This contradicts the assumption on 0 to be a cluster point of $\{t_k\}$. ■

It is convenient to discuss separately the case of finitely and infinitely many Serious Steps.

Proposition 3.8 *Let* (3.39) *hold and suppose one makes infinitely many Serious Steps in* (3.36). *Then for a suitable subsequence*

$$\lim_{i\to\infty} z_{k(i)} = 0 \quad \textit{and} \quad \lim_{i\to\infty} \sigma_{k(i)} = 0.$$

Proof. We may assume $t_k \geq \underline{t} > 0$ for all k since, otherwise, the assertion follows from Proposition 3.7. Now let $\{x_{k(i)}\}_{i\in\mathrm{N}}$ be a subsequence resulting in Serious Steps, i.e.

$$f(x_{k(i)+1}) - f(x_{k(i)}) < m_1 v_{k(i)}.$$

Hence for $l \geq 1$ (note that $x_{k+1} = x_k$ for Null Steps)

$$f(x_{k(l)+1}) - f(x_{k(1)}) < m_1 \sum_{i=1}^{l} v_{k(i)}.$$

We conclude

$$f(\bar{x}) - f(x_{k(1)}) \leq m_1 \sum_{i=1}^{\infty} v_{k(i)}$$

and thus $0 \geq \sum_{i=1}^{\infty} v_{k(i)} > -\infty$. The assertion follows from (3.37) since by assumption $t_k \geq \underline{t} > 0$ for all k. ■

Proposition 3.9 *Suppose* (3.39) *holds and one makes only finitely many Serious Steps. Then for a suitable subsequence*

$$\lim_{i\to\infty} z_{k(i)} = 0, \ \lim_{i\to\infty} \sigma_{k(i)} = 0.$$

Proof. Because of Proposition 3.7 we can assume again

$$t_k \geq \underline{t} > 0 \quad \text{for all } k. \tag{3.46}$$

Further there exists by assumption some $\bar{k}$ with

$$x_k = x_{\bar{k}} \quad \text{for } k \geq \bar{k}.$$

In step (i) we will discuss the change in the minimal value w_k of (3.29) for $w_k \to w_{k+1}$ and $k \geq \bar{k}$. This will be used in (ii) to show $w_k \to 0$, which proves the assertion because of (3.38).

Ad (i): We fix some $k \geq \bar{k}$ and consider the function of $\nu \in [0, 1]$

$$Q(\nu) := \frac{1}{2}\|(1-\nu)z_k + \nu g_{k+1}\|^2 + (1-\nu)\frac{1}{t_{k+1}}\sigma_k + \nu\frac{1}{t_{k+1}}\alpha_{k+1,k+1}.$$

A glance at (3.29) tells us that

$$w_{k+1} \leq \min\{Q(\nu) \mid 0 \leq \nu \leq 1\} =: \tilde{w}. \tag{3.47}$$

To unburden the notation we put

$$\Delta_k := \frac{1}{t_{k+1}} - \frac{1}{t_k}$$

and skip the subscript k in the rest of part (i) and write $+$ for $k+1$. Simple arithmetic shows

$$\begin{aligned} Q(\nu) &= \frac{1}{2}\nu^2\|z - g_+\|^2 + \nu(z^T g_+ - \|z\|^2) + \frac{1}{2}\|z\|^2 + \frac{1}{t_+}\sigma + \frac{1}{t_+}\nu(\alpha_{+,+} - \sigma) \\ &= \frac{1}{2}\nu^2\|z - g_+\|^2 + \nu(z^T g_+ - \|z\|^2) + w + \Delta\sigma + \frac{1}{t_+}\nu(\alpha_{+,+} - \sigma). \end{aligned} \tag{3.48}$$

Since we only make Null Steps for $k \geq \bar{k}$, one has as consequence from **NS**(i) (or from (3.33))

$$g_+^T(-tz) - \alpha_{+,+} \geq m_1 v > m_2 v = m_2(-t\|z\|^2 - \sigma) \tag{3.49}$$

and thus

$$g_+^T z \leq -\frac{1}{t}\alpha_{+,+} + m_2(\|z\|^2 + \frac{1}{t}\sigma). \tag{3.50}$$

This inequality allows us to continue (3.48) for $\nu \in [0,1]$

$$\begin{aligned}
Q(\nu) &\leq \frac{1}{2}\nu^2\|z - g_+\| + \nu(-\frac{1}{t}\alpha_{+,+} + m_2\|z\|^2 + m_2\frac{1}{t}\sigma - \|z\|^2) + w \\
&\quad +\Delta\sigma + \frac{1}{t_+}\nu(\alpha_{+,+} - \sigma) \\
&= \frac{1}{2}\nu^2\|z - g_+\|^2 - \nu(1 - m_2)(\frac{1}{t}\sigma + \|z\|^2) + w \\
&\quad -\frac{1}{t}\nu(\alpha_{+,+} - \sigma) + \Delta\sigma + \frac{1}{t_+}\nu(\alpha_{+,+} - \sigma) \\
&\leq \frac{1}{2}\nu^2\|z - g_+\|^2 - \nu(1 - m_2)w + w + \nu\Delta(\alpha_{+,+} - \sigma) + \Delta\sigma \\
&=: q(\nu).
\end{aligned} \tag{3.51}$$

With

$$(C_k =)C := \max\{\|z\|, \|g_+\|, \frac{1}{t}\sigma, 1\} \tag{3.52}$$

we can go on:

$$\begin{aligned}
Q(\nu) &\leq q(\nu) \\
&\leq 2\nu^2C^2 - \nu(1 - m_2)w + w + \nu\Delta(\alpha_{+,+} - \sigma) + \Delta\sigma \\
&=: \bar{q}(\nu).
\end{aligned}$$

For the special $\bar{\nu} := (1 - m_2)w/4C^2$ we obtain from (3.47) and the last inequality (note that $\bar{\nu} \in [0,1]$ since $\bar{\nu} \leq (1 - m_2)(\frac{1}{2}C^2 + C)/4C^2 < 1$):

$$w_+ \leq \tilde{w} \leq \bar{q}(\bar{\nu}) = w - (1 - m_2)^2\frac{w^2}{8C^2} + (1 - m_2)\frac{w}{4C^2}\Delta(\alpha_{+,+} - \sigma) + \Delta\sigma. \tag{3.53}$$

Inequality (3.53) compares w_{k+1} with w_k for $k \geq \bar{k}$. This will be used in the next step to show $w_k \to 0$.

Ad (ii): We add again the index $k(\geq \bar{k})$ to w, σ, α, t, Δ and C from (3.52). Since we only make Null Steps for $k \geq \bar{k}$, the t_k are monotonically decreasing from $\bar{k}$ on (see Figure 3.1) and we conclude from (3.46) that

$$\Delta_k \to 0 \quad \text{as } k \to \infty. \tag{3.54}$$

By Proposition 3.7 the terms z_k, g_{k+1}, σ_k in (3.52) are bounded; this together with (3.46) implies the existence of $\bar{C}$ with $\bar{C} \geq C_k$ for all k. Combining all this, (3.53) is simplified to

$$\begin{aligned}
w_{k+1} &\leq w_k - (1 - m_2)^2\frac{w_k^2}{8}\bar{C}^{-2} \\
&\quad +(1 - m_2)\frac{w_k}{4}C_k^{-2}\Delta_k(\alpha_{k+1,k+1} - \sigma_k) + \Delta_k\sigma_k \quad \text{for } k \geq \bar{k}.
\end{aligned} \tag{3.55}$$

We use once more Lemma 3.6 to see that $\{w_k\}_{k\in\mathbb{N}}$ is bounded. Let a be the greatest cluster point and assume

$$w_{k(i)+1} \to a \quad \text{for } i \to \infty.$$

Now let b be any other cluster point of the sequence $w_{k(i)}$, i.e., for a further subsequence we have

$$w_{k(i(j))} \to b \quad \text{for } j \to \infty.$$

From (3.54) and (3.55) we obtain for $j \to \infty$

$$a \leq b - [(1 - m_2)^2 \tfrac{1}{8} \bar{C}^{-2}] b^2 + 0.$$

Since, by choice, $b \leq a$ this can hold only if $a = b = 0$. This proves $w_k \to 0$ and the assertion follows from (3.38) and the boundedness of the t_k. ∎

Our key convergence results for convex f follow now easily.

Theorem 3.10 *If $X^* \neq \emptyset$ then x_k converges to some $x^* \in X^*$ as $k \to \infty$.*

Proof. Obviously (3.39) holds and the x_k converge to some $\tilde{x}$ (Lemma 3.5). From (3.24) we get for each k and with z_k, σ_k from (3.31)

$$z_k^T (x - x_k) \leq f(x) - f(x_k) + \sigma_k \quad \text{for all } x.$$

If we fix x and choose a subsequence as in Proposition 3.7, 3.8, 3.9, then we obtain for $k \to \infty$

$$0 \leq f(x) - f(\tilde{x}).$$

Hence $x^* := \tilde{x} \in X^*$. ∎

The above result can be supplemented as follows:

Theorem 3.11 *If $X^* = \emptyset$ then $f(x_k)$ converges to* $\inf\{f(x) \mid x \in X\} \in [-\infty, \infty)$.

Proof. By construction the $f(x_k)$ are monotonically decreasing. Now suppose the assertion not to be true, i.e. for some $\bar{x}$ one has $f(\bar{x}) \leq f(x_k)$ for all k. Just as above, we conclude that $x_k \to \tilde{x} \in X^*$, which contradicts $X^* = \emptyset$. ∎

3.3 BT-ALGORITHM: THE NONCONVEX CASE

We shortly discuss the modifications necessary for nonconvex f. All we assume on f in Section 3.3 is to be locally Lipschitz and weakly semismooth on $\mathbb{R}^n$.

3.3.1 Model and algorithm

For nonconvex f the subgradient inequality (3.5) does not hold, i.e., the linearization $f(y_i) + g_i^T(\cdot - x_i)$ can lie above $f(\cdot)$ at x_k and the linearization errors $\alpha_{k,i}$ become negative. As a consequence the cutting plane model $f_{\text{CP}}(x_k;\cdot)$ is no longer an approximation of $f(x_k + \cdot) - f(x_k)$ from below; in particular, we may have $f_{\text{CP}}(x_k; 0) = \max\{-\alpha_{k,i}\} > f(x_k + 0) - f(x_k)$. To cope with this difficulty we follow a standard strategy in the bundle context and replace $\alpha_{k,i}$ by

$$\beta_{k,i} := \beta(x_k, y_i) := \max\{\alpha_{k,i}, c_0 \|x_k - y_i\|^2\}$$

with fixed small positive real c_0. By construction, $\beta_{k,i} \geq 0$ and the modified model

$$f_{\text{CP}}(x_k; d) := \max_{i \in J_k} \{g_i^T d - \beta_{k,i}\} \quad \text{for } d \in \mathbb{R}^n \tag{3.56}$$

coincides with $f(x_k+d)-f(x_k)$ at least at $d=0$. The $\beta_{k,i}$ copy the penalty role of the $\alpha_{k,i}$ in Section 3.2: Whenever $\alpha_{k,i}<0$ and y_i is "far away" from the current iterate x_k then $\beta_{k,i}$ is large and thus g_i plays but a minor role in (3.56). We have to admit, however, that the above f_{CP} is a much less satisfactory model than the f_{CP} in (3.13) for convex f.

Now replace in Section 3.2 the $\alpha_{k,i}$ by the new weights $\beta_{k,i}$. This does not change the character of (3.15) and (3.29), and thus Lemma 3.1 and the duality between (3.15) and (3.29) remain true. For inequality (3.24), however, the subgradient inequality and thus convexity was essential; hence (3.26) and the corresponding criterion in (3.30)(1) does no longer imply the ε-optimality (3.27) for x_k. All we get for nonconvex f from

$$\|\sum_{i\in J_k}\lambda^i g_i\|\le\varepsilon \quad \text{and} \quad \sum_{i\in J_k}\lambda^i\beta_{k,i}\le\varepsilon$$

is, that 0 "lies up to ε" in the convex hull of certain g_i's from $\partial f(y_i)$ for which the "y_i are not far away from x_k" (since $\sum_{i\in J_k}\lambda^i\beta_{k,i}\le\varepsilon$). In view of Proposition 2.10, this corresponds to "almost" stationarity in smooth optimization.

We come to the inner iteration, which is the heart of our algorithm, and discuss the modifications required by nonconvexity. The decision for making a Serious Step or a Null Step, respectively, is now based on the criteria

SS(i) $f(y^j)-f(x_k)<m_1v^j$,

NS(i) $f(y^j)-f(x_k)\ge m_1v^j$,

NS(ii) $(a)\ \alpha(x_k,y^j)\le m_3\sigma_{k-1}$ or (b) $|f(x_k)-f(y^j)|\le\|z_{k-1}\|+\sigma_{k-1}$,

NS(iii) $(g^j)^Td^j-\beta_{k,j}\ge m_2v^j$.

Note that **SS(i)**, **NS(i)** and **NS(ii)** copy the corresponding conditions for convex f from Section 3.2. The difference is that **SS(ii)** has disappeared and **NS(iii)** appears as additional criterion. These two changes take care that the convergence analysis from 3.2 can be adapted to the nonconvex case.

For the following discussion it will be helpful to recall the arguments from the previous section after the introduction of **SS(i)** - **NS(ii)**.

Why SS(ii) is dropped: The first part of **SS(ii)**

$$(g^j)^Td^j\ge m_2v^j \tag{3.57}$$

is actually a present from convexity. For convex f and small enough t inequality (3.57) follows from **SS(i)** and the subgradient inequality (see e.g. part (iii) in the proof of Theorem 3.3). The condition is however not needed in the convergence analysis of Section 3.2 and only has been added for numerical reasons. Since for nonconvex f the relation (3.57) does not follow anymore from **SS(i)**, we simply skip **SS(ii)**; as already said this does not touch the convergence analysis. This explains that and why (3.30)(2)(a) & (b) shrink to one step (3.63)(2)(a) below.

Why NS(iii) is added: The second change is not of such a simple nature and is indeed deeply rooted in the nonconvexity. Suppose we are on the right branch of Figure 3.1, i.e., **NS(i)** holds. For convex f this implies inequality (3.33), which corresponds to **NS(iii)**. This inequality ensures that, after a Null Step, the updated model provides a direction d, which differs sufficiently from the the unsuccessful present one, to let us escape from x_k or detect optimality of x_k. In the convergence analysis inequality (3.33) (or **NS(iii)**, respectively) leads to the key inequality (3.50) in the proof of Proposition 3.9, which allows to compare w_{k+1} to w_k for $k\ge\bar{k}$, thus establishing convergence of the method. For nonconvex f, however, **NS(i)** and **NS(ii)** may hold without **NS(iii)** being true. So we have to introduce **NS(iii)** as an additional precondition for making a Null Step. This explains step (2)(b) in (3.63) below, which for convex f is reduced to (3.30)(2)(c).

It remains to answer what to do in the new situation

$$\textbf{NS(i)} \text{ and } \textbf{NS(ii)} \text{ hold but } \textbf{NS(iii)} \text{ does not,} \tag{3.58}$$

in which case a Null Step is not justified. Here a more sophisticated argument is needed. Recall the role of **NS(ii)(b)** in the convergence analysis for convex f. It was convenient to treat the case $t_k \to 0$ separately and this was done in Proposition 3.7. Our argument in the proof was indirect. Namely, suppose (3.44) does not hold. Then, whenever we were on the right branch of Figure 3.1 and t was smaller than some $\tilde{T}$, the condition **NS(ii)(b)** (and thus **NS(ii)**) was satisfied and a Null Step was made. Hence there was no reason for t_k to go to 0, in contradiction to the assumption on $\{t_k\}$. This proved the convergence of the method for the special case where $t_k \to 0$. Now, for nonconvex f and in situation (3.58), **NS(ii)(b)** may hold (i.e. (3.45) is satisfied) but, different to the convex case, we are not allowed to make a Null Step. The alternative, to make t_k smaller, is also excluded, since this would suspend our contradiction argument. So what to do?

To escape this deadlock—where (3.58) together with **NS(ii)(b)** holds—we leave our trust region philosophy and use as emergency exit a *line search*, which ends up with a situation, ready for a Serious Step or a Null Step, respectively. This line search for nonsmooth nonconvex f is described in detail in Lemaréchal, 1981. We do not go into the technical details, since we consider such a line search as an emergency exit only, which is against the spirit of our trust-region-approach. And, indeed, when our method runs into numerical troubles, then usually this is because we have to switch to a line search which ends in a collapse.

In Lemaréchal, 1981 it is shown that for weakly semismooth f this line search provides in finitely many steps a stepsize $s_k \in (0,1)$ such that at $y_{k+1} := x_k + s_k d_k$ and with $g_{k+1} \in \partial f(y_{k+1})$ either

$$f(y_{k+1}) - f(x_k) < m_1 s_k v_k, \quad g_{k+1}^T d_k \geq m_2 v_k \tag{3.59}$$

holds, or (alternatively)

$$\begin{aligned} f(y_{k+1}) - f(x_k) \geq m_1 s_k v_k, \quad \beta(x_k, y_{k+1}) \leq m_3 \sigma_{k-1} \\ g_{k+1}^T d_k - \beta(x_k, y_{k+1}) \geq m_2 v_k \end{aligned} \tag{3.60}$$

is satisfied. According to the above discussion, (3.59) describes the situation in which we should make a Serious Step, whereas (3.60) fits precisely to a Null Step. Following this philosophy we make a so-called *Short Serious Step* if (3.59) holds,

$$\text{put } x_{k+1} := y_{k+1} \text{ and add } g_{k+1} \text{ to the bundle,} \tag{3.61}$$

or a *Null Step* in case (3.60) is satisfied,

$$\text{put } x_{k+1} := x_k \text{ and add } g_{k+1} \text{ to the bundle.} \tag{3.62}$$

This leads to the following adaption of (3.30) to nonconvex f:

Inner Iteration $x_k \to x_{k+1}$**:** (3.63)

(0) *Choose* $t^1 := t_{k-1}$. *Set* $l^1 := 0$, $u^1 := T$ *and* $j := 1$.

(1) *Compute the solution* $(v^j, d^j) = (v(t^j), d(t^j))$ *of* (3.15) *with* $\alpha_{k,i}$ *replaced by* $\beta_{k,i}$. *If* $\frac{1}{t^j}\|d^j\| \leq \varepsilon$ *and* $-\frac{1}{t^j}\|d^j\|^2 - v^j \leq \varepsilon$, *then stop*: x_k *is "almost stationary". Otherwise put* $y^j := x_k + d^j$ *and compute* $g^j \in \partial f(y^j)$.

(2) (a) *If* **SS**(i) *holds, then make a* **Serious Step**: *Put* $x_{k+1} := y_{k+1} := y^j$, $g_{k+1} := g^j$ *and stop.*

(b) *If* **NS**(i), **NS**(ii) *and* **NS**(iii) *hold, then make a* **Null Step**: *Put* $x_{k+1} := x_k$, $y_{k+1} := y^j$, $g_{k+1} := g^j$ *and stop.*

(c) *If* **NS**(i), **NS**(ii) *hold but not* **NS**(iii), *then*:

(i) *if* **NS**(ii)(b) *holds, then put* $d_k := d^j$, $v_k := v^j$ *and make a line search along* $x_k + sd_k$, $s \geq 0$ *(see above)*,

(ii) *otherwise put* $u^{j+1} := t^j$, $l^{j+1} := l^j$, $t^{j+1} := \frac{1}{2}(u^{j+1} + l^{j+1})$, $j := j + 1$ *and go back to* (1).

(d) *If* **NS**(i) *holds but not* **NS**(ii), *then put* $u^{j+1} := t^j$, $l^{j+1} := l^j$, $t^{j+1} := \frac{1}{2}(u^{j+1} + l^{j+1})$, $j := j + 1$ *and go back to* (1).

We summarize (3.63) in a flow chart (see Figure 3.2).

In the overall algorithm the updating of the $\beta_{k,i}$ together with the reset strategy has to be adapted to the new situation. This is done just as in Kiwiel's aggregate subgradient method, where one avoids again the storing of the previous x_i and y_i; these technicalities are skipped here.

We further mention that linear constraints can be added without major difficulties.

3.3.2 Convergence analysis

Suppose we use the above definition of $\beta_{k,i}$ and do not apply a reset strategy. A look at Figure 3.2 tells us that the inner iteration (3.63) is again a finite process. Lemmas 3.4 and 3.5 rely decisively on the subgradient inequality and do not carry over to nonconvex f. Hence the statement of Lemma 3.5 becomes now an assumption:

$$\{x_k\}_{k \in \mathbb{N}} \textit{ is bounded.} \tag{3.64}$$

Usually one will ensure (3.64) by requiring the starting point x_0 to be chosen such that the level set $\{x \mid f(x) \leq f(x_0)\}$ is bounded.

Now replace the technical assumption (3.39) by the boundedness assumption (3.64). For Lipschitz f the map $x \mapsto \partial f(x)$ is again closed and upper semicontinuous (Theorem 2.9) and the proof of Lemma 3.6 carries over word-by-word to the nonconvex situation. The same holds for Propositions 3.7 and 3.8. Only the proof of Proposition 3.9 requires some straightforward technical modifications, which take into account Short Serious Steps. With these changes one reaches again an inequality which corresponds to (3.50) and which is the key to the convergence proof.

Summarizing, we get for nonconvex f and for $\varepsilon = 0$ the following convergence result (for details see Schramm, 1989):

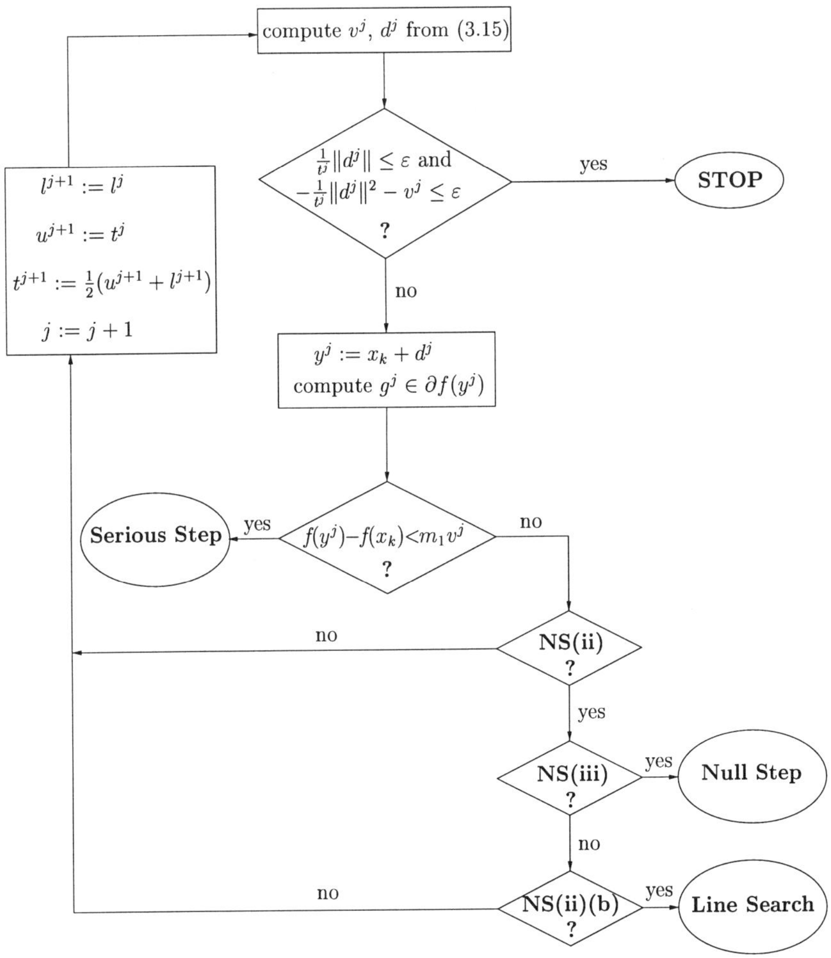

Figure 3.2. Flow chart for inner iteration (nonconvex f)

Theorem 3.12 *Let f be weakly semismooth and locally Lipschitz on $\mathbb{R}^n$. If f is bounded below and* (3.64) *holds, then there exists a cluster point $\bar{x}$ of the sequence $\{x_k\}_{k\in\mathrm{N}}$ such that* $0 \in \partial f(\bar{x})$.

3.4 NONSMOOTH NEWTON'S METHOD

In Chapters 4 and 6 we will see that the considered equilibrium problems can also be written in the form of an equation

$$E(x, z) = 0,$$

where for each $x \in U_{\text{ad}}$ the function $E(x, \cdot)$ $[\mathbb{R}^k \to \mathbb{R}^k]$ is semismooth (cf. Definition 2.13). This allows to compute the equilibria for given controls x by a "nonsmooth" variant of the

Newton's method. Naturally such Newton's methods are based on 1st-order approximations of $E(x, \cdot)$ (directional derivatives, generalized Jacobians, limit sets of Thibault); see e.g. Kummer, 1992; Qi and Sun, 1993; Qi, 1993; Pang and Qi, 1993. We shortly sketch the variant due to Qi, 1993 which is used in our mechanical applications (Chapters 9–11).

To shorten the notation, we put with fixed $x \in U_{\rm ad}$

$$F(z) := E(x, z),$$

and consider the equation

$$F(z) = 0. \tag{3.65}$$

We assume throughout this section that F *is semismooth on* $\mathbb{R}^k$ which implies that F is locally Lipschitz and directionally differentiable on $\mathbb{R}^k$; cf. Proposition 2.21. Let $z \in \mathbb{R}^k$ be fixed and h be a variable vector from $\mathbb{R}^k$. By the mentioned properties of F,

$$F(z + h) - F(z) - F'(z; h) = o(\|h\|), \tag{3.66}$$

i.e., the positively homogeneous map $F'(z; \cdot)$ has the same approximation property as the Fréchet derivative; cf. Shapiro, 1990. We speak of *Bouligand differentiability* and make use of (3.66) in the main convergence statement (Theorem 3.14 below). Besides semismoothness, we will require from F still another property.

Definition 3.1 *We say that F is* strongly BD-regular *at z if all matrices* $V \in \partial_B F(z)$ *are nonsingular.*

Lemma 3.13 *Assume that F is strongly BD-regular at z. Then there exist a neighbourhood* $\mathcal{U}$ *of z and a constant* $c \geq 0$ *such that all matrices* $V \in \partial_B F(v)$ *with* $v \in \mathcal{U}$ *are nonsingular and that*

$$\|V^{-1}\| \leq c. \tag{3.67}$$

Proof. First we show the existence of a neighbourhood $\mathcal{U}$ and a constant c such that the Jacobians $\mathcal{J}F(v)$ are nonsingular for all $v \in \mathcal{U} \setminus \Omega_F$ (the set where F is not differentiable) and

$$\|\mathcal{J}F(v)\|^{-1} \leq c. \tag{3.68}$$

If this claim is not true then there is a sequence $z_k \to z$, $z_k \notin \Omega_F$ such that either all Jacobians $\mathcal{J}F(z_k)$ are singular or $\|(\mathcal{J}F(z_k))^{-1}\| \to \infty$. Since F is Lipschitz near z, the Jacobians $\mathcal{J}F(z_k)$ are bounded in a neighbourhood of z. Hence, there is a subsequence of $\{z_k\}$, say $\{z_{k'}\}$, such that $\{\mathcal{J}F(z_{k'})\}$ converges to a matrix V. This V must be singular. By the definition, $V \in \partial_B F(z)$ and this contradicts the strong BD-regularity of F at z. Hence, inequality (3.68) holds for all $v \in \mathcal{U} \setminus \Omega_F$. At the points $v \in \mathcal{U} \cap \Omega_F$ it suffices to observe that each $V \in \partial_B F(v)$ is the limit of a sequence $\mathcal{J}F(z_i)$ with $z_i \to v$ and $z_i \in \mathcal{U} \setminus \Omega_F$ for all i. ∎

The Newton iteration based on generalized Jacobians is defined by

$$z_{k+1} = z_k - V_k^{-1} F(z_k), \quad k = 0, 1, \ldots \tag{3.69}$$

where $V_k \in \partial F(z_k)$. A local convergence result was given in Kummer, 1992 and Qi and Sun, 1993. In Qi, 1993 a variant is studied, where the V_k in (3.69) is taken from $\partial_B F(z_k)$.

This allows to weaken the assumptions ensuring local superlinear convergence. Recall that a sequence $\{x_k\}$ is said to converge *superlinearly* to x^*, if

$$\|x_{k+1} - x^*\| = o\left(\|x_k - x^*\|\right).$$

The resulting algorithm can be written in the form:

Nonsmooth Newton's Method: *Choose a starting point* $z_0 \in \mathbb{R}^m, \varepsilon \geq 0$ *and put* $k := 0$. (3.70)

(1) *If* $\|F(z_k)\| \leq \varepsilon$, *then STOP.*

(2) *Compute an arbitrary matrix* $V_k \in \partial_B F(z_k)$ *and put*

$$z_{k+1} = z_k - V_k^{-1} F(z_k).$$

(3) *Put* $k := k+1$ *and go back to (1).*

The next theorem gives a convergence result for the above method in the case $\varepsilon = 0$.

Theorem 3.14 (Qi, 1993) *Suppose that* z^* *is a solution of (3.65) and* F *is strongly BD-regular at* z^*. *Then the Newton's method (3.70) is well-defined and converges locally superlinearly to* z^*.

Proof. Let $\mathcal{U}$ be a neighbourhood of z^* specified by Lemma 3.13 (with $z = z^*$) and assume that $z_k \in \mathcal{U}$. Then, by Lemma 3.13, the kth Newton iteration in (2) is well defined, and

$$\begin{aligned}
\|z_{k+1} - z^*\| &= \|z_k - z^* - V_k^{-1} F(z_k)\| \\
&= \left\|V_k^{-1}\left[V_k(z_k - z^*) - F(z_k) + F(z^*) + F'(z^*; z_k - z^*) - F'(z^*; z_k - z^*)\right]\right\| \\
&\leq \left\|V_k^{-1}\left[F(z_k) - F(z^*) - F'(z^*; z_k - z^*)\right]\right\| \\
&\qquad + \left\|V_k^{-1}\left[V_k(z_k - z^*) - F'(z^*; z_k - z^*)\right]\right\| \\
&\leq \|V_k^{-1}\| \left[\|F(z_k) - F(z^*) - F'(z^*; z_k - z^*)\| + \|V_k(z_k - z^*) - F'(z^*; z_k - z^*)\|\right].
\end{aligned}$$

By (3.66)

$$F(z_k) - F(z^*) - F'(z^*; z_k - z^*) = o(\|z_k - z^*\|)$$

and, due to Theorem 2.22(iv), also

$$V_k(z_k - z^*) - F'(z^*; z_k - z^*) = o(\|z_k - z^*\|).$$

Consequently, there is a function $\varphi(t) = o(t)$ (not dependent on k) such that

$$\|z_{k+1} - z^*\| \leq \varphi(\|z_k - z^*\|). \tag{3.71}$$

This implies the superlinear convergence of $\{z_k\}$ to z^*, provided z_0 is sufficiently close to z^* and all iterates z_k stay in $\mathcal{U}$. To this purpose, however, it suffices to start in such a ball $\mathcal{B} := z^* + \rho\mathbb{B}$, which satisfies the following conditions:

(i) $\mathcal{B} \subset \mathcal{U}$;

(ii) for all $\nu \in [0, \rho]$ one has $\varphi(\nu) \leq k\nu$ with a $k \in (0, 1)$.

The statement has been proved. ■

Under additional assumptions on F, even local quadratic convergence can be proved for (3.70) (Qi, 1993). Although we had no difficulties with the convergence of this method in our mechanical problems, the method may not converge at all, similarly as the classic Newton's method in the smooth case. To achieve global convergence, various modifications of the presented method (and also other nonsmooth Newton variants) are available (Pang, 1990a; Qi, 1993; Pang and Qi, 1993). These questions go, however, beyond the scope of this book.

Bibliographical notes

For a general description of the subgradient methods, we refer to the monograph by Shor, 1985.

The cutting plane method is independently due to Cheney and Goldstein, 1959 and Kelley, 1960. The step from the pure cutting plane model to the bundle concept was done by Lemaréchal, 1974; Lemaréchal, 1975 and Wolfe, 1975, and the algorithmic realization M1FC1 is due to Lemaréchal and Imbert, 1985. The idea to combine the bundle concept with the trust region technique was around for some time before it was studied in detail in Schramm, 1989 and Schramm and Zowe, 1992 and implemented as code BT; see Schramm and Zowe, 1991.

Closely related concepts like the proximal point idea were studied, e.g., by Kiwiel, 1990.

Recent extensions of bundle methods which try to improve the numerical behaviour by incorporating second order information are due to Lemaréchal and Sagastizábal, 1997; Lukšan and Vlček, 1996; Mifflin, 1996 and Mifflin et al., 1996.

The monographs Kiwiel, 1985 and Hiriart-Urruty and Lemaréchal, 1993 are excellent references for a complete and detailed setting of the bundle concept and related algorithmic ideas.

Starting with a technical report by Josephy, 1979, much attention has been paid to suitable modifications of the Newton's method to the solution of nonsmooth equations. In the early papers (e.g. Pang, 1990b), the Newton iteration was constructed by the so-called $B-$derivative; cf. (3.66). In Kummer, 1992, another possibilities were proposed, among other also (3.69). This variant of the Newton's method was then studied in many papers (e.g. Qi and Sun, 1993; Qi, 1993) and led to effective implementations for both NCPs (DeLuca et al., 1996) and VIs (Facchinei et al., 1995).

4 GENERALIZED EQUATIONS

As we will see later, variational inequalities (and complementarity problems) provide a convenient and elegant tool for characterizing manifold equilibria. The aim of this chapter is to spell out how these models can be brought into the equally useful form of a generalized equation

$$0 \in C(z) + N_Q(z), \tag{4.1}$$

where $C[\mathbb{R}^k \to \mathbb{R}^k]$ is a continuous mapping, Q a nonempty, closed, convex subset of $\mathbb{R}^k$ and $N_Q(z)$ its normal cone to Q at z; cf. Definition 2.6. Q is called the *feasible set* of the GE (4.1). Oftentimes, the rewriting as a "nonsmooth equation" is not only possible but very helpful. While proceeding, we also collect several basic results on existence and uniqueness needed in the later chapters. Our objective is to prepare for subsequent analysis and computations.

As said, we deal with equilibria described by variational inequalities, nonlinear complementarity problems and a special subclass of quasi-variational inequalities, called implicit complementarity problems. Section 4.1 gives the definitions and exhibits the corresponding generalized and nonsmooth equations. These will substantially simplify the analysis in the next chapters. The approach originates pretty much from the work of Robinson (Robinson, 1980; Robinson, 1991). We mention, though, that the conversion of a variational inequality to an equivalent nonsmooth equation was given much earlier; cf. the proof of Theorem 4.1 below, which goes back to Hartman and Stampacchia, 1966.

Section 4.2 deals with existence and uniqueness results, common to the extensive literature on this subject. Concerning existence, only basic theorems are presented which rely on compactness and coercivity (monotonicity) arguments. The uniqueness results depend (except in Theorem 4.7) on the concept of strong monotonicity.

4.1 EQUIVALENT FORMULATIONS

Equilibrium problems can often be stated in the form of a *variational inequality* (VI):

$$\left.\begin{array}{l}\text{Find } y \in \Omega \text{ such that}\\ \qquad \langle F(y), v - y\rangle \geq 0 \quad \text{for all } v \in \Omega.\end{array}\right\} \tag{4.2}$$

Here $F[\mathbb{R}^m \to \mathbb{R}^m]$ is a continuous mapping and Ω a nonempty closed convex subset of $\mathbb{R}^m$. Frequently, Ω is given by a system of equations and inequalities

$$\Omega = \{v \in \mathbb{R}^m \mid h^i(v) = 0,\ i = 1, 2, \ldots, \ell,\ g^j(v) \leq 0,\ j = 1, 2, \ldots, s\}, \tag{4.3}$$

where the functions $h^i[\mathbb{R}^m \to \mathbb{R}]$, $i = 1, 2, \ldots, \ell$, are affine and $g^j[\mathbb{R}^m \to \mathbb{R}]$, $j = 1, 2, \ldots, s$, are convex and continuously differentiable.

Two examples follow next:

Convex programming. Given an objective function $f[\mathbb{R}^m \to \mathbb{R}]$ which is Gâteaux differentiable and convex on $\mathbb{R}^m$, the program

$$\begin{array}{ll}\text{minimize} & f(y)\\ \text{subject to} & \\ & y \in \Omega\end{array} \tag{4.4}$$

is intimately tied to (4.2). Indeed, letting $F(y) = \nabla f(y)$, (4.2) gives necessary and sufficient optimality conditions for (4.4). In this sense, we can characterize optimality for any convex program with differentiable objectives by variational inequalities.

Complementarity problems. Let $\Psi := (\psi^1, \psi^2, \ldots, \psi^m)^T \in \mathbb{R}^m$ be given. If the set Ω is the shifted nonnegative orthant, i.e.,

$$\left.\begin{array}{ll} & \Omega = \{v \in \mathbb{R}^m \mid g^j(v) \leq 0,\ j = 1, 2, \ldots, s\}\\ \text{with} & \\ & g^j(v) = \psi^j - v^j, \quad j = 1, 2, \ldots, m,\end{array}\right\} \tag{4.5}$$

then (4.2) is reduced to the *nonlinear complementarity problem* (NCP):

$$\left.\begin{array}{l}\text{Find } y \in \mathbb{R}^m \text{ such that}\\ \qquad F(y) \geq 0,\ y - \Psi \geq 0,\ \langle F(y), y - \Psi\rangle = 0.\end{array}\right\} \tag{4.6}$$

In the literature this name is commonly reserved for the case $\Omega = \mathbb{R}^m_+$. If in addition F happens to be affine (and $\Omega = \mathbb{R}^m_+$), one speaks of *linear complementarity problem* (LCP).

Another class of equilibria is described by the *quasi-variational inequality* (QVI):

$$\left.\begin{array}{l}\text{Find } y \in \Gamma(y) \text{ such that}\\ \qquad \langle F(y), v - y\rangle \geq 0 \quad \text{for all } v \in \Gamma(y).\end{array}\right\} \tag{4.7}$$

Here $\Gamma[\mathbb{R}^m \rightsquigarrow \mathbb{R}^m]$ is a closed- and convex-valued multifunction. The implicit nature of the constraint $y \in \Gamma(y)$ makes this problem substantially more difficult than (4.2). We will mainly deal with a special case of (4.7), where

$$\Gamma(y) = \{v \in \mathbb{R}^m \mid v^i \geq \varphi^i(y),\ i = 1, 2, \ldots, m\} \tag{4.8}$$

figuring continuously differentiable functions $\varphi^i[\mathbb{R}^m \to \mathbb{R}]$, $i = 1, 2, \ldots, m$. Then for each y the set $\Gamma(y)$ has the structure (4.5) and problem (4.7) is reduced to the *implicit complementarity problem* (ICP):

$$\left.\begin{array}{l}\text{Find } y \in \mathbb{R}^m \text{ such that} \\ \qquad F(y) \geq 0,\ y - \Phi(y) \geq 0,\ \langle F(y), y - \Phi(y)\rangle = 0\end{array}\right\} \tag{4.9}$$

with $\Phi(y) := (\varphi^1(y), \varphi^2(y), \ldots, \varphi^m(y))^T$.

We now explain how the above variational and quasi-variational inequalities can be cast in the form (4.1). Put in (4.1)

$$k := m, \quad z := y, \quad C := F \quad \text{and} \quad Q := \Omega.$$

Then a look at the definition of $N_\Omega(y)$ shows that the GE

$$0 \in F(y) + N_\Omega(y) \tag{4.10}$$

is just a condensed form of writing (4.2). Note that for $y \notin \Omega$ the set $N_\Omega(y) = \emptyset$ (so $F(y) + N_\Omega(y) = \emptyset$), which ensures the feasibility of y.

Next suppose that Ω is given by (4.3). To make good use of (4.10), we need a handy representation of $N_\Omega(y)$. Assume that the *Slater constraint qualification* holds for (4.3), i.e.,

(SCQ): There exists $\overline{y} \in \mathbb{R}^m$ such that $g^j(\overline{y}) < 0$, $j = 1, 2, \ldots, s$, and $h^i(\overline{y}) = 0$, $i = 1, 2, \ldots, \ell$.

Then, as shown in Corollary 2.25, for each $y \in \Omega$ one has

$$N_\Omega(y) = \left\{ \sum_{i=1}^{\ell} \mu^i \nabla h^i(y) + \sum_{i=1}^{s} \lambda^i \nabla g^i(y) \mid \mu \in \mathbb{R}^\ell, \lambda \in \mathbb{R}^s_+, \right.$$
$$\left. \lambda^i g^i(y) = 0,\ i = 1, 2, \ldots, s \right\},$$

and the GE (4.10) becomes the following system of equations and inequalities in $(y, \mu, \lambda) \in \mathbb{R}^m \times \mathbb{R}^\ell \times \mathbb{R}^s$:

$$\left.\begin{array}{ll} F(y) + \sum_{i=1}^{\ell} \mu^i \nabla h^i(y) + \sum_{i=1}^{s} \lambda^i \nabla g^i(y) = 0, & \\ h^i(y) = 0, & i = 1, \ldots, \ell, \\ g^i(y) \leq 0, & i = 1, 2, \ldots, s, \\ \lambda^i \geq 0, & i = 1, 2, \ldots, s, \\ \lambda^i g^i(y) = 0, & i = 1, 2, \ldots, s. \end{array}\right\} \tag{4.11}$$

In case $F(y) = \nabla f(y)$ for some function $f[\mathbb{R}^m \to \mathbb{R}]$ relations (4.11) are the well-known Karush-Kuhn-Tucker conditions for the *mathematical program*

$$\begin{array}{lll} \text{minimize} & f(y) & \\ \text{subject to} & & \\ & h^i(y) = 0, & i = 1, 2 \ldots, \ell \\ & g^i(y) \leq 0, & i = 1, 2, \ldots, s. \end{array} \tag{4.12}$$

Hence (4.11) is also said to be the *Karush-Kuhn-Tucker* (KKT) *system* for (4.10) with Ω given by (4.3). Further, the map

$$\mathcal{L}(y,\mu,\lambda) := F(y) + \sum_{i=1}^{\ell} \mu^i \nabla h^i(y) + \sum_{i=1}^{s} \lambda^i \nabla g^i(y) \tag{4.13}$$

from $\mathbb{R}^m \times \mathbb{R}^\ell \times \mathbb{R}^s_+$ into $\mathbb{R}^m$ is called the *D-Lagrangian* of (4.10). Note that (4.13) is the derivative of the standard Lagrangian of (4.12) in mathematical programming.

One easily verifies that

$$N_{\mathbb{R}^m}(y) = 0_{\mathbb{R}^m} \quad \text{for all } y \in \mathbb{R}^m$$

and

$$N_{\mathbb{R}^s_+}(\lambda) = \begin{cases} \{u \in \mathbb{R}^s \,|\, u^i = 0 \text{ if } \lambda^i > 0 \text{ and } u^i \le 0 \text{ if } \lambda^i = 0\} & \text{for } \lambda \in \mathbb{R}^s_+ \\ \emptyset & \text{otherwise.} \end{cases}$$

The KKT system (4.11) can thus be written as the GE (4.1) in the variable $z = (y,\mu,\lambda) \in \mathbb{R}^m \times \mathbb{R}^\ell \times \mathbb{R}^s$

$$0 \in \begin{bmatrix} \mathcal{L}(y,\mu,\lambda) \\ H(y) \\ -G(y) \end{bmatrix} + \begin{bmatrix} N_{\mathbb{R}^m}(y) \\ N_{\mathbb{R}^\ell}(\mu) \\ N_{\mathbb{R}^s_+}(\lambda) \end{bmatrix}$$

with $H(y) := [h^1(y), h^2(y), \ldots, h^\ell(y)]^T$ and $G(y) := [g^1(y), g^2(y), \ldots, g^s(y)]^T$. This is equivalent (cf. Aubin and Frankowska, 1990) to

$$0 \in \begin{bmatrix} \mathcal{L}(y,\mu,\lambda) \\ H(y) \\ -G(y) \end{bmatrix} + N_{\mathbb{R}^m \times \mathbb{R}^\ell \times \mathbb{R}^s_+}(y,\mu,\lambda). \tag{4.14}$$

Note that in (4.14) the feasible set $\mathbb{R}^m \times \mathbb{R}^\ell \times \mathbb{R}^s_+$ is of simple structure. Therefore, in spite of the additional variables μ, λ, the GE (4.14) is a convenient tool and will extensively be used further on.

Before we come to the ICP, we recall that, by convexity of Ω,

$$w \in N_\Omega(y) \text{ if and only if } y = \mathrm{Proj}_\Omega(y + w); \tag{4.15}$$

cf. Figure 2.2. Hence, the GE (4.10) (and thus also (4.2)) can also be written as a nonsmooth equation in y:

$$y = \mathrm{Proj}_\Omega(y - F(y)). \tag{4.16}$$

Sometimes Ω in (4.16) is replaced by a multifunction of y, e.g., $\Gamma(y)$ defined in (4.8). In all these cases we will speak of *nonsmooth equation* (NSE). Relation (4.16) is especially useful, if we have an explicit formula for $\mathrm{Proj}_\Omega(\cdot)$. This situation occurs, e.g., when Ω is of form (4.5); then (4.16) attains a particularly simple form

$$\min{}_c\{F(y), y - \Psi\} = 0, \tag{4.17}$$

where $\min_c$ means that the minimum is taken componentwise. From NSE (4.17) one immediately gets back the original form of the NCP (4.6).

The GE (4.14) amounts to a system of equations

$$\begin{aligned} \mathcal{L}(y,\mu,\lambda) &= 0 \\ H(y) &= 0 \end{aligned}$$

together with the GE

$$0 \in -G(y) + N_{\mathbb{R}^s_+}(\lambda).$$

By (4.15), the latter is equivalent to the NSE

$$\lambda = \operatorname{Proj}_{\mathbb{R}^s_+}(\lambda + G(y)).$$

In this way, we get a useful NSE which is equivalent to the VI with Ω from (4.3):

$$\begin{bmatrix} \mathcal{L}(y,\mu,\lambda) \\ H(y) \\ \min_c\{-G(y),\lambda\} \end{bmatrix} = 0. \tag{4.18}$$

Just as (4.10) characterizes solutions of the VI (4.2), we realize that a vector y solves the QVI (4.7) if and only if

$$0 \in F(y) + N_{\Gamma(y)}(y). \tag{4.19}$$

Relation (4.19) is not a GE in the sense of (4.1) ($Q = \Gamma(y)$ is now the image of y in multifunction Γ). We can, however, write (4.19) with the help of (4.15) as a nonsmooth equation

$$y = \operatorname{Proj}_{\Gamma(y)}(y - F(y)). \tag{4.20}$$

In the special case of ICP when Γ is given by (4.8), we can rewrite (4.20) as a GE of the form (4.1). For this Γ, the operator $\operatorname{Proj}_{\Gamma(\cdot)}(\cdot - F(\cdot))$ assigns $y \in \mathbb{R}^m$ the unique solution of the optimization problem in v

$$\begin{aligned} &\text{minimize} && \tfrac{1}{2}\|v - y + F(y)\|^2 \\ &\text{subject to} && \\ & && v^i \geq \varphi^i(y), \quad i = 1,2,\ldots,m, \end{aligned}$$

or, equivalently, the unique v-component of the solution to the GE (in variables (v,λ))

$$0 \in \begin{bmatrix} v - y + F(y) - \lambda \\ v - \Phi(y) \end{bmatrix} + N_{\mathbb{R}^m \times \mathbb{R}^m_+}(v,\lambda), \tag{4.21}$$

where $\Phi(y) = (\varphi^1(y), \varphi^2(y), \ldots, \varphi^m(y))^T$, cf. (4.9). A comparison of (4.20) and (4.21) shows that y is a solution of the ICP if and only if $(y,\lambda) \in \mathbb{R}^m \times \mathbb{R}^m_+$ solves the GE

$$0 \in \begin{bmatrix} F(y) - \lambda \\ y - \Phi(y) \end{bmatrix} + N_{\mathbb{R}^m \times \mathbb{R}^m_+}(y,\lambda). \tag{4.22}$$

This is the wanted GE of the type (4.1) with $z = (y,\lambda)$ and $Q = \mathbb{R}^m \times \mathbb{R}^m_+$ for ICP.

Due to the specific structure of Γ, we have again an easy explicit formula for $\operatorname{Proj}_{\Gamma(\cdot)}(\cdot)$. Equation (4.20) thus attains a very convenient form

$$\min_c\{F(y), y - \Phi(y)\} = 0, \tag{4.23}$$

which is only a slight generalization of (4.17). Therefore, in the framework of NSEs, we can treat NCP and ICP together. As in the case of NCP, from (4.23) one immediately deduces the standard form (4.9) of ICP.

Table 4.1 provides an overview of the used equilibrium models and their equivalent formulations as GEs and NSEs.

Table 4.1.

Model	GE	NSE
VI (4.2)	$0 \in F(y) + N_\Omega(y)$	$y = \mathrm{Proj}_\Omega(y - F(y))$
NCP (4.6)	$0 \in F(y) + N_{\Psi + \mathbb{R}^m_+}(y)$	$\min_c\{F(y), y - \Psi\} = 0$
VI with Ω given by (4.3)	$0 \in \begin{bmatrix} \mathcal{L}(y,\mu,\lambda) \\ H(y) \\ -G(y) \end{bmatrix} + N_{\mathbb{R}^m \times \mathbb{R}^\ell \times \mathbb{R}^s_+}(y,\mu,\lambda)$	$\begin{bmatrix} \mathcal{L}(y,\mu,\lambda) \\ H(y) \\ \min_c\{-G(y),\lambda\} \end{bmatrix} = 0$
ICP (4.9)	$0 \in \begin{bmatrix} F(y) - \lambda \\ y - \Phi(y) \end{bmatrix} + N_{\mathbb{R}^m \times \mathbb{R}^m_+}(y,\lambda)$	$\min_c\{F(y), y - \Phi(y)\} = 0$

We add that there are more ways to write (4.1) as an equation; cf., e.g., Robinson, 1991.

We illustrate some of the above formulations by simple examples which will be used for illustration in the following text.

Example 4.1 Consider the convex optimization problem

$$\text{minimize } (y^1)^2 + (y^2)^2 - y^1 - \frac{1}{2}y^2$$

subject to

$$y \in \Omega := \left\{ v \in \mathbb{R}^2 \mid (v^i - 1)^2 \le \frac{1}{4},\ i = 1,2 \right\}$$

with the unique solution $\hat{y} = \left(\frac{1}{2}, \frac{1}{2}\right)$. Clearly, this problem is equivalent to the GE (4.10) with

$$F(y) = \begin{bmatrix} 2y^1 - 1 \\ 2y^2 - \frac{1}{2} \end{bmatrix}.$$

Further, with D-Lagrangian

$$\mathcal{L}(y,\lambda) = F(y) + \lambda^1 \begin{bmatrix} 2(y^1 - 1) \\ 0 \end{bmatrix} + \lambda^2 \begin{bmatrix} 0 \\ 2(y^2 - 1) \end{bmatrix},$$

the corresponding GE (4.14) attains the form

$$0 \in \begin{bmatrix} \mathcal{L}(y,\lambda) \\ \frac{1}{4} - (y^1-1)^2 \\ \frac{1}{4} - (y^2-1)^2 \end{bmatrix} + N_{\mathbb{R}^2 \times \mathbb{R}^2_+}(y,\lambda).$$

This GE has the solution $(\hat{y},\hat{\lambda}) = \left(\frac{1}{2},\frac{1}{2},0,\frac{1}{2}\right)$. △

Example 4.2 Consider the ICP (4.9) with $m = 4$,

$$F(y) := \begin{bmatrix} 2 & -1 & 0 & 0 \\ -1 & 2 & -1 & 0 \\ 0 & -1 & 2 & -1 \\ 0 & 0 & -1 & 2 \end{bmatrix} y + \begin{bmatrix} 1 \\ 1 \\ 1 \\ 1 \end{bmatrix},$$

and

$$\varphi^i(y) := \left\langle \begin{array}{ll} -2.5 - F^i(y) & \text{for } i = 1,4 \\ -3 - F^i(y) & \text{for } i = 2,3. \end{array} \right.$$

One easily verifies that $\hat{y} = (-2,-3,-3,-2)$ is a solution of this problem. The GE (4.22) attains the form

$$0 \in \begin{bmatrix} 2y^1 - y^2 - \lambda^1 + 1 \\ -y^1 + 2y^2 - y^3 - \lambda^2 + 1 \\ -y^2 + 2y^3 - y^4 - \lambda^3 + 1 \\ -y^3 + 2y^4 - \lambda^4 + 1 \\ 3y^1 - y^2 + 3.5 \\ -y^1 + 3y^2 - y^3 + 4 \\ -y^2 + 3y^3 - y^4 + 4 \\ -y^3 + 3y^4 + 3.5 \end{bmatrix} + N_{\mathbb{R}^4 \times \mathbb{R}^4_+}(y,\lambda),$$

with the solution $(\hat{y},0,0,0,0)$ and the NSE (4.23) reads as follows:

$$\min \left\{ \begin{array}{ll} 2y^1 - y^2 + 1, & 3y^1 - y^2 + 3.5 \\ -y^1 + 2y^2 - y^3 + 1, & -y^1 + 3y^2 - y^3 + 4 \\ -y^2 + 2y^3 - y^4 + 1, & -y^2 + 3y^3 - y^4 + 4 \\ -y^3 + 2y^4 + 1, & -y^3 + 3y^4 + 3.5 \end{array} \right\} = 0.$$

△

Example 4.3 Let us go through the above formulations of a concrete problem from mechanics. A rigorous introduction of this problem will be given in Chapter 9.

We consider a linear model of an elastic string on an interval $(0,1)$ fixed at the two end-points and subject to a perpendicular load b. The deflection y of such a string can be computed as a solution of the second-order elliptic differential equation:

$$\begin{aligned} -\Delta y &= b \qquad \text{in } (0,1) \\ y &= 0 \qquad \text{at 0 and 1.} \end{aligned}$$

Assuming the existence of a rigid obstacle given by a function ψ as illustrated in Figure 4.1, this equation becomes an (infinite-dimensional) variational inequality

$$\begin{aligned}&\text{Find } y \in \Omega := \{z \in H_0^1((0,1)) | z \geq \psi \text{ on } (0,1)\} \text{ such that}\\ &(-\Delta y - b, v - y) \geq 0 \quad \text{for all } v \in \Omega.\end{aligned}$$

After discretization by the finite element method the functions become vectors in $\mathbb{R}^m$

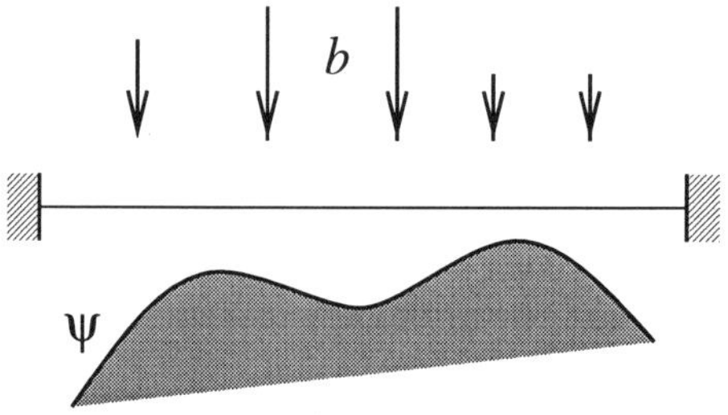

Figure 4.1. String with a rigid obstacle

(denoted by the same symbols), and we get the variational inequality formulation of our problem in $\mathbb{R}^m$:

$$\begin{aligned}&\text{Find } y \in \Omega \text{ such that}\\ &\langle Ay - b, v - y \rangle \geq 0 \quad \text{for all } v \in \Omega\end{aligned}$$

with

$$\Omega := \{v \in \mathbb{R}^m | v^i \geq \psi^i\}$$

and with a symmetric and positive definite "stiffness" matrix A.

The GE associated with the discretized problem reads:

$$0 \in Ay - b + N_\Omega(y).$$

In words, when the string does not touch the obstacle at any discretization node, then y lies in the interior of Ω and the associated normal cone shrinks to zero; the GE is reduced to a system of linear equations. Now assume for clarity that $\psi \leq 0$ in $(0,1)$, that b is only nonzero in one component (say k), and that the value of $b^k := b_{\text{eq}}$ is such that the string just touches the obstacle at the corresponding discretization node. Then, obviously, after increasing the value of b_{eq} by some b_{add}, the solution y does not change—the "additional" force is absorbed by the obstacle, see Figure 4.2. Negative value of this surplus force b_{add} is called reaction. And the normal cone $N_\Omega(y)$ is just a set of the surplus forces $b_{\text{add}} = b_{\text{add}} + b_{\text{eq}} - Ay = b - Ay$.

Because of the simple structure of Ω, the above GE in fact amounts to the complementarity problem:

$$Ay \geq b, \qquad y \geq \psi, \qquad \langle Ay - b, y - \psi \rangle = 0.$$

In words, for any component (node), either the (non-negative) reaction $(Ay - b)^i$ is zero or the (non-negative) gap $y^i - \psi^i$ is zero (or both). By a standard trick we transform the above problem to an LCP on $\mathbb{R}^m_+$. With $z := y - \psi$ the problem becomes an LCP in z

$$Az + A\psi - b \geq 0, \qquad z \geq 0, \qquad \langle Az + A\psi - b, z \rangle = 0.$$

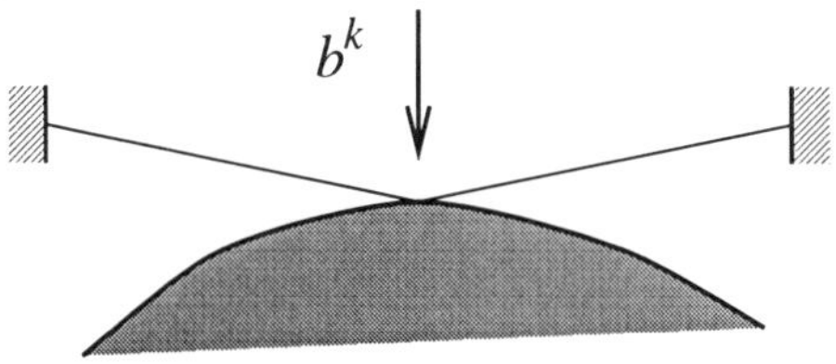

Figure 4.2. Deflection under b^k

Finally, the formulation of the string problem as a nonsmooth equation becomes

$$y = \mathrm{Proj}_{\mathbb{R}^m_+}(y - Ay + b),$$

which can be simplified to

$$\min{}_c\{Ay - b, y - \psi\}.$$

△

Example 4.4 Let us again consider the string problem of Example 4.3 but with a *compliant* obstacle instead of a rigid one. The original shape of the obstacle is again given by a function ψ but the *actual* shape depends on the surplus force $b - Ay$. Now, this force is only partly absorbed by the obstacle and the rest of the force deforms its shape as depicted in Figure 4.3(b). So the constraints depend on the surplus force and hence on the solution.

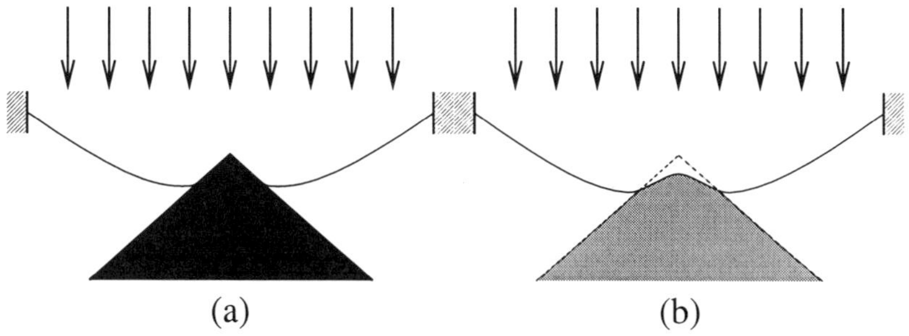

Figure 4.3. String with (a) rigid and (b) compliant obstacle

Thus we deal with an implicit complementarity problem

$$Ay \geq b, \qquad y \geq \varphi(y), \qquad \langle Ay - b, y - \varphi(y)\rangle = 0,$$

where A and b have the same meaning as in Example 4.3 and the function φ is defined as

$$\varphi(y) := k(Ay - b) + \psi, \qquad k \geq 0.$$

Here $1/k$ is the so-called coefficient of compliance. Note that $k = 0$ gives the LCP from Example 4.3. For this ICP we can easily derive the corresponding generalized equation in

(y, λ)

$$0 \in \begin{bmatrix} Ay - b - \lambda \\ y - k(Ay - b) - \psi \end{bmatrix} + N_{\mathbb{R}^m \times \mathbb{R}^m_+}(y, \lambda)$$

and the nonsmooth equation in y

$$\min{}_c\{Ay - b, y - k(Ay - b) - \psi\} = 0.$$

△

4.2 EXISTENCE AND UNIQUENESS

In this section we present some basic existence and uniqueness results for the GE (4.10), (4.14) and for the ICP (4.9). Let us start with the GE (4.10)

$$0 \in F(y) + N_\Omega(y), \tag{4.10}$$

where $F[\mathbb{R}^m \to \mathbb{R}^m]$ is a continuous mapping and, as before, Ω is a nonempty closed convex subset of $\mathbb{R}^m$. The classic existence results rely on compactness or coercivity arguments or on a combination of both (Kinderlehrer and Stampaccia, 1980; Baiocchi and Capelo, 1984).

Theorem 4.1 *The GE (4.10) has a solution provided the feasible set Ω is compact.*

Proof. The projection map $\text{Proj}_\Omega(\cdot)$ is continuous (see Zarantonello, 1971) and thus Brouwer's Fixed Point Theorem implies the existence of a solution for NSE (4.16) if Ω is compact. This proves the claim since (4.16) is just a rewriting of (4.10). ■

In many applications, however, Ω is unbounded (e.g. in complementarity problems). Here helps the following trick. We replace Ω by a compact set

$$\Omega_r = \Omega \cap r\mathbb{B}$$

with a positive real r. Then the following GE

$$0 \in F(y) + N_{\Omega_r}(y) \tag{4.24}$$

possesses a solution and it remains to look for conditions guaranteeing that this solution also solves (4.10). Here comes such result:

Theorem 4.2 *For the existence of a solution of the GE (4.10) it is necessary and sufficient that, for some $r > 0$, there exists a solution y_r of the GE (4.24) with $\|y_r\| < r$.*

Proof. If y solves (4.10), then it also solves (4.24) with arbitrary $r > \|y\|$. Conversely, if y_r solves (4.24) and $\|y_r\| < r$ then, by the calculus rules for normal cones of convex sets (e.g. Rockafellar, 1970, Cor. 23.8.1),

$$N_{\Omega_r}(y_r) = N_\Omega(y_r) + N_{r\mathbb{B}}(y_r) = N_\Omega(y_r) + \{0\} = N_\Omega(y_r)$$

and, consequently, y_r is a solution of (4.10). ■

The above result applies, e.g., to the class of coercive functions.

Definition 4.1 *We say that $F[\mathbb{R}^m \to \mathbb{R}^m]$ is* coercive *on Ω, if there exists a point $y_0 \in \mathbb{R}^m$ such that*

$$\frac{\langle F(y) - F(y_0), y - y_0 \rangle}{\|y - y_0\|} \to \infty \quad \textit{for } \|y\| \to \infty,\ y \in \Omega. \tag{4.25}$$

Corollary 4.3 *Assume that F in (4.10) is coercive on Ω. Then the GE (4.10) possesses a solution.*

Proof. Because of Theorem 4.1 we can assume without loss of generality that Ω is unbounded. Suppose that (4.25) holds with some y_0 and choose reals $H > \|F(y_0)\|$ and $r > \|y_0\|$ such that

$$\langle F(y) - F(y_0), y - y_0 \rangle \geq H\|y - y_0\| \quad \text{for } \|y\| \geq r,\ y \in \Omega.$$

Then

$$\begin{aligned} \langle F(y), y - y_0 \rangle &\geq H\|y - y_0\| + \langle F(y_0), y - y_0 \rangle \\ &\geq H\|y - y_0\| - \|F(y_0)\|\|y - y_0\| \\ &\geq (H - \|F(y_0)\|)(\|y\| - \|y_0\|) \\ &> 0 \end{aligned} \tag{4.26}$$

for $\|y\| \geq r$, $y \in \Omega$. By construction, the GE (4.24) has a solution, say y_r. Using the equivalence of (4.2) and (4.10) with Ω replaced by Ω_r, we get for this y_r

$$\langle F(y_r), y_r - y_0 \rangle = -\langle F(y_r), y_0 - y_r \rangle \leq 0.$$

Thus, due to (4.26), $\|y_r\| < r$ and we are done. ■

Monotonicity properties are also helpful to guarantee the existence and uniqueness of solutions to GEs.

Definition 4.2 *The mapping $F[\mathbb{R}^m \to \mathbb{R}^m]$ is said to be* strictly monotone *on Ω, if*

$$\langle F(v) - F(w), v - w \rangle > 0 \quad \textit{for all } v, w \in \Omega,\ v \neq w. \tag{4.27}$$

F is said to be strongly monotone *on Ω, if there exists $\alpha > 0$ such that*

$$\langle F(v) - F(w), v - w \rangle \geq \alpha\|v - w\|^2 \quad \textit{for all } v, w \in \Omega. \tag{4.28}$$

Theorem 4.4 *(i) If F in (4.10) is strictly monotone on Ω, then the GE (4.10) has at most one solution.*
(ii) If F in (4.10) is strongly monotone on Ω, then the GE (4.10) has exactly one solution.

Proof. (i) Assume that $y_1, y_2 \in \Omega$ are two different solutions of the GE (4.10). The equivalence of (4.2) and (4.10) yields

$$\langle F(y_1), y_2 - y_1 \rangle \geq 0 \quad \text{and} \quad \langle F(y_2), y_1 - y_2 \rangle \geq 0.$$

Adding these two inequalities and multiplying by -1, one gets

$$\langle F(y_1) - F(y_2), y_1 - y_2 \rangle \leq 0,$$

and thus $y_1 = y_2$ because of (4.27).
(ii) It suffices to note that strong monotonicity implies strict monotonicity as well as the coercivity condition (4.25), so that the existence is guaranteed by Corollary 4.3. ∎

In most applications, F is continuously differentiable on $\mathbb{R}^m$; therefore one can ensure the above monotonicity properties by assumptions on the Jacobian $\mathcal{J}F$ of F. In this connection we use the notion of uniform positive definiteness.

Definition 4.3 *Let O be a map from $\mathbb{R}^n$ into $\mathbb{R}^{m \times m}$ and let Ξ be a subset of $\mathbb{R}^n$. We say that O is* positive definite *on Ξ, if*

$$\langle d, O(z)d \rangle > 0 \quad \textit{for all } z \in \Xi \textit{ and all } d \in \mathbb{R}^m \setminus \{0\}.$$

We say that O is uniformly positive definite *on Ξ, if there exists $\alpha > 0$ such that*

$$\langle d, O(z)d \rangle \geq \alpha \|d\|^2 \quad \textit{for all } z \in \Xi \textit{ and all } d \in \mathbb{R}^m.$$

Proposition 4.5 *Let F be continuously differentiable on $\mathbb{R}^m$.*

(i) If $\mathcal{J}F$ is positive definite on Ω, then F is strictly monotone on Ω.

(ii) If $\mathcal{J}F$ is uniformly positive definite on Ω, then F is strongly monotone on Ω.

(iii) If F is strongly monotone on an open set $\widetilde{\Omega}$, then $\mathcal{J}F$ is uniformly positive definite on $\widetilde{\Omega}$.

Proof. (i) By the mean-value theorem, one has for arbitrary $v, w \in \Omega$, $v \neq w$,

$$\langle F(v) - F(w), v - w \rangle = \int_0^1 \langle \mathcal{J}F(w + t(v - w))(v - w), v - w \rangle dt. \tag{4.29}$$

As Ω is convex, the integrand in (4.29) is positive for all $t \in [0, 1]$ and so F is strictly monotone on Ω.
(ii) For uniformly positive definite $\mathcal{J}F$ on Ω, we get from (4.29) for all $v, w \in \Omega$

$$\langle F(v) - F(w), v - w \rangle \geq \alpha \|v - w\|^2.$$

(iii) Suppose that (4.28) holds with $\alpha > 0$ and Ω replaced by $\widetilde{\Omega}$. Then for each $y \in \widetilde{\Omega}$ and $d \in \mathbb{R}^m$ one has

$$\langle d, \mathcal{J}F(y)d \rangle = \langle d, \lim_{t \to 0} \frac{F(y + td) - F(y)}{t} \rangle \geq \lim_{t \to 0} \frac{1}{t^2} \alpha \|td\|^2 = \alpha \|d\|^2.$$

∎

For affine F, strong monotonicity amounts to positive definiteness of the defining matrix, which is unduly strong for some GEs. In Theorem 4.6 below we show how this requirement can be weakened.

Definition 4.4 *Let A be an $m \times m$ matrix and K be a cone in $\mathbb{R}^m$ with vertex at the origin. We say that A is* strictly copositive with respect to K, *if*

$$\langle d, Ad \rangle > 0 \quad \textit{for all } d \in K,\ d \neq 0. \tag{4.30}$$

Theorem 4.6 *Let $F(y) = Ay + b$ with an $m \times m$ matrix A and $b \in \mathbb{R}^m$. Assume that Ω is a cone and A is strictly copositive with respect to the cone $\Omega - \Omega$. Then the GE (4.10) has a unique solution.*

Proof. Note first that

$$\Omega - \Omega = \mathrm{lin}\Omega,$$

and thus $\Omega - \Omega$ is closed. Hence, using a standard compactness argument, we get from (4.30)

$$\alpha := \min\{\langle \frac{d}{\|d\|}, A\frac{d}{\|d\|}\rangle | d \in \Omega - \Omega, d \neq 0\} > 0,$$

and thus

$$\langle d, Ad \rangle \geq \alpha \|d\|^2 \quad \text{for all } d \in \Omega - \Omega. \tag{4.31}$$

It follows for all $v, w \in \Omega$,

$$\langle F(v) - F(w), v - w \rangle = \langle A(v - w), v - w \rangle \geq \alpha \|v - w\|^2,$$

which proves the strong monotonicity of F on Ω and thus the claim because of Theorem 4.4(ii). ∎

The conditions of Theorem 4.4 and Theorem 4.6 are only *sufficient* for existence and uniqueness of solutions to the GE (4.10). If, however, (4.10) stands for an LCP, then one can establish a stronger result which is both *sufficient* and *necessary*.

Definition 4.5 *A square matrix M is called a P-*matrix *if all its principal subdeterminants are positive.*

Theorem 4.7 *Consider the LCP*

$$Ay + b \geq 0, \quad y \geq 0, \quad \langle (Ay + b), y \rangle = 0 \tag{4.32}$$

with an $m \times m$ matrix A and $b \in \mathbb{R}^m$. This problem has a unique solution for each $b \in \mathbb{R}^m$ if and only if A is a P-matrix.

The proof of this assertion is quite long and technical; therefore it is postponed to Appendix C.1.

Remark. From Theorem 4.4 and Proposition 4.5 we deduce that the positive definiteness of the matrix A in an LCP ensures existence and uniqueness of the solution. A symmetric matrix is positive definite if and only if it is a P-matrix, cf. Fiedler, 1986. Thus the condition in Theorem 4.4(ii) is not only sufficient but also necessary for the LCP (4.32) with symmetric A to have a solution for each $b \in \mathbb{R}^m$.

Remark. If we restrict b to $\mathrm{int}\mathbb{R}^m_+$ or to $\mathbb{R}^m_+$, then the assumptions on A can be further weakened. This leads to the class of semi-monotone and strictly semi-monotone matrices, respectively. For these definitions and resulting properties the reader is referred to Murty, 1988.

We now turn our attention to the GE (4.14)

$$0 \in \begin{bmatrix} \mathcal{L}(y,\mu,\lambda) \\ H(y) \\ -G(y) \end{bmatrix} + N_{\mathbb{R}^m \times \mathbb{R}^\ell \times \mathbb{R}^s_+}(y,\mu,\lambda), \tag{4.33}$$

We say that the *linear independence constraint qualification* (LICQ) holds at $\overline{y} \in \Omega$, if

(LICQ): The gradients $\nabla h^i(\overline{y})$ for $i = 1, 2, \ldots, \ell$ and $\nabla g^j(\overline{y})$ for $j \in I(\overline{y}) := \{k \in \{1, 2, \ldots, s\} \mid g^k(\overline{y}) = 0\}$ are linearly independent.

Under (LICQ) we immediately get the following result.

Theorem 4.8 *Let F be strongly monotone on Ω given by (4.3) and $\overline{y}$ be the unique solution of the corresponding GE (4.10). Assume that (LICQ) holds at $\overline{y}$. Then the GE (4.14) possesses a unique solution $(\hat{y}, \hat{\mu}, \hat{\lambda})$ with $\hat{y} = \overline{y}$.*

Proof. Note that (LICQ) implies (SCQ) and so the step from (4.10) to (4.14) is legal. The uniqueness of $\hat{y} = \overline{y}$ is due to the strong monotonicity of F and the uniqueness of the associated KKT pair $(\hat{\mu}, \hat{\lambda})$ follows from the equation

$$\mathcal{L}(\hat{y},\mu,\lambda) := F(\hat{y}) + \sum_{i=1}^{\ell} \mu^i \nabla h^i(\hat{y}) + \sum_{i \in I(\hat{y})} \lambda^i \nabla g^i(\hat{y}) = 0$$

and the linear independence assumption. ■

We close this section with an existence and uniqueness result for the ICP (4.9). In QVIs (which include ICPs), existence results are again based on compactness and/or coercivity (monotonicity) assumptions (Chan and Pang, 1982; Kočvara and Outrata, 1995a). Uniqueness statements, however, are rare and require additional technical assumptions. The following result gives such example.

Theorem 4.9 *Let F be continuously differentiable, strongly monotone and (globally) Lipschitz on $\mathbb{R}^m$. Further assume that Φ is continuously differentiable on $\mathbb{R}^m$ and for all $d \in \mathbb{R}^m$ and $y := F^{-1}(w)$ with $w \in \mathbb{R}^m_+$, one has*

$$\langle d, \mathcal{J}\Phi(y)(\mathcal{J}F(y))^{-1} d\rangle \le 0. \tag{4.34}$$

Then there exists a unique solution to the ICP (4.9).

Proof. Due to Ortega and Rheinboldt, 1970, Prop. 5.4.5, F is a homeomorphism of $\mathbb{R}^m$ onto $\mathbb{R}^m$ and its inverse is continuously differentiable and strongly monotone on $\mathbb{R}^m$; cf. Appendix C.2. Now let $y \in \mathbb{R}^m$ and put $w := F(y)$. In the variable w the ICP (4.9) attains the form

$$w \ge 0, \quad F^{-1}(w) - \Phi(F^{-1}(w)) \ge 0, \quad \langle w, F^{-1}(w) - \Phi(F^{-1}(w))\rangle = 0,$$

which is a standard NCP (no implicit structure anymore) and is equivalent to the VI:

$$\left.\begin{array}{l}\text{Find } w \in \mathbb{R}^m_+ \text{ such that} \\ \left\langle F^{-1}(w) - \Phi \circ F^{-1}(w), v - w \right\rangle \geq 0 \quad \text{for all } v \in \mathbb{R}^m_+ . \end{array}\right\} \tag{4.35}$$

As explained in Appendix C.2, F^{-1} is strongly monotone on $\mathbb{R}^m$ and thus $\mathcal{J}F^{-1}$ is uniformly positive definite on $\mathbb{R}^m$ (Proposition 4.5(iii)). Hence $\mathcal{J}F^{-1} - \mathcal{J}(\Phi \circ F^{-1})$ will be uniformly positive definite on $\mathbb{R}^m_+$ (and thus $F^{-1} - \Phi \circ F^{-1}$ strongly monotone on $\mathbb{R}^m_+$ because of Proposition 4.5(ii)), provided we can ensure positive semidefiniteness of $-\mathcal{J}(\Phi \circ F^{-1})(w)$ on $\mathbb{R}^m_+$. Now

$$\mathcal{J}(\Phi \circ F^{-1})(w) = \mathcal{J}\Phi(y)\mathcal{J}F^{-1}(w) = \mathcal{J}\Phi(y)(\mathcal{J}F(y))^{-1},$$

which completes the proof because of (4.34). ∎

We come back to the examples of Section 4.1 and check which from the above results can be applied. Concerning existence in Example 4.1, both Theorems 4.1 and 4.2 apply; further, for all $y \in \mathbb{R}^2$

$$\mathcal{J}F(y) = \begin{bmatrix} 2 & 0 \\ 0 & 2 \end{bmatrix},$$

and so $\mathcal{J}F$ is uniformly positive definite on $\mathbb{R}^2$. Hence, by Proposition 4.5, F is strongly monotone on $\mathbb{R}^2$ and thus the respective optimization problem even has a unique solution (Theorem 4.4). As (LICQ) holds at $\hat{y}$, Theorem 4.8 shows in addition that the corresponding GE (4.14) possesses a unique solution pair $(\hat{y}, \hat{\lambda})$.

The linear map F in Example 4.2 is of course Lipschitz on $\mathbb{R}^4$. Furthermore, for all $y \in \mathbb{R}^4$ one has

$$\mathcal{J}F(y) = \begin{bmatrix} 2 & -1 & 0 & 0 \\ -1 & 2 & -1 & 0 \\ 0 & -1 & 2 & -1 \\ 0 & 0 & -1 & 2 \end{bmatrix},$$

which is a positive definite matrix. Therefore, F is again strongly monotone on the whole space. To apply Theorem 4.9 it remains to verify inequality (4.34). For an arbitrary $y \in \mathbb{R}^4$ one has

$$\mathcal{J}\Phi(y) = -\mathcal{J}F(y)$$

and thus

$$\mathcal{J}\Phi(y)(\mathcal{J}F(y))^{-1} = -E.$$

Therefore this ICP has a unique solution by Theorem 4.9.

Bibliographical notes

The variational inequality problem was introduced by Hartman and Stampacchia in 1966 and subsequently expanded in several classic papers. These early studies were motivated by boundary value problems posed in the form of partial differential equations, cf. Hartman and Stampacchia, 1966 and Kinderlehrer and Stampaccia, 1980. The possibility to model various economic equilibria (Nash equilibrium, Wardrop equilibrium, etc.) as finite-dimensional VIs was recognized only much later; cf. Harker and Pang, 1990 and the references therein. The nonlinear complementarity problem first appeared in 1966 (Cottle, 1966), but only a few years later it was recognized as a special case of the VI (Karamardian, 1971). Quasi-variational inequalities were introduced in the seventies by Bensoussan and Lions in connection with stochastic impulse control problems (e.g. Bensoussan and Lions, 1973). They also turned out to be very convenient for modelling various equilibria in both mechanics (Mosco, 1976; Baiocchi and Capelo, 1984) and mathematical economy (Harker, 1991).

The generalized equations first appeared in the cited works by Robinson in the form (4.1), i.e., with the normal cone mapping. However, in the recent works on this subject no special structure of the set-valued part is assumed (Dontchev and Hager, 1994; Dontchev, 1995).

Theorem 4.1 is the famous result from Hartman and Stampacchia, 1966 (in the finite-dimensional setting) and also Theorem 4.2 comes from this paper. Proposition 4.5 is extracted from Ortega and Rheinboldt, 1970, where various monotonicity properties of an operator are related to positive definiteness or positive semi-definiteness of the respective Jacobians. The existence and uniqueness questions in LCPs are fairly well understood and the underlying theory goes far beyond the scope of this book. The interested reader is referred to Murty, 1988 or Cottle et al., 1992. Finally, Theorem 4.9 is a generalization of a result from Pang, 1981, concerning the linear implicit complementarity problem and comes from Kočvara and Outrata, 1994c.

5 STABILITY OF SOLUTIONS TO PERTURBED GENERALIZED EQUATIONS

Over the past twenty years a comprehensive theory has been developed on the stability of solutions to perturbed generalized equations, in particular on the Lipschitz behaviour of these solutions. In this chapter we present the part of this theory relevant to our later applications. We assume that the generalized equations from the previous chapter are now subjected to perturbations and we analyse the local behaviour of the associated solution maps S. Basically, we follow Robinson's work.

Section 5.1 considers the GE (1.1) and contains the essential result of Robinson, 1980, which directly leads to the concept of *strong regularity* (Theorem 5.2).

The frequent case of a polyhedral feasible set Q is studied in Section 5.2. In this case, the directional differentiability of the projection onto polyhedral sets (Theorem 2.31) together with a result on polyhedral multifunctions (Theorem 2.4) allow to characterize the strong regularity in a very neat way.

Section 5.3 applies the results of the preceding sections to the GEs (4.14),(4.22) with maps F, H, G and Φ which now depend on a perturbation parameter x. The special structure of these GEs enables to get conditions ensuring that the respective solution maps S possess Lipschitz selections.

5.1 ANALYSIS OF THE IMPLICIT MAP

Let x_0 be a certain reference value of the perturbation parameter, $\mathcal{A}$ an open set in $\mathbb{R}^n$ containing x_0 and consider the perturbed GE

$$0 \in C(x, z) + N_Q(z), \tag{5.1}$$

where C is now a continuously differentiable map from $\mathcal{A} \times \mathbb{R}^k$ into $\mathbb{R}^k$ and Q is again a nonempty closed and convex subset of $\mathbb{R}^k$. We further assume that $\mathcal{O} \subset \mathcal{A}$ is a neighbourhood of x_0 and that z_0 solves (5.1) for $x = x_0$. The *local stability analysis* studies the behaviour of the *solution map* $S[\mathcal{A} \rightsquigarrow \mathbb{R}^k]$ defined by

$$S(x) = \{z \in \mathbb{R}^k \,|\, 0 \in C(x,z) + N_Q(z)\} \tag{5.2}$$

close to the *reference pair* (x_0, z_0). More precisely, we ask for conditions which guarantee the existence of a single-valued Lipschitz map $\sigma : x \mapsto z$ on the neighbourhood $\mathcal{O}$ of x_0 such that

$$\sigma(x_0) = z_0 \quad \text{and} \quad \sigma(x) \in S(x) \quad \text{for } x \in \mathcal{O}, \tag{5.3a}$$

or even,

$$\sigma(x) = S(x) \cap V \quad \text{for } x \in \mathcal{O} \tag{5.3b}$$

with some suitable neighbourhood V of z_0.

Along with $S(x)$ we will work with the multivalued map $\Sigma[\mathbb{R}^k \rightsquigarrow \mathbb{R}^k]$ given by

$$\Sigma(\xi) := \{z \in \mathbb{R}^m \mid \xi \in C(x_0, z_0) + \mathcal{J}_z C(x_0, z_0)(z - z_0) + N_Q(z)\}, \tag{5.4}$$

which is generated by partial linearization of $C(x, z)$ in (5.1) with respect to z at (x_0, z_0). Let us introduce the notation

$$r(x,z) := C(x_0, z_0) + \mathcal{J}_z C(x_0, z_0)(z - z_0) - C(x, z).$$

We can use $\Sigma(\cdot)$ and $r(x, \cdot)$ to characterize $S(x)$.

Proposition 5.1 *It holds that*

$$z \in S(x) \quad \textit{if and only if} \quad z \in \Sigma(r(x,z)).$$

Proof. Going back to the definitions of Σ and r, we see that

$$z \in S(x) \quad \text{if and only if} \quad 0 \in C(x,z) + N_Q(z) \tag{5.5}$$

and

$$\begin{aligned} &z \in \Sigma(r(x,z)) \quad \text{if and only if} \\ &\qquad r(x,z) \in C(x_0, z_0) + \mathcal{J}_z C(x_0, z_0)(z - z_0) + N_Q(z). \end{aligned} \tag{5.6}$$

Simple manipulation shows that the GEs in (5.5) and (5.6) are equivalent. ■

Let us add a simple observation before we come to the announced result. Since C is continuously differentiable on $\mathcal{A} \times \mathbb{R}^k$, we can choose neighbourhoods $\widetilde{U}$ of x_0, $\widetilde{V}$ of z_0 and a positive real L such that

$$\|C(x_1, z) - C(x_2, z)\| \leq L\|x_1 - x_2\| \quad \text{for all } x_1, x_2 \in \widetilde{U},\ z \in \widetilde{V}. \tag{5.7}$$

In what follows the (uniform) *Lipschitz modulus* L from (5.7) plays an important role.

Theorem 5.2 *(a) Assume there exists a single-valued Lipschitz function ϕ with modulus γ from a neighbourhood W of $0 \in \mathbb{R}^k$ into $\mathbb{R}^k$ such that*

$$\phi(0) = z_0 \quad \textit{and} \quad \phi(\xi) \in \Sigma(\xi) \quad \textit{for all } \xi \in W. \tag{5.8}$$

Then for each $\varepsilon > 0$ there exist neighbourhoods U_ε and V_ε of x_0 and z_0, respectively, and a single-valued map $\sigma : x \mapsto z$ from U_ε into V_ε with

$$\sigma(x_0) = z_0 \quad \textit{and} \quad \sigma(x) \in S(x) \quad \textit{for all } x \in U_\varepsilon. \tag{5.9}$$

The map σ is Lipschitz on U_ε with Lipschitz modulus $(\gamma+\varepsilon)L$ and L from (5.7).

(b) If, in addition, there exists a neighbourhood V of z_0 such that

$$\phi(\xi) = \Sigma(\xi) \cap V \quad \textit{for all } \xi \in W, \tag{5.10}$$

then

$$\sigma(x) = S(x) \cap V_\varepsilon \quad \textit{for all } x \in U_\varepsilon. \tag{5.11}$$

Remark. The assumptions made in (a) and (b) are satisfied if, for example, $Q = \mathbb{R}^k$ and $\mathcal{J}_z C(x_0, z_0)$ has full rank. Hence Theorem 5.2 contains the classic Implicit Function Theorem as a special case.

Proof of Theorem 5.2.. **(a)** For arbitrary but fixed $\varepsilon > 0$ we choose $\delta = \delta(\varepsilon) > 0$, $\rho = \rho(\varepsilon) > 0$ and a neighbourhood U_ε of x_0 such that with $V_\varepsilon := z_0 + \rho B$ one has:

$$\begin{aligned} &\gamma\delta < \varepsilon/(\gamma+\varepsilon) && \\ &r(x,z) \in W && \text{for all } (x,z) \in U_\varepsilon \times V_\varepsilon \\ &\|\mathcal{J}_z C(x_0,z_0) - \mathcal{J}_z C(x,z)\| \le \delta && \text{for all } (x,z) \in U_\varepsilon \times V_\varepsilon \\ &\|C(x_0,z_0) - C(x,z_0)\| \le (1-\gamma\delta)\rho/\gamma && \text{for all } x \in U_\varepsilon. \end{aligned} \tag{5.12}$$

The relations (5.12) will be used below without further notice. We split the proof of (a) into two parts:

(i) construction of σ;

(ii) verification of the Lipschitz continuity of σ.

Ad (i) *Construction of σ*. For each *fixed* $\overline{x} \in U_\varepsilon$ we define a map $\Phi_{\overline{x}}$ from $\mathbb{R}^k$ into $\mathbb{R}^k$ by

$$\Phi_{\overline{x}}(\cdot) := \phi(r(\overline{x}, \cdot)). \tag{5.13}$$

Below we will show that

$$\Phi_{\overline{x}} \text{ is a contraction on } V_\varepsilon \text{ which maps } V_\varepsilon \text{ into } V_\varepsilon \tag{5.14}$$

(see (5.15) and (5.16) below). Hence Banach's Fixed Point Theorem applies and there is some $\overline{z}$ in V_ε with

$$\overline{z} = \Phi_{\overline{x}}(\overline{z}) = \phi(r(\overline{x}, \overline{z}))$$

and thus, by (5.8),

$$\overline{z} \in \Sigma(r(\overline{x}, \overline{z})).$$

Proposition 5.1 implies $\overline{z} \in S(\overline{x})$ and, since $\overline{x}$ was arbitrary in U_ε, the existence of a map

$$\sigma : x \mapsto z \in S(x)$$

on U_ε follows. From

$$\Phi_{x_0}(z_0) = \phi(r(x_0, z_0)) = \phi(0) = z_0$$

we get for this map $\sigma(x_0) = z_0$, which taken together proves (5.9). It remains to verify (5.14).

To check the contraction property of $\Phi_{\overline{x}}$, let $z_1, z_2 \in V_\varepsilon$. The Lipschitz continuity of ϕ on W yields

$$\begin{aligned}\|\Phi_{\overline{x}}(z_1) - \Phi_{\overline{x}}(z_2)\| &\leq \gamma \,\|r(\overline{x}, z_1) - r(\overline{x}, z_2)\| \\ &\leq \gamma \cdot \sup\{\|\mathcal{J}_z r(\overline{x}, (1-\mu)z_1 + \mu z_2)\| \mid \mu \in (0,1)\} \cdot \|z_1 - z_2\|.\end{aligned}$$

Now $\mathcal{J}_z r(\overline{x}, z) = \mathcal{J}_z C(x_0, z_0) - \mathcal{J}_z C(\overline{x}, z)$ and, because of (5.12), we can continue

$$\|\Phi_{\overline{x}}(z_1) - \Phi_{\overline{x}}(z_2)\| \leq \gamma\delta\|z_1 - z_2\| \quad \text{for all } z_1, z_2 \in V_\varepsilon. \tag{5.15}$$

Since $\gamma\delta < 1$ by choice of δ, the map $\Phi_{\overline{x}}$ is indeed a contraction.

Further we have

$$\begin{aligned}\|\Phi_{\overline{x}}(z_0) - z_0\| &= \|\phi(r(\overline{x}, z_0)) - \phi(0)\| \\ &\leq \gamma\|r(\overline{x}, z_0) - 0\| \\ &= \gamma\|C(x_0, z_0) - C(\overline{x}, z_0)\| \\ &\leq (1-\gamma\delta)\rho.\end{aligned}$$

This implies for $z \in V_\varepsilon (= z_0 + \rho I\!B)$

$$\begin{aligned}\|\Phi_{\overline{x}}(z) - z_0\| &\leq \|\Phi_{\overline{x}}(z) - \Phi_{\overline{x}}(z_0)\| + \|\Phi_{\overline{x}}(z_0) - z_0\| \\ &\leq \gamma\delta\|z - z_0\| + (1-\gamma\delta)\rho \qquad (5.16) \\ &\leq \rho,\end{aligned}$$

i.e., $\Phi_{\overline{x}}$ maps V_ε onto itself. The inequalities (5.15) and (5.16) justify the use of Banach's Fixed Point Theorem and thus guarantee the existence of the map σ.

Ad (ii) *Lipschitz continuity of* σ. Next we prove that σ is Lipschitz on U_ε with modulus $(\gamma + \varepsilon)L$. We can assume without loss of generality that $U_\varepsilon \times V_\varepsilon \subset \widetilde{U} \times \widetilde{V}$ with $\widetilde{U}, \widetilde{V}$ from (5.7). Then for arbitrary $x_1, x_2 \in U_\varepsilon$

$$\begin{aligned}\|\sigma(x_1) - \sigma(x_2)\| &= \|\Phi_{x_1}(\sigma(x_1)) - \Phi_{x_2}(\sigma(x_2))\| \\ &\leq \|\Phi_{x_1}(\sigma(x_1)) - \Phi_{x_1}(\sigma(x_2))\| + \|\Phi_{x_1}(\sigma(x_2)) - \Phi_{x_2}(\sigma(x_2))\|.\end{aligned}$$

For the first term of the last expression we get from (5.15) the upper bound (note: $\overline{x} \in U_\varepsilon$ was arbitrary)

$$\gamma\delta\|\sigma(x_1) - \sigma(x_2)\|.$$

The Lipschitz continuity of ϕ yields the following upper bound for the second term:

$$\begin{aligned}\|\Phi_{x_1}(\sigma(x_2)) - \Phi_{x_2}(\sigma(x_2))\| &= \|\phi(r(x_1, \sigma(x_2))) - \phi(r(x_2, \sigma(x_2)))\| \\ &\leq \gamma\|C(x_1, \sigma(x_2)) - C(x_2, \sigma(x_2))\|.\end{aligned}$$

Combining these estimates and using (5.7) we see that

$$\begin{aligned}\|\sigma(x_1) - \sigma(x_2)\| &\leq \gamma\delta\|\sigma(x_1) - \sigma(x_2)\| + \gamma\|C(x_1, \sigma(x_2)) - C(x_2, \sigma(x_2))\| \\ &\leq \gamma\delta\|\sigma(x_1) - \sigma(x_2)\| + \gamma L\|x_1 - x_2\|,\end{aligned}$$

which implies the claimed Lipschitz property for σ on U_ε:

$$\|\sigma(x_1) - \sigma(x_2)\| \leq \frac{\gamma L}{1-\gamma\delta}\|x_1 - x_2\| < (\gamma + \varepsilon)L\|x_1 - x_2\|.$$

(b) By making ρ in (5.12) smaller, if necessary, we can assume $V_\varepsilon \subset V$. Now fix some $x \in U_\varepsilon$ and let z be an arbitrary element from $S(x) \cap V_\varepsilon$. To prove (5.11) we will show that $z = \sigma(x)$. By Proposition 5.1, $z \in \Sigma(r(x,z)) \cap V_\varepsilon$. From (5.12) we know that $r(x,z) \in W$ and thus by assumption (5.10) and definition (5.13)

$$z = \phi(r(x,z)) = \Phi_x(z).$$

Since $\Phi_x(\cdot)$ has only one fixed point in V_ε, z must be this uniquely determined fixed point $\sigma(x)$ from (a), which shows

$$\sigma(x) = S(x) \cap V_\varepsilon \quad \text{for } x \in U_\varepsilon.$$

■

Remark. The above proof did not use any special properties of the set-valued part in (5.1). Hence Theorem 5.2 remains true if we replace the map $N_Q(\cdot)$ in (5.1) by any other set-valued map. The main result of Dontchev and Hager, 1994, which contains Theorem 5.2 as a special case, appears in this more general form. Also the continuous differentiability of C can be weakened. In fact, the statement remains true if C possesses property (5.7) and if we replace the GE in (5.4) by

$$\xi \in \mathcal{G}(z) + N_Q(z),$$

where $\mathcal{G}(z_0) = C(x_0, z_0)$ and $\mathcal{G}$ "strongly" approximates C in z at (x_0, z_0); cf. Robinson, 1991.

In this book we mainly use part (b) of Theorem 5.2.
It is convenient to introduce the following definition.

Definition 5.1 (Strong regularity condition) *Let $z_0 \in S(x_0)$ with S given by (5.2). Suppose there exist neighbourhoods W of $0 \in \mathbb{R}^k$ and V of z_0 such that the map $\xi \mapsto \Sigma(\xi) \cap V$ is single-valued and Lipschitz on W with modulus γ. Then we call (5.1)* strongly regular *at (x_0, z_0) (with Lipschitz modulus γ) or say that (5.1) satisfies the* strong regularity condition *(SRC) at (x_0, z_0).*

From Theorem 5.2 we know that (SRC) at (x_0, z_0) implies for each $\varepsilon > 0$ the existence of another pair of neighbourhoods U_ε of x_0 and V_ε of z_0, such that the map $x \mapsto \sigma(x) := S(x) \cap V_\varepsilon$ is single-valued and Lipschitz on U_ε with modulus $(\gamma + \varepsilon)L$.

The next two sections will give more convenient criteria which ensure (SRC).

5.2 GENERALIZED EQUATIONS WITH POLYHEDRAL FEASIBLE SETS

For polyhedral feasible sets Q the strong regularity admits a useful characterization. Let us rewrite the GE (5.1) in the NSE form

$$z = \operatorname{Proj}_Q(z - C(x,z)).$$

As in Section 5.1, let (x_0, z_0) be a reference pair for (5.1) and denote by K the critical cone of Q with respect to z_0 and $-C(x_0, z_0)[= (z_0 - C(x_0, z_0)) - z_0]$, i.e., (see (2.50))

$$K = T_Q(z_0) \cap \{C(x_0, z_0)\}^{\perp}. \tag{5.17}$$

Then Theorem 2.31 and Corollary 2.5 provide a condition equivalent to (SRC) at (x_0, z_0).

Theorem 5.3 *Let Q in (5.1) be polyhedral. Then the following statements are equivalent:*

(i) The GE (5.1) is strongly regular at (x_0, z_0).

(ii) The map

$$\Lambda : \xi \mapsto \{\eta \in \mathbb{R}^k \mid \xi \in \mathcal{J}_z C(x_0, z_0)\eta + N_K(\eta)\} \tag{5.18}$$

is single-valued on $\mathbb{R}^k$.

Proof. We start by showing the existence of neighbourhoods V and W of z_0 and $0 \in \mathbb{R}^k$, respectively, such that

$$\Sigma(\xi) \cap V = z_0 + \Lambda(\xi) \cap (V - z_0) \qquad \text{for all} \qquad \xi \in W. \tag{5.19}$$

For this purpose we fix ξ and consider the generalized equation from the right-hand side of (5.4)

$$0 \in -\xi + C(x_0, z_0) + \mathcal{J}_z C(x_0, z_0)(z - z_0) + N_Q(z).$$

This GE can equivalently be written as a NSE of type (4.16)

$$z = \text{Proj}_Q[z - (-\xi + C(x_0, z_0) + \mathcal{J}_z C(x_0, z_0)(z - z_0))],$$

which shows that

$$\Sigma(\xi) = \{z \in \mathbb{R}^k \mid \text{Proj}_Q[\xi - C(x_0, z_0) - \mathcal{J}_z C(x_0, z_0)(z - z_0) + z] = z\}. \tag{5.20}$$

It is clear that the neighbourhoods V and W can be chosen in such a way that the norm of the vector

$$\xi - \mathcal{J}_z C(x_0, z_0)(z - z_0) + z - z_0$$

becomes arbitrarily small for all $z \in V$ and $\xi \in W$. Since Q is polyhedral and $0 \in C(x_0, z_0) + N_Q(z_0)$, i.e.,

$$\text{Proj}_Q(z_0 - C(x_0, z_0)) = z_0$$

(cf. (4.16)), Theorem 2.31 applies. If we put there

$$\Omega := Q, y := z_0, x := z_0 - C(x_0, z_0), k := \xi - \mathcal{J}_z C(x_0, z_0)(z - z_0) + z - z_0,$$

then we get for the projection map in (5.20)

$$\begin{aligned} &\text{Proj}_Q\,(\xi - C(x_0, z_0) - \mathcal{J}_z C(x_0, z_0)(z - z_0) + z) \\ &\qquad = z_0 + \text{Proj}_K\,(\xi - \mathcal{J}_z C(x_0, z_0)(z - z_0) + z - z_0) \quad \text{for all } z \in V,\ \xi \in W, \end{aligned}$$

with K from (5.17). For $\xi \in W$ the sets $\Sigma(\xi) \cap V$ can thus be characterized by

$$\Sigma(\xi) \cap V = \{z \in V \mid z - z_0 = \text{Proj}_K(\xi - \mathcal{J}_z C(x_0, z_0)(z - z_0) + z - z_0)\}$$

or, going back from the NSE to the corresponding GE, by

$$\Sigma(\xi) \cap V = \{z \in V \mid \xi \in \mathcal{J}_z C(x_0, z_0)(z - z_0) + N_K(z - z_0)\} \quad \text{for } \xi \in W.$$

With $\eta := z - z_0$ this becomes

$$\Sigma(\xi) \cap V = \{z_0 + \eta \mid \xi \in \mathcal{J}_z C(x_0, z_0)\eta + N_K(\eta)\} \cap V \tag{5.21}$$

and finally we get with Λ from (5.18),

$$\Sigma(\xi) \cap V = z_0 + \Lambda(\xi) \cap (V - z_0).$$

This proves (5.19) and it remains to verify the equivalence of the statements:

(i)' The map $\xi \mapsto \Lambda(\xi) \cap (V - z_0)$ is single-valued and Lipschitz on W;

(ii) Λ is single-valued on $\mathbb{R}^m$.

Λ is the inverse of the map $\eta \mapsto \mathcal{J}_z C(x_0, z_0)\eta + N_K(\eta)$ which is polyhedral. Hence, by Corollary 2.5, Λ is Lipschitz on a convex subset of $\mathbb{R}^k$, whenever Λ is single-valued on this set. Moreover, Λ is positively homogeneous so that single-valuedness on a neighbourhood of $0 \in \mathbb{R}^k$ is equivalent to single-valuedness on the whole $\mathbb{R}^k$. This completes the proof. ∎

Combining Theorem 4.6 and Theorem 5.3 one gets the following criterion which ensures the single-valuedness of the map Λ from (5.18) and thus the strong regularity at (x_0, z_0). We prefer to give a more self-contained proof which is not based on the equivalence of the above statements (i)',(ii).

Theorem 5.4 *Let Q in (5.1) be polyhedral and suppose that $\mathcal{J}_z C(x_0, z_0)$ is strictly copositive with respect to $K - K$. Then the GE (5.1) is strongly regular at (x_0, z_0).*

Proof. As explained in the first part of the proof of Theorem 5.3, we can find neighbourhoods V and W of z_0 and 0, respectively, such that

$$\Sigma(\xi) \cap V = z_0 + \Lambda(\xi) \cap (V - z_0) \quad \text{for all } \xi \in W.$$

Therefore, by Theorem 5.2, it suffices to show that Λ is single-valued and Lipschitz on a neighbourhood of 0, which is equivalent to the single-valuedness and the Lipschitz continuity of Λ on the whole $\mathbb{R}^k$. K is a polyhedral cone with vertex at the origin and so, by Theorem 4.6, the strict copositivity of $\mathcal{J}_z C(x_0, z_0)$ with respect to $K - K$ implies the single-valuedness of Λ. To establish the Lipschitz continuity of Λ, we observe that, as in the proof of Theorem 4.6, there exists an $\alpha > 0$ such that

$$\langle z_1 - z_2, \mathcal{J}_z C(x_0, z_0)z_1 - \mathcal{J}_z C(x_0, z_0)z_2\rangle \geq \alpha\|z_1 - z_2\|^2 \text{ for all } z_1, z_2 \in K. \tag{5.22}$$

Now let ξ_1, ξ_2 be any elements from $\mathbb{R}^k$ and put $\eta_1 = \Lambda(\xi_1)$ and $\eta_2 = \Lambda(\xi_2)$. Then the corresponding VI formulation yields

$$\begin{aligned} \langle \mathcal{J}_z C(x_0, z_0)\eta_1 - \xi_1, \eta_2 - \eta_1\rangle &\geq 0 \\ \langle \mathcal{J}_z C(x_0, z_0)\eta_2 - \xi_2, \eta_1 - \eta_2\rangle &\geq 0, \end{aligned}$$

and consequently

$$\langle \mathcal{J}_z C(x_0, z_0)\eta_2 - \xi_2, \eta_2 - \eta_1\rangle \leq \langle \mathcal{J}_z C(x_0, z_0)\eta_1 - \xi_1, \eta_2 - \eta_1\rangle\,. \tag{5.23}$$

By combining (5.22) and (5.23) we arrive at

$$\begin{aligned}\alpha\|\eta_2-\eta_1\|^2 &\le \langle \mathcal{J}_z C(x_0,z_0)\eta_2 - \mathcal{J}_z C(x_0,z_0)\eta_1, \eta_2-\eta_1\rangle \\ &= \langle \mathcal{J}_z C(x_0,z_0)\eta_2 - \xi_2, \eta_2-\eta_1\rangle - \langle \mathcal{J}_z C(x_0,z_0)\eta_1 - \xi_1, \eta_2-\eta_1\rangle \\ &\qquad + \langle \xi_2-\xi_1, \eta_2-\eta_1\rangle \le \langle \xi_2-\xi_1, \eta_2-\eta_1\rangle,\end{aligned}$$

and thus

$$\|\Lambda(\xi_1)-\Lambda(\xi_2)\| = \|\eta_2-\eta_1\| \le \frac{1}{\alpha}\|\xi_2-\xi_1\|.$$

This proves the Lipschitz continuity of $\Lambda(\cdot)$ on $\mathbb{R}^k$. ■

In the next section we will become more specific with respect to Q.

5.3 ADMISSIBLE SETS OF PARTICULAR INTEREST

We first consider the *perturbed* version of the GE (4.14)

$$0 \in \begin{bmatrix} \mathcal{L}(x,y,\mu,\lambda,) \\ H(x,y) \\ -G(x,y) \end{bmatrix} + N_{\mathbb{R}^m\times\mathbb{R}^\ell\times\mathbb{R}^s_+}(y,\mu,\lambda), \tag{5.24}$$

in which the functions F, $G=(g^1,\dots,g^s)^T$ and $H=(h^1,\dots,h^\ell)^T$ now also depend on a perturbation parameter x which runs over an open set $\mathcal{A}\subset\mathbb{R}^n$. Let $\mathcal{L}$ denote the resulting perturbed D-Lagrangian (cf. (4.13)

$$\mathcal{L}(x,y,\mu,\lambda) = F(x,y) + \sum_{i=1}^{\ell}\mu^i\nabla_y h^i(x,y) + \sum_{i=1}^{s}\lambda^i\nabla_y g^i(x,y).$$

Henceforth we assume that

(i) F is continuously differentiable on $\mathcal{A}\times\mathbb{R}^m$;

(ii) H and G are twice continuously differentiable on $\mathcal{A}\times\mathbb{R}^m$;

(iii) for each $x\in\mathcal{A}$, the map $H(x,\cdot)$ is affine and $g^i(x,\cdot)$ is convex on $\mathbb{R}^m$, $i=1,2,\dots,s$.

For the perturbation $x\in\mathcal{A}$ we denote by

$$\Delta(x) := \{y\in\mathbb{R}^m \mid h^i(x,y)=0,\ i=1,2,\dots,\ell,\ g^j(x,y)\le 0,\ j=1,2,\dots,s\} \tag{5.25}$$

the corresponding feasible set. Now fix some $\overline{x}\in\mathcal{A}$ and assume that the following *extended Slater constraint qualification* holds at $\overline{x}$:

(ESCQ): There exists $\overline{y}:=y(\overline{x})\in\mathbb{R}^m$ such that $g^i(\overline{x},\overline{y})<0$ for $i=1,2,\dots,s$ and $h^j(\overline{x},\overline{y})=0$ for $j=1,2,\dots,\ell$.

We know from Section 4.1 that in this situation the GE (5.24) (with $x=\overline{x}$) is equivalent to the VI:

$$\left.\begin{array}{l}\text{Find } y\in\Delta(\overline{x}) \text{ such that}\\ \qquad\langle F(\overline{x},y), v-y\rangle \ge 0 \quad \text{for all } v\in\Delta(\overline{x}).\end{array}\right\} \tag{5.26}$$

The reference point for (5.24) now becomes a quadruple $(x_0, y_0, \mu_0, \lambda_0)$ (which satisfies (5.24)). For such $(x_0, y_0, \mu_0, \lambda_0)$ we denote by

$$I(x_0, y_0) := \{i \in \{1, 2, \dots, s\} \mid g^i(x_0, y_0) = 0\}$$

the index set of *active* inequality constraints and by

$$L(x_0, y_0) := \{1, 2, \dots, s\} \setminus I(x_0, y_0)$$

the index set of *inactive* inequalities. For technical reasons we further decompose the set $I(x_0, y_0)$ for a given λ_0 into the set of *strongly active* inequalities

$$I^+(x_0, y_0, \lambda_0) := \left\{i \in I(x_0, y_0) \mid \lambda_0^i > 0\right\}$$

and the set of *weakly active* inequalities

$$I^0(x_0, y_0, \lambda_0) := I(x_0, y_0) \setminus I^+(x_0, y_0, \lambda_0).$$

The cardinalities of $I(x_0, y_0)$, $I^+(x_0, y_0, \lambda_0)$ and $I^0(x_0, y_0, \lambda_0)$ will be denoted by a, a^+ and a^0, respectively, i.e.,

$$a = |I(x_0, y_0)|, \qquad a^+ = |I^+(x_0, y_0, \lambda_0)|, \qquad a^0 = |I^0(x_0, y_0, \lambda_0)|.$$

Later we will often use the above index sets as vector and matrix subscripts and skip the arguments. Finally, let us introduce the $(m + l + s) \times (m + l + s)$ matrix

$$D(x_0, y_0, \mu_0, \lambda_0) := \begin{bmatrix} \mathcal{J}_y \mathcal{L}(x_0, y_0, \mu_0, \lambda_0) & (\mathcal{J}_y H(x_0, y_0))^T & (\mathcal{J}_y G(x_0, y_0))^T \\ \mathcal{J}_y H(x_0, y_0) & 0 & 0 \\ -\mathcal{J}_y G(x_0, y_0) & 0 & 0 \end{bmatrix}$$

and, for a subset N of $I(x_0, y_0)$, the square matrix

$$D_{(N)}(x_0, y_0, \mu_0, \lambda_0) := \begin{bmatrix} \mathcal{J}_y \mathcal{L}(x_0, y_0, \mu_0, \lambda_0) & (\mathcal{J}_y H(x_0, y_0))^T & (\mathcal{J}_y G_N(x_0, y_0))^T \\ \mathcal{J}_y H(x_0, y_0) & 0 & 0 \\ -\mathcal{J}_y G_N(x_0, y_0) & 0 & 0 \end{bmatrix}, \tag{5.27}$$

where G_N consists of the functions g^i, $i \in N$.

Proposition 5.5 *The GE (5.24) is strongly regular at* $(x_0, y_0, \mu_0, \lambda_0)$ *if and only if the map*

$$\widetilde{\Lambda} : \xi \mapsto \left\{\tilde{\eta} \in \mathbb{R}^{m+\ell+a^++a^0} \mid \xi \in D_{(I)}(x_0, y_0, \mu_0, \lambda_0)\tilde{\eta} + N_{\mathbb{R}^m \times \mathbb{R}^\ell \times \mathbb{R}^{a^+} \times \mathbb{R}_+^{a^0}}(\tilde{\eta})\right\} \tag{5.28}$$

is single-valued on $\mathbb{R}^{m+\ell+a^++a^0}$.

Proof. The GE (5.24) has a polyhedral feasible set $\mathbb{R}^m \times \mathbb{R}^\ell \times \mathbb{R}_+^s$ so that Theorem 5.3 applies. A closer look at (5.24) shows that the GE associated with the map Λ in (5.18) now attains the form

$$\xi \in D(x_0, y_0, \mu_0, \lambda_0)\eta + N_K(\eta), \tag{5.29}$$

with

$$\begin{aligned} K &= \{\eta = (v, w, u) \in \mathbb{R}^m \times \mathbb{R}^\ell \times \mathbb{R}^s \mid u^i \geq 0 \text{ for } i \in L(x_0, y_0) \cup I^0(x_0, y_0, \lambda_0)\} \\ &\quad \cap \{\eta = (v, w, u) \in \mathbb{R}^m \times \mathbb{R}^\ell \times \mathbb{R}^s \mid u^i = 0 \text{ for } i \in L(x_0, y_0)\} \\ &= \{\eta = (v, w, u) \in \mathbb{R}^m \times \mathbb{R}^\ell \times \mathbb{R}^s \mid u_{I^0} \geq 0,\ u_L = 0\}. \end{aligned} \tag{5.30}$$

Consequently, for $\eta = (v, w, u) \in K$,

$$\begin{aligned} N_K(\eta) = \{\eta^* = (v^*, w^*, u^*) \in \mathbb{R}^m \times \mathbb{R}^\ell \times \mathbb{R}^s \mid v^* = 0,\ w^* = 0,\ u^*_{I^+} = 0 \\ \text{and } u^*_{I^0} \leq 0 \text{ with } (u^*)^i u^i \doteq 0 \text{ for } i \in I^0(x_0, y_0, \lambda_0)\}. \end{aligned} \tag{5.31}$$

If we take (5.31) into account, then we observe that the GE (5.29) can be simplified:

(i) the columns of the matrix $D(x_0, y_0, \mu_0, \lambda_0)$ corresponding to inactive inequality constraints can be omitted because, due to (5.30), the components u^i are zero for $i \in L(x_0, y_0)$;

(ii) the rows of (5.29) corresponding to inactive constraints can be omitted, because the components $(u^*)^i$ of $N_K(\eta)$ are free for $i \in L(x_0, y_0)$. Thus these rows do not present any restrictions for the variable η.

On the basis of these simplifications we conclude that it suffices to guarantee the single-valuedness of the corresponding reduced map $\widetilde{\Lambda}$ instead of the original map Λ given by (5.29). ■

Remark. The shrinking of Λ to $\widetilde{\Lambda}$ is the well-known *reduction procedure* due to Robinson, explained in Robinson, 1980 in a more general framework. This procedure "identifies that portion of the problem to which we have to attach conditions in order for the original GE to be strongly regular at $(x_0, y_0, \mu_0, \lambda_0)$".

The next lemma will help to combine Theorem 4.7 and Proposition 5.5.

Lemma 5.6 *Let n_1 and n_2 be positive integers and A be an $(n_1 + n_2) \times (n_1 + n_2)$ matrix*

$$A = \begin{bmatrix} A_{11} & A_{12} \\ A_{21} & A_{22} \end{bmatrix},$$

where A_{11} is $n_1 \times n_1$. Denote by T the map which assigns the (set of) solutions $p = (p_1, p_2) \in \mathbb{R}^{n_1} \times \mathbb{R}^{n_2}$ of the GE

$$\xi \in Ap + N_{\mathbb{R}^{n_1} \times \mathbb{R}^{n_2}_+}(p_1, p_2)$$

to $\xi = (\xi_1, \xi_2) \in \mathbb{R}^{n_1} \times \mathbb{R}^{n_2}$. Then the following statements are equivalent:

(i) T is single-valued and Lipschitz on $\mathbb{R}^{n_1} \times \mathbb{R}^{n_2}$;

(ii) A_{11} is nonsingular and the Schur complement A/A_{11} of A_{11} in A (i.e., the matrix $A_{22} - A_{21}A_{11}^{-1}A_{12}$) is a P-matrix.

Proof. We start with the more straightforward part (ii) $\Rightarrow$ (i). For $\xi = (\xi_1, \xi_2) \in \mathbb{R}^{n_1} \times \mathbb{R}^{n_2}$ the GE defining T splits into the linear equation

$$A_{11}p_1 + A_{12}p_2 = \xi_1$$

and the LCP

$$(A/A_{11})p_2 - b \geq 0, \quad p_2 \geq 0, \quad \langle (A/A_{11})p_2 - b, p_2 \rangle = 0, \tag{5.32}$$

where

$$b := \xi_2 - A_{21}A_{11}^{-1}\xi_1.$$

For a P-matrix A/A_{11} we know from Theorem 4.7 that (5.32) possesses a unique solution p_2 for each $b \in \mathbb{R}^{n_2}$ (i.e., for each ξ) and, since A_{11} is nonsingular, also the component p_1 is uniquely determined. This shows that T is single-valued on $\mathbb{R}^{n_1} \times \mathbb{R}^{n_2}$. Now T is the inverse of the polyhedral map $p \mapsto Ap + N_{\mathbb{R}^{n_1} \times \mathbb{R}^{n_2}_+}(p_1, p_2)$ and thus Lipschitz by Corollary 2.5.

To prove the implication (i) $\Rightarrow$ (ii), we suppose by contradiction that A_{11} is singular and choose any $u \in \mathbb{R}^{n_1}$ not in the range of A_{11}. With this u and some $v \in \mathbb{R}^{n_2}$ with strictly negative components we define

$$\xi_\vartheta := \begin{bmatrix} \vartheta u \\ v \end{bmatrix} \quad \text{for } \vartheta \geq 0.$$

Then $0 \in T(\xi_0)$ and from the Lipschitz continuity of T we conclude that the point $p_\vartheta := T(\xi_\vartheta)$ will be close to zero for all $\vartheta > 0$ sufficiently small. Let $p_{\vartheta 1} \in \mathbb{R}^{n_1}$ and $p_{\vartheta 2} \in \mathbb{R}^{n_2}$ be the decomposition of p_ϑ and observe that

$$v - A_{21}p_{\vartheta 1} - A_{22}p_{\vartheta 2} \in N_{\mathbb{R}^{n_2}_+}(p_{\vartheta 2}). \tag{5.33}$$

For small ϑ the left-hand side of (5.33) will be strictly negative in all components, which implies $p_{\vartheta 2} = 0$. Hence

$$\vartheta u = A_{11}p_{\vartheta 1} + A_{12}p_{\vartheta 2} = A_{11}p_{\vartheta 1}$$

and this contradicts the choice of u. Thus A_{11} must be nonsingular and, consequently, A/A_{11} is well defined. Since $T(\xi)$ is a singleton for each $\xi \in \mathbb{R}^{n_1} \times \mathbb{R}^{n_2}$, the LCP (5.32) has a unique solution for each $b \in \mathbb{R}^{n_2}$ which, due to Proposition 4.7, implies that A/A_{11} is a P-matrix. ∎

By combining Proposition 5.5 and Lemma 5.6 we get a characterization of the strong regularity of the GE (5.24) at $(x_0, y_0, \mu_0, \lambda_0)$.

Theorem 5.7 *The following statements are equivalent:*

(i) The GE (5.24) is strongly regular at $(x_0, y_0, \mu_0, \lambda_0)$.

(ii) The matrix $D_{(I+)}(x_0, y_0, \mu_0, \lambda_0)$ *is nonsingular and its Schur complement in* $D_{(I)}(x_0, y_0, \mu_0, \lambda_0)$ *is a P-matrix.*

Proof. The claim follows from Lemma 5.6 if we put there

$$n_1 := m + \ell + a^+, \qquad n_2 := a^0$$

and (with $\widetilde{\Lambda}$ from (5.28))

$$T := \widetilde{\Lambda}, \quad A := D_{(I)}(x_0, y_0, \mu_0, \lambda_0), \quad A_{11} := D_{(I+)}(x_0, y_0, \mu_0, \lambda_0).$$ ∎

We say that the *extended linear independence constraint qualification* (ELICQ) holds at $(\overline{x}, \overline{y})$ provided

(ELICQ): The gradients $\nabla_y h^i(\overline{x}, \overline{y})$ for $i = 1, 2, \ldots, \ell$ and $\nabla_y g^j(\overline{x}, \overline{y})$ for $j \in I(\overline{x}, \overline{y})$ are linearly independent.

It follows from Theorem 5.7 that the strong regularity of the GE (5.24) at $(x_0, y_0, \mu_0, \lambda_0)$ implies (ELICQ) at (x_0, y_0). Indeed, since each P-matrix is nonsingular, condition (ii) of Theorem 5.7 implies that $D_{(I)}(x_0, y_0, \mu_0, \lambda_0)$ is nonsingular. If, however, (ELICQ) does not hold at (x_0, y_0) then $D_{(I)}(x_0, y_0, \mu_0, \lambda_0)$ is necessarily singular and thus condition (ii) cannot hold.

We conclude that, whenever (SRC) holds at $(x_0, y_0, \mu_0, \lambda_0)$, then the pair (μ_0, λ_0) is the only KKT vector associated with (x_0, y_0). Hence, we can skip the argument λ_0 in $I^+(x_0, y_0, \lambda_0)$ and $I^0(x_0, y_0, \lambda_0)$ in this situation.

The next result shows that (ELICQ) together with an additional monotonicity assumption on $\mathcal{J}_y \mathcal{L}(x_0, y_0, \mu_0, \lambda_0)$ implies the strong regularity of (5.24).

Theorem 5.8 *Suppose that (ELICQ) holds at (x_0, y_0) and that*

$$\left.\begin{array}{r} \mathcal{J}_y \mathcal{L}(x_0, y_0, \mu_0, \lambda_0) \textit{ is strictly copositive with respect to} \\ \operatorname{Ker}(\mathcal{J}_y H(x_0, y_0)) \cap \operatorname{Ker}(\mathcal{J}_y G_{I^+}(x_0, y_0)). \end{array}\right\} \quad (5.34)$$

Then the GE (5.24) is strongly regular at $(x_0, y_0, \mu_0, \lambda_0)$.

Proof. According to Proposition 5.5 it suffices to prove that the map $\widetilde{\Lambda}$ from (5.28) is single-valued on $\mathbb{R}^{m+\ell+a^+ + a^0}$. To this purpose fix ξ and $\tilde{\eta} \in \widetilde{\Lambda}(\xi)$ with

$$\tilde{\eta} = (\tilde{\eta}_1, \tilde{\eta}_2, \tilde{\eta}_3, \tilde{\eta}_4) \in \mathbb{R}^m \times \mathbb{R}^\ell \times \mathbb{R}^{a^+} \times \mathbb{R}^{a^0}$$

and

$$\xi = (\xi_1, \xi_2, \xi_3, \xi_4) \in \mathbb{R}^m \times \mathbb{R}^\ell \times \mathbb{R}^{a^+} \times \mathbb{R}^{a^0}.$$

From the analysis in Section 4.1 we know that $\tilde{\eta}_1$ is a solution and $(\tilde{\eta}_2, \tilde{\eta}_3, \tilde{\eta}_4)$ a KKT vector of the GE

$$\xi_1 \in \mathcal{J}_y \mathcal{L}(x_0, y_0, \mu_0, \lambda_0)\tilde{\eta}_1 + N_{Q_\xi}(\tilde{\eta}_1)$$

where Q_ξ is the cone

$$\begin{aligned} Q_\xi \quad := \quad & \{v \in \mathbb{R}^m \mid \mathcal{J}_y H(x_0, y_0)v = \xi_2, \\ & \mathcal{J}_y G_{I^+}(x_0, y_0)v + \xi_3 = 0, \quad \mathcal{J}_y G_{I^0}(x_0, y_0)v + \xi_4 \le 0\}. \end{aligned}$$

Due to Theorem 4.6 the solution $\tilde{\eta}$ of this GE is unique if

$$\langle v_1 - v_2, \mathcal{J}_y \mathcal{L}(x_0, y_0, \mu_0, \lambda_0)(v_1 - v_2)\rangle > 0 \quad \text{for all } v_1, v_2 \in Q_\xi, v_1 \neq v_2.$$

But this holds by the assumption on $\mathcal{J}_y \mathcal{L}(x_0, y_0, \mu_0, \lambda_0)$ since

$$v_1 - v_2 \in \operatorname{Ker}(\mathcal{J}_y H(x_0, y_0)) \cap \operatorname{Ker}(\mathcal{J}_y G_{I^+}(x_0, y_0)) \text{ for } v_1, v_2 \in Q_\xi.$$

Since (ELICQ) holds at (x_0, y_0) also the KKT vector $(\tilde{\eta}_2, \tilde{\eta}_3, \tilde{\eta}_4)$ is unique. ■

Remark. Due to Theorem 5.7, (ELICQ) at (x_0, y_0) together with the condition (5.34) imply the statement (ii) of Theorem 5.7.

If the GE (5.24) represents first-order optimality conditions of an optimization problem, then (5.34) in Theorem 5.8 is just the *strong second-order sufficient condition*, well-known in mathematical programming (e.g., Fiacco and McCormick, 1968). Therefore we will use the same name and refer to (5.34) as (SSOSC).

We now turn our attention to the GE (5.1) with $k = m$, $z = y$, and

$$Q = \mathbb{R}^m_+ + \Psi, \tag{5.35}$$

where $\Psi := (\psi_1, \psi_2, \ldots, \psi_m)^T$ is a given vector in $\mathbb{R}^m$. This GE is equivalent to the perturbed NCP (cf. (4.6)):

$$\left.\begin{array}{l}\text{Find } y \in \mathbb{R}^m \text{ such that} \\ \quad F(x,y) \geq 0,\ y - \Psi \geq 0,\ \langle F(x,y), y - \Psi\rangle = 0.\end{array}\right\} \tag{5.36}$$

As (4.5) is a special case of (4.3), the corresponding GE (5.24) attains the form

$$0 \in \begin{bmatrix} F(x,y) - \lambda \\ y - \Psi \end{bmatrix} + N_{\mathbb{R}^m \times \mathbb{R}^m_+}(y, \lambda). \tag{5.37}$$

We observe that (ELICQ) holds at all (x_0, y_0) which satisfy (5.37) (with $\lambda_0 = F(x_0, y_0)$), and that

$$\begin{aligned} I^+(x_0, y_0) &= \left\{i \in \{1, 2, \ldots, m\} \mid F^i(x_0, y_0) > 0\right\}, \\ L(x_0, y_0) &= \left\{i \in \{1, 2, \ldots, m\} \mid y_0^i > \psi^i\right\}, \\ I^0(x_0, y_0) &= \left\{i \in \{1, 2, \ldots, m\} \mid F^i(x_0, y_0) = 0,\ y_0^i = \psi^i\right\}. \end{aligned}$$

Theorem 5.9 *The following statements are equivalent:*

(i) The GE (5.1) with $k = m$, $z = y$ and Q given by (5.35) is strongly regular at (x_0, y_0).

(ii) The matrix $\mathcal{J}_y F_{L,L}(x_0, y_0)$ is nonsingular and its Schur complement in the matrix

$$\begin{bmatrix} \mathcal{J}_y F_{L,L}(x_0, y_0) & \mathcal{J}_y F_{L,I^0}(x_0, y_0) \\ \mathcal{J}_y F_{I^0,L}(x_0, y_0) & \mathcal{J}_y F_{I^0,I^0}(x_0, y_0) \end{bmatrix} \tag{5.38}$$

is a P-matrix.

Proof. By Theorem 5.3, statement (i) is equivalent to the unique solvability of the GE

$$\xi \in \mathcal{J}_y F(x_0, y_0)\eta + N_K(\eta) \tag{5.39}$$

for each fixed $\xi \in \mathbb{R}^m$. In (5.39) the critical cone K from (5.17) is reduced to

$$\begin{aligned} K &= T_Q(y_0) \cap \{F(x_0, y_0)\}^\perp \\ &= \{\eta \in \mathbb{R}^m \mid \eta_I \geq 0\} \cap \{\eta \in \mathbb{R}^m \mid \eta_{I^+} = 0\} \\ &= \{\eta \in \mathbb{R}^m \mid \eta_{I^+} = 0,\ \eta_{I^0} \geq 0\} \end{aligned}$$

and thus for $\eta \in K$

$$N_K(\eta) = \left\{\eta^* \in \mathbb{R}^m \mid \eta^*_L = 0 \text{ and } \eta^*_{I^0} \leq 0 \text{ with } (\eta^*)^i \eta^i = 0 \text{ for } i \in I^0(x_0, y_0)\right\}.$$

Hence (5.39) is reduced to

$$\begin{bmatrix} \xi_L \\ \xi_{I^0} \end{bmatrix} \in \begin{bmatrix} \mathcal{J}_y F_{L,L}(x_0, y_0) & \mathcal{J}_y F_{L,I^0}(x_0, y_0) \\ \mathcal{J}_y F_{I^0,L}(x_0, y_0) & \mathcal{J}_y F_{I^0,I^0}(x_0, y_0) \end{bmatrix} \begin{bmatrix} \eta_L \\ \eta_{I^0} \end{bmatrix} + N_{\mathbb{R}^{m-a^+-a^0} \times \mathbb{R}^{a^0}_+}(\eta_L, \eta_{I^0}), \tag{5.40}$$

where, as before, a^+ and a^0 denote the cardinalities of $I^+(x_0, y_0)$ and $I^0(x_0, y_0)$, respectively. It remains to apply Lemma 5.6 to the GE (5.40). ■

Remark. The positive definiteness of the matrix (5.38) guarantees uniqueness of the solution to (5.40) for all left-hand side vectors (ξ_L, ξ_{I^0}) and thus both statements of Theorem 5.9.

We close this section with several considerations on the strong regularity of the perturbed version of the GE (4.22) in which now also Φ depends on the perturbation parameter x:

$$0 \in \begin{bmatrix} F(x, y) - \lambda \\ y - \Phi(x, y) \end{bmatrix} + N_{\mathbb{R}^m \times \mathbb{R}^m_+}(y, \lambda). \tag{5.41}$$

We assume that both, F and Φ, are continuously differentiable on $\mathcal{A} \times \mathbb{R}^m$. This GE corresponds to the perturbed ICP (4.9):

$$\left.\begin{array}{l} \text{Find } y \in \mathbb{R}^m \text{ such that} \\ F(x, y) \geq 0,\ y - \Phi(x, y) \geq 0,\ \langle F(x, y), y - \Phi(x, y)\rangle = 0. \end{array}\right\} \tag{5.42}$$

We note again that the strong regularity of the GE (5.41) at (x_0, y_0, λ_0) $(\lambda_0 = F(x_0, y_0))$ implies that (ELICQ) holds at (x_0, y_0) and that

$$\begin{aligned} I^+(x_0, y_0) &= \left\{i \in \{1, 2, \ldots, m\} \mid F^i(x_0, y_0) > 0\right\}, \\ L(x_0, y_0) &= \left\{i \in \{1, 2, \ldots, m\} \mid y^i > \varphi^i(x_0, y_0)\right\}, \\ I^0(x_0, y_0) &= \left\{i \in \{1, 2, \ldots, m\} \mid F^i(x_0, y_0) = 0,\ y^i = \varphi^i(x_0, y_0)\right\}. \end{aligned}$$

The GE (5.41) has almost the same structure as the GE (5.24). Therefore, when computing the critical cone K, we can repeat the argument of (5.30) and obtain

$$K = \{\eta = (v, u) \in \mathbb{R}^m \times \mathbb{R}^m \mid u_{I^0} \geq 0,\ u_L = 0\}.$$

So we can proceed as in the proof of Proposition 5.5.With the $(m+a^++a^0) \times (m+a^++a^0)$ matrix

$$Z(x_0, y_0) = \begin{bmatrix} \mathcal{J}_y F(x_0, y_0) & -E^T_{I^+} & -E^T_{I^0} \\ E_{I^+} - \mathcal{J}_y \Phi_{I^+}(x_0, y_0) & 0 & 0 \\ E_{I^0} - \mathcal{J}_y \Phi_{I^0}(x_0, y_0) & 0 & 0 \end{bmatrix}$$

we get that the GE (5.41) is strongly regular at (x_0, y_0, λ_0) if and only if the map

$$\widetilde{\Lambda} : \xi \mapsto \left\{\hat{\eta} \in \mathbb{R}^{m+a^++a^0} \mid \xi \in Z(x_0, y_0)\hat{\eta} + N_{\mathbb{R}^{m+a^+} \times \mathbb{R}^{a^0}_+}(\hat{\eta})\right\}$$

is single-valued on $\mathbb{R}^{m+a^++a^0}$. By Lemma 5.6 this leads to the following counterpart of Theorem 5.7.

Theorem 5.10 *The following statements are equivalent:*

(i) The GE (5.41) is strongly regular at (x_0, y_0, λ_0).

(ii) The matrix

$$\begin{bmatrix} \mathcal{J}_y F(x_0, y_0) & -E_{I+}^T \\ E_{I+} - \mathcal{J}_y \Phi_{I+}(x_0, y_0) & 0 \end{bmatrix}$$

is nonsingular and its Schur complement in $Z(x_0, y_0)$ *is a* P*-matrix.*

Unfortunately, condition (ii) is not easy to verify in terms of the original problem data. Simple monotonicity assumptions generally do not imply uniqueness results for QVIs, hence there is no counterpart to the easy-to-use Theorem 5.8. At least in some cases, the following sufficient condition may be useful.

Proposition 5.11 *Assume that* $\mathcal{J}_y F(x_0, y_0)$ *is positive definite and* $\mathcal{J}_y \Phi(x_0, y_0)\ (\mathcal{J}_y F(x_0, y_0))^{-1}$ *is negative semidefinite. Then the GE (5.41) is strongly regular at* (x_0, y_0, λ_0).

Proof. The map Σ from (5.4) assigns the solutions (y, λ) of the GE

$$\begin{bmatrix} \xi_1 \\ \xi_2 \end{bmatrix} \in \begin{bmatrix} 0 \\ y_0 - \Phi(x_0, y_0) \end{bmatrix}$$

$$+ \begin{bmatrix} \mathcal{J}_y F(x_0, y_0) & -E \\ E - \mathcal{J}_y \Phi(x_0, y_0) & 0 \end{bmatrix} \begin{bmatrix} y - y_0 \\ \lambda - \lambda_0 \end{bmatrix} + N_{\mathbb{R}^m \times \mathbb{R}^m_+}(y, \lambda)$$

to a vector $\xi = (\xi_1, \xi_2) \in \mathbb{R}^m \times \mathbb{R}^m$. Note that Σ is the inverse of a polyhedral multifunction and hence also polyhedral. Thus the single-valuedness of Σ on $\mathbb{R}^m \times \mathbb{R}^m$ implies the Lipschitz continuity of Σ on $\mathbb{R}^m \times \mathbb{R}^m$ (Corollary 2.5) and, by Theorem 5.2, it suffices to prove that the GE

$$\begin{bmatrix} \xi_1 \\ \xi_2 \end{bmatrix} \in \begin{bmatrix} \mathcal{J}_y F(x_0, y_0) & -E \\ E - \mathcal{J}_y \Phi(x_0, y_0) & 0 \end{bmatrix} \begin{bmatrix} y \\ \lambda \end{bmatrix} + N_{\mathbb{R}^m \times \mathbb{R}^m_+}(y, \lambda) \tag{5.43}$$

has a unique solution (y, λ) for each ξ.

Since $\mathcal{J}_y F(x_0, y_0)$ is positive definite, it is nonsingular and its inverse is also positive definite. Hence, (5.43) splits to the equation

$$y = (\mathcal{J}_y F(x_0, y_0))^{-1} (\xi_1 + \lambda)$$

and the LCP

$$\begin{aligned} &\xi_2 - (\mathcal{J}_y \Phi(x_0, y_0) - E)(\mathcal{J}_y F(x_0, y_0))^{-1} \xi_1 \\ &\quad \in (E - \mathcal{J}_y \Phi(x_0, y_0))(\mathcal{J}_y F(x_0, y_0))^{-1} \lambda + N_{\mathbb{R}^m_+}(\lambda). \end{aligned}$$

Under the assumptions, this LCP has a unique solution λ for each ξ and then also the y-component of the solution to (5.43) is uniquely determined. ∎

Proposition 5.11 is closely related to Theorem 4.9 although the proof techniques differ. For $\Phi(x, y) = \Psi$, the ICP (5.42) is reduced to the NCP (5.36) and the assumptions of Proposition 5.11 are reduced to the positive definiteness of $\mathcal{J}_y F(x_0, y_0)$.

Remark. Alternatively to the above proof, one can show that the assumptions of Proposition 5.11 imply property (ii) of Theorem 5.10.

At the end of this section we perturb the problems from Examples 4.1, 4.2 and 4.3 and apply results of this section to verify the strong regularity at the considered reference points.

Example 5.1 (Example 4.1 continued) Consider the perturbed optimization problem

$$\begin{aligned} &\text{minimize} \quad (y^1)^2 + (xy^2)^2 - xy^1 - \tfrac{1}{2}y^2 \\ &\text{subject to} \\ &\qquad (y^1 - x)^2 \le \tfrac{1}{4}, \\ &\qquad (y^2 - x)^2 \le \tfrac{5}{4} - x. \end{aligned} \tag{5.44}$$

For $x := x_0 = 1$ we get the convex optimization problem from Example 4.1. (ESCQ) holds at all x from a neighborhood of x_0 and so problem (5.44) is equivalent (on a neighbourhood of x_0) to the GE of type (5.24)

$$0 \in \begin{bmatrix} \mathcal{L}(x,y,\lambda) \\ \frac{1}{4} - (y^1 - x)^2 \\ \frac{5}{4} - x - (y^2 - x)^2 \end{bmatrix} + N_{\mathbb{R}^2 \times \mathbb{R}^2_+}(y,\lambda), \tag{5.45}$$

with

$$\mathcal{L}(x,y,\lambda) = \begin{bmatrix} 2y^1 - x \\ 2xy^2 - \frac{1}{2} \end{bmatrix} + \lambda^1 \begin{bmatrix} 2(y^1 - x) \\ 0 \end{bmatrix} + \lambda^2 \begin{bmatrix} 0 \\ 2(y^2 - x) \end{bmatrix}.$$

For $x := x_0$ this GE has a solution $(y_0, \lambda_0) = \left(\frac{1}{2}, \frac{1}{2}, 0, \frac{1}{2}\right)$. (ELICQ) holds at (x_0, y_0) and the matrix

$$\mathcal{J}_y\mathcal{L}(x_0,y_0,\lambda_0) = \begin{bmatrix} 2 + 2\lambda_0^1 & 0 \\ 0 & 2x_0 + 2\lambda_0^2 \end{bmatrix} = \begin{bmatrix} 2 & 0 \\ 0 & 3 \end{bmatrix}$$

is positive definite. Hence, Theorem 5.8 provides the information that the GE (5.45) is strongly regular at (x_0, y_0, λ_0). △

Example 5.2 (Example 4.2 continued) Let $x \in \mathbb{R}$ and consider the GE (5.41) with

$$F(x,y) = \begin{bmatrix} 2x & -1 & 0 & 0 \\ -1 & 2x & -1 & 0 \\ 0 & -1 & 2x & -1 \\ 0 & 0 & -1 & 2x \end{bmatrix} y + \begin{bmatrix} 1 \\ 1 \\ 1 \\ 1 \end{bmatrix} x$$

and

$$\varphi^i(x,y) = \begin{cases} -2.5x - F^i(x,y) & \text{for} \quad i = 1,4, \\ -3x - F^i(x,y) & \text{for} \quad i = 2,3. \end{cases}$$

This GE corresponds to the appropriately perturbed ICP from Example 4.2 and, for $x := x_0 = 1$, has a solution $(y_0, \lambda_0) = (-2, -3, -3, -2, 0, 0, 0, 0)$. One easily checks that

$$\mathcal{J}_y F(x_0, y_0) = \begin{bmatrix} 2 & -1 & 0 & 0 \\ -1 & 2 & -1 & 0 \\ 0 & -1 & 2 & -1 \\ 0 & 0 & -1 & 2 \end{bmatrix}$$

is positive definite. Since further

$$\mathcal{J}_y \Phi(x_0, y_0) = -\mathcal{J}_y F(x_0, y_0),$$

Proposition 5.11 guarantees that the above GE is strongly regular at (x_0, y_0, λ_0). △

Example 5.3 (Example 4.3 continued) Let us introduce a control parameter in the string problem with a rigid obstacle from Example 4.3. In the second part of the book we mostly deal with optimum shape design problems, so the control parameter in our model example is the length $x \in \mathbb{R}$ of the string. Obviously, the displacement $y \in \mathbb{R}^m$ (solution of the variational inequality from Example 4.3) depends on this length; cf. Figure 5.1. We assume

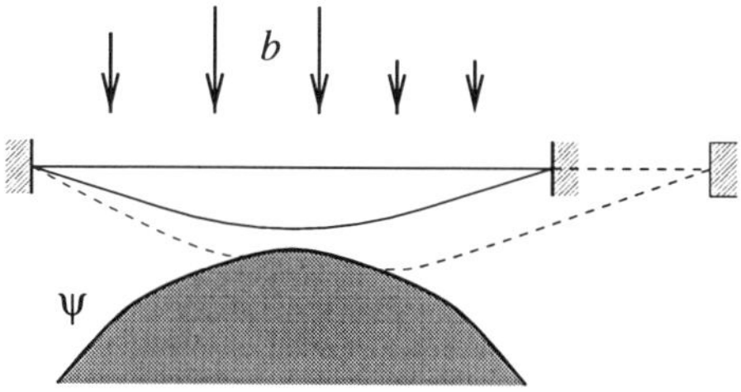

Figure 5.1. String of variable length with a rigid obstacle

that the number $m+1$ of finite elements used for discretization of the string will be the same for all values of x and that the discretization nodes are uniformly distributed, hence all the finite elements have the same length $x/(m+1)$. Then the stiffness matrix A depends on x in the following simple way:

$$A(x) = \frac{m+1}{x} \begin{pmatrix} 2 & -1 & & \\ -1 & 2 & -1 & \\ & & \ddots & \\ & & -1 & 2 \end{pmatrix}.$$

For $0 \le x_L \le x \le x_U$, the matrix $A(x)$ is symmetric and positive definite, hence all its principal submatrices are nonsingular. Further, the Schur complement of each principal submatrix in each larger principal submatrix is a P-matrix. Thus, according to Theorem 5.9, the corresponding generalized equation

$$0 \in A(x)y - b(x) + N_\Omega(y), \quad \Omega := \{v \in \mathbb{R}^m | v^i \ge \psi^i\} \tag{5.46}$$

is strongly regular at each $(\overline{x}, \overline{y})$, provided $x_L \le \overline{x} \le x_U$ and $\overline{y}$ solves (5.46) for $x := \overline{x}$. △

Bibliographical notes

The central stability results for variational inequalities and generalized equations go back to S.M. Robinson. We make extensive use of the papers Robinson, 1979; Robinson, 1980; Robinson, 1991. Robinson, 1979 studies general stability in connection with GEs and, in particular, stability of linear GEs. The concept of strong regularity was introduced and studied in Robinson, 1980. Robinson, 1991 deals with implicit maps defined by nonsmooth equations. The original version of Theorem 5.2 comes from Robinson, 1980; further studies connected with the concept of strong regularity can be found in Dontchev and Hager, 1994 and Dontchev, 1995. The results of Mordukhovich, 1994 allow to derive efficient criteria even for the pseudo-Lipschitz behaviour of the solutions to GEs. The class of not strongly regular GEs of type (5.1) with a pseudo-Lipschitz solution map S is not large, though; cf. Dontchev and Rockafellar, 1996; Kummer, 1997. There are many further results on stability of generalized equations, above all in the context of parametric programming (e.g. Kojima, 1980; Jittorntrum, 1984).

Theorem 5.3 presents an equivalent definition of strong regularity for polyhedral feasible sets; it comes from Robinson, 1980 and Robinson, 1985, see also Kyparisis, 1990. In our presentation, Theorem 5.4 is a direct consequence of Theorem 5.3 and Theorem 4.6. The present proof is taken from Jiang, 1997. Proposition 5.5 is just a special case of the reduction procedure, explained in Robinson, 1980 and Lemma 5.6 goes back to Robinson, 1980. Also Theorems 5.7, 5.9 and Theorem 5.8 can be found in a slightly different form in Robinson, 1980. An alternative approach to these results, based on stability theory for LCPs, was given in Mangasarian and Shiau, 1987. Finally, Proposition 5.11 originates from Outrata, 1995.

6 DERIVATIVES OF SOLUTIONS TO PERTURBED GENERALIZED EQUATIONS

In this chapter we use directional derivatives (Section 6.1) and generalized Jacobians (Section 6.2) to describe the local behaviour of the solution maps S for the perturbed generalized equations from Chapter 5. We restrict ourselves to the case of polyhedral feasible sets and assume further that the strong regularity condition holds. Similarly as in the preceding chapter, we successively analyze the GEs (5.1), (5.24) and (5.41). The transformation of the GEs into the equivalent nonsmooth equations will be a decisive tool in our approach. Further we will prove that under strong regularity the respective selections of the solution maps of the studied GEs are semismooth (Section 6.3). This semismoothness is a crucial prerequisite to apply the numerical methods of Chapter 3.

6.1 DIRECTIONAL DERIVATIVES

We consider the GE (5.1) and the solution map S from (5.2). Assume that (SRC) holds at a reference point (x_0, z_0). Then, by Theorem 5.2, there exist neighbourhoods $\mathcal{U}$ of x_0 and $\mathcal{V}$ of z_0 together with a Lipschitz map $\sigma[\mathcal{U} \to \mathbb{R}^k]$ such that $\sigma(x_0) = z_0$ and

$$\sigma(x) = S(x) \cap \mathcal{V} \quad \text{for all } x \in \mathcal{U}. \tag{6.1}$$

To characterize the behaviour of σ close to x_0, we study its differentiability properties at x_0. From the very beginning we assume that Q is *polyhedral* so that Theorem 2.31 (directional differentiability of the projection map) can be applied. Beside this, our main tool will be a lemma due to Kummer, 1992 dealing with locally Lipschitz homeomorphisms.

Definition 6.1 *A function $f[\mathbb{R}^n \to \mathbb{R}^n]$ is called a* Lipschitz homeomorphism near $x_0 \in \mathbb{R}^n$ *if there are open neighbourhoods $\mathcal{O}$ of x_0 and $\mathcal{N}$ of $f(x_0)$ such that*

(i) f is a bijection of $\mathcal{O}$ and $\mathcal{N}$;

(ii) f *and* $f^{-1} \cap \mathcal{O}$ *are Lipschitz on* $\mathcal{O}$ *and* $\mathcal{N}$*, respectively.*

Lemma 6.1 *Let* $f[\mathbb{R}^n \to \mathbb{R}^n]$ *be a Lipschitz homeomorphism near* x_0 *and* $\mathcal{O}, \mathcal{N}$ *be neighbourhoods as specified in Definition 6.1. Assume that* f *is directionally differentiable at* $x \in \mathcal{O}$*. Then the local inverse function* $g = f^{-1} \cap \mathcal{O}$ *is directionally differentiable at* $f(x) \in \mathcal{N}$ *and, for each direction* $h \in \mathbb{R}^n$,

$$g'(f(x); h) = v, \quad \textit{where } v \textit{ is such that} \quad h = f'(x; v).$$

Proof. Let $x \in \mathcal{O}$ and $h \in \mathbb{R}^n$ be fixed. Put $y := f(x)$ and

$$v(t) := \frac{g(y+th) - g(y)}{t} \quad \text{for } t > 0 \text{ sufficiently small.}$$

Since f is a Lipschitz homeomorphism, there exist some v and a sequence $t_i \downarrow 0$ with $v(t_i) \to v$ for $i \to \infty$. With this sequence $\{t_i\}$ we get for the directional derivative of f in the direction v

$$\begin{aligned} f'(x; v) &= \lim_{i \to 0} \frac{f(x + t_i v) - f(x)}{t_i} \\ &= \lim_{i \to 0} \left\{ \frac{f(x + t_i v(t_i)) - f(x)}{t_i} + \frac{f(x + t_i v) - f(x + t_i v(t_i))}{t_i} \right\} \\ &= \lim_{i \to 0} \left\{ \frac{y + t_i h - y}{t_i} + \frac{f(x + t_i v) - f(x + t_i v(t_i))}{t_i} \right\}. \end{aligned}$$

For the second term in the bracket we get by assumptions on f and $v(t_i)$

$$\left| \frac{f(x + t_i v) - f(x + t_i v(t_i))}{t_i} \right| \le L \|v - v(t_i)\| \to 0 \text{ for } i \to \infty,$$

where L is the Lipschitz modulus of f. Thus

$$f'(x; v) = h. \tag{6.2}$$

Now consider another sequence $\hat{t}_i \downarrow 0$ and assume that $v(\hat{t}_i) \to \hat{v}$ for $i \to \infty$. If we can show $v = \hat{v}$, then $g'(y; h)$ exists and, using (6.2), we are done. By (6.2) one has $f'(x; v) = f'(x; \hat{v})$, and thus

$$\|f(x + tv) - f(x + t\hat{v})\| \le o(t).$$

This means

$$\|tv - t\hat{v}\| \le o(t)$$

due to the Lipschitz continuity of g. Hence, $v = \hat{v}$ and the lemma has been proved. ■

Let us now come to the GE (5.1) which is equivalent to the NSE (cf. (4.16))

$$z = \operatorname{Proj}_Q(z - C(x, z)), \tag{6.3}$$

where the right-hand side maps $\mathbb{R}^n \times \mathbb{R}^k$ into $\mathbb{R}^k$. For an application of Lemma 6.1 we need a mapping between spaces of the same dimension. Hence we blow (6.3) up and define with a dummy variable $w \in \mathbb{R}^n$ a function $\mathcal{F}[\mathbb{R}^n \times \mathbb{R}^k \to \mathbb{R}^n \times \mathbb{R}^k]$ by

$$\mathcal{F}(w, z) := \begin{bmatrix} w \\ \operatorname{Proj}_Q(z - C(w, z)) - z \end{bmatrix}. \tag{6.4}$$

In terms of $\mathcal{F}$ our NSE (6.3) becomes

$$\mathcal{F}(w,z) = \begin{bmatrix} x \\ 0 \end{bmatrix}. \tag{6.5}$$

Lemma 6.2 *Assume that the GE (5.1) satisfies (SRC) at (x_0, z_0). Then the map $\mathcal{F}[\mathbb{R}^n \times \mathbb{R}^k \to \mathbb{R}^n \times \mathbb{R}^k]$ from (6.4) is a Lipschitz homeomorphism near (w_0, z_0), where $w_0 = x_0$.*

Proof. First note that $\mathcal{F}$ is Lipschitz on each bounded neighbourhood of (w_0, z_0), since the projection operator onto a convex set is nonexpansive (cf. Zarantonello, 1971) and C is continuously differentiable by assumption. Let (x, s) be the image of (w, z), i.e.,

$$\mathcal{F}(w,z) = \begin{bmatrix} x \\ s \end{bmatrix}. \tag{6.6}$$

To prove the statement, we find open neighbourhoods $\mathcal{O}_1$ of $x_0 = w_0$, $\mathcal{N}_2$ of $s_0 := 0_{\mathbb{R}^k}$ and $\mathcal{O}_2$ of z_0 such that $\mathcal{F}^{-1}(\cdot,\cdot) \cap (\mathcal{O}_1 \times \mathcal{O}_2)$ is single-valued and Lipschitz on $\mathcal{O}_1 \times \mathcal{N}_2$. Then, automatically, $\mathcal{F}^{-1}(\mathcal{O}_1 \times \mathcal{N}_2) \cap (\mathcal{O}_1 \times \mathcal{O}_2)$ is a neighbourhood of (w_0, z_0), due to the continuity of $\mathcal{F}$.

Let us come back to the reduced form of (6.6) (note that $w = x$)

$$\operatorname{Proj}_Q(z - C(x,z)) = z + s \tag{6.7}$$

which, after introducing the new variable $r := z + s$, becomes the GE

$$0 \in C(x, r - s) + s + N_Q(r). \tag{6.8}$$

Now consider $x \in \mathbb{R}^n$ and $s \in \mathbb{R}^k$ in (6.8) to be perturbation parameters. Clearly,

$$\mathcal{F}^{-1}(x,s) = \{(w,z) \in \mathbb{R}^n \times \mathbb{R}^k \mid w = x, z = r - s, \text{ where } r \text{ is a solution of (6.8)}\}.$$

Write $\widetilde{C}(x,s,r)$ for $C(x, r-s)$. With this function $\widetilde{C}[\mathbb{R}^n \times \mathbb{R}^k \times \mathbb{R}^k \to \mathbb{R}^k]$ relation (6.8) becomes,

$$0 \in \widetilde{C}(x,s,r) + s + N_Q(r). \tag{6.9}$$

For the linearization of $\widetilde{C}$ at $(x_0, s_0, r_0) = (x_0, 0, z_0)$ with respect to r, we get

$$\widetilde{C}(x_0, 0, z_0) + \mathcal{J}_r\widetilde{C}(x_0, 0, z_0)(r - r_0) = C(x_0, z_0) + \mathcal{J}_z C(x_0, z_0)(z - z_0)$$

and thus C and $\widetilde{C}$ yield the same multi-valued map Σ in (5.4). Hence (SRC) for (5.1) at (x_0, z_0) implies (SRC) for (6.9) at $(x_0, 0, z_0)$ and from Theorem 5.2 we get the existence of neighbourhoods $\tilde{\mathcal{O}}_1$ of x_0, $\tilde{\mathcal{N}}_2$ of $0_{\mathbb{R}^k}$ and $\tilde{\mathcal{O}}_2$ of z_0 and of a Lipschitz map $\kappa[\tilde{\mathcal{O}}_1 \times \tilde{\mathcal{N}}_2 \to \mathbb{R}^k]$ such that $\kappa(x_0, 0) = z_0$ and $\kappa(x, s)$ is unique solution of (6.9) in $\tilde{\mathcal{O}}_2$ for all $(x, s) \in \tilde{\mathcal{O}}_1 \times \tilde{\mathcal{N}}_2$. This implies that we can find open neighbourhoods $\mathcal{O}_1$ of x_0, $\mathcal{N}_2$ of $0_{\mathbb{R}^k}$ and $\mathcal{O}_2$ of z_0 with $\mathcal{O}_1 \subset \tilde{\mathcal{O}}_1$, $\mathcal{N}_2 \subset \tilde{\mathcal{N}}_2$ and $\mathcal{O}_2 + \mathcal{N}_2 \subset \tilde{\mathcal{O}}_2$ such that

$$\mathcal{F}^{-1}(x,s) \cap (\mathcal{O}_1 \times \mathcal{O}_2) = \{(w,z) \in \mathcal{O}_1 \times \mathcal{O}_2 \mid w = x,\ z = \kappa(x,s) - s\}$$

for all $(x, s) \in \mathcal{O}_1 \times \mathcal{N}_2$. Thus $\mathcal{F}$ is a bijection of $\mathcal{F}^{-1}(\mathcal{O}_1 \times \mathcal{N}_2) \cap (\mathcal{O}_1 \times \mathcal{O}_2)$ and $(\mathcal{O}_1 \times \mathcal{N}_2)$. The Lipschitz continuity of $F^{-1}(\cdot,\cdot) \cap (\mathcal{O}_1 \times \mathcal{O}_2)$ on $(\mathcal{O}_1 \times \mathcal{N}_2)$ follows from the properties of κ. ■

The above lemmas help to compute the directional derivatives of σ at x_0 for the case of polyhedral feasible sets Q.

Theorem 6.3 *Assume that the GE (5.1) satisfies (SRC) at (x_0, z_0). Then the map σ is directionally differentiable at x_0. For each direction h, the directional derivative $\sigma'(x_0; h)$ is the unique solution of the GE in the variable v*

$$0 \in \mathcal{J}_x C(x_0, z_0)h + \mathcal{J}_z C(x_0, z_0)v + N_K(v), \tag{6.10}$$

where K is the critical cone of Q corresponding to z_0 and $-C(x_0, z_0)$ (given by (5.17)).

Proof. By (6.5) and Lemmas 6.1, 6.2, σ is directionally differentiable at x_0 and the directional derivative $\sigma'(x_0; h)$ equals v, where v (together with a vector $k \in \mathbb{R}^n$) solves the equation

$$\mathcal{F}'(x_0, z_0; k, v) = \begin{bmatrix} h \\ 0 \end{bmatrix}.$$

This implies $k = h$ and, due to Theorem 2.31,

$$\operatorname{Proj}_K(v - \mathcal{J}_x C(x_0, z_0)k - \mathcal{J}_z C(x_0, z_0)v) = v.$$

The last equation is exactly the GE (6.10); cf. (4.16). ■

Next consider the GE (5.24) and assume that (SRC) holds at the reference quadruple $(x_0, y_0, \mu_0, \lambda_0)$. This implies in particular that (μ_0, λ_0) is the unique KKT vector associated with the pair (x_0, y_0). Therefore, we will simplify the notation and write $I^+(x_0, y_0)$ and $I^0(x_0, y_0)$ instead of $I^+(x_0, y_0, \lambda_0)$ and $I^0(x_0, y_0, \lambda_0)$, respectively. For our purpose, it will be convenient to split the unique Lipschitz selection σ of S passing through $(x_0, y_0, \mu_0, \lambda_0)$ into three Lipschitz operators $\sigma_1, \sigma_2, \sigma_3$ which assign x from a neighbourhood of x_0 the y-, μ- and λ-component of the solution to (5.24).

Theorem 6.4 *Consider the GE (5.24) and assume that (SRC) holds at $(x_0, y_0, \mu_0, \lambda_0)$. Then the operators $\sigma_1, \sigma_2, \sigma_3$ (whose existence is guaranteed by Theorem 5.2) are directionally differentiable at x_0. For each direction $h \in \mathbb{R}^n$, the composed directional derivative $(\sigma_1'(x_0; h),\ \sigma_2'(x_0; h),\ \sigma_3'(x_0; h))$ is the unique solution of the system in (v, w, u)*

$$\begin{aligned} 0 \in &\begin{bmatrix} \mathcal{J}_x \mathcal{L}(x_0, y_0, \mu_0, \lambda_0) \\ \mathcal{J}_x H(x_0, y_0) \\ -\mathcal{J}_x G_{I^+}(x_0, y_0) \\ -\mathcal{J}_x G_{I^0}(x_0, y_0) \end{bmatrix} h + D_{(I)}(x_0, y_0, \mu_0, \lambda_0) \begin{bmatrix} v \\ w \\ u_{I^+} \\ u_{I^0} \end{bmatrix} + \begin{bmatrix} 0_{\mathbb{R}^m} \\ 0_{\mathbb{R}^\ell} \\ 0_{\mathbb{R}^{a+}} \\ N_{\mathbb{R}_+^{a^0}}(u_{I^0}) \end{bmatrix} \\ u_L = &\ 0, \end{aligned} \tag{6.11}$$

where $D_{(I)}(x_0, y_0, \mu_0, \lambda_0)$ is given by (5.27).

Proof. By Theorem 6.3, the corresponding Lipschitz selection σ is directionally differentiable at x_0 and, for $h \in \mathbb{R}^n$, one has

$$\sigma_1'(x_0; h) = v_h,\ \sigma_2'(x_0; h) = w_h,\ \sigma_3'(x_0; h) = u_h, \text{ where } (v_h, w_h, u_h)$$

is the unique solution of the GE

$$0 \in \begin{bmatrix} \mathcal{J}_x\mathcal{L}(x_0, y_0, \mu_0, \lambda_0) \\ \mathcal{J}_x H(x_0, y_0) \\ -\mathcal{J}_x G(x_0, y_0) \end{bmatrix} h + D(x_0, y_0, \mu_0, \lambda_0) \begin{bmatrix} v \\ w \\ u \end{bmatrix} + N_K(v, w, u) \tag{6.12}$$

with

$$K = \{(v, w, u) \in \mathbb{R}^m \times \mathbb{R}^\ell \times \mathbb{R}^s \mid u_{I^0} \geq 0,\ u_L = 0\};$$

cf. (5.30). Therefore, $N_K(v, w, u)$ is given by (5.31) and for all indices $i \in L(x_0, y_0)$ one has $(\sigma_3')^i(x_0; h) = 0$. The claim follows if we apply to (6.12) the reduction procedure, explained in the proof of Proposition 5.5. ■

Corollary 6.5 *Let the assumptions of Theorem 6.4 hold and suppose that $I^0(x_0, y_0) = \emptyset$. Then σ is Fréchet differentiable at x_0.*

Proof. A closer look at (6.11) shows that, in absence of weakly active constraints (i.e., $I^0(x_0, y_0) = \emptyset$), the map $\sigma'(x_0; \cdot)$ is linear and thus σ is Gâteaux differentiable at x_0. However, σ is Lipschitz near x_0 and so, in fact, it is Fréchet differentiable at x_0. ■

To illustrate Theorem 6.4, let us go back to the GE from Example 5.1 and compute the corresponding $\sigma'(x_0; h)$ for $h = \pm 1$.

Example 6.1 (Example 5.1 continued) Consider again the GE (5.45) at $(x_0, y_0, \lambda_0) = \left(1, \frac{1}{2}, \frac{1}{2}, 0, \frac{1}{2}\right)$ and recall that (SRC) holds at this point. Then (6.11) becomes

$$0 \in \begin{bmatrix} -1 \\ 0 \\ -1 \\ -2 \end{bmatrix} h + \begin{bmatrix} 2 & 0 & -1 & 0 \\ 0 & 3 & 0 & -1 \\ 1 & 0 & 0 & 0 \\ 0 & 1 & 0 & 0 \end{bmatrix} \begin{bmatrix} v^1 \\ v^2 \\ u^1 \\ u^2 \end{bmatrix} + \begin{bmatrix} 0 \\ 0 \\ N_{\mathbb{R}_+}(u^1) \\ 0 \end{bmatrix}.$$

By testing the alternatives $u^1 = 0$ and $u^1 > 0$, we get for $h = 1$ the solution

$$\sigma'(1; 1) = (1, 2, 1, 6).$$

Analogously, for $h = -1$ we obtain

$$\sigma'(1; -1) = \left(-\frac{1}{2}, -2, 0, -6\right)$$

and thus σ is not differentiable at x_0 $(= 1)$. △

Consider now the GE (5.41) and assume again that strong regularity holds at the reference point (x_0, y_0, λ_0). Then the map σ splits into two Lipschitz operators σ_1, σ_2 which assign all x, close to x_0, the y- and the λ-component of the solution to (5.41). From (5.41), however, one immediately deduces that $\sigma_2(x) = \lambda(x) = F(x, \sigma_1(x))$ on a neighbourhood of x_0. Hence it suffices to deal with σ_1 and Lemma 6.1 can be directly applied to the equivalent perturbed NSE of type (4.23):

$$\min{}_c\{F(x, y), y - \Phi(x, y)\} = 0. \tag{6.13}$$

This leads to our next result.

Theorem 6.6 *Consider the GE (5.41) and assume that (SRC) holds at* (x_0, y_0, λ_0). *Then the operator* σ_1 *is directionally differentiable at* x_0 *and for each direction* $h \in \mathbb{R}^n$ *the derivative* $\sigma_1'(x_0; h)$ *is the unique solution of the following equation in the variable* v*:*

$$\begin{aligned} \mathcal{J}_x F_L(x_0, y_0)h + \mathcal{J}_y F_L(x_0, y_0)v &= 0 \\ v_{I^+} - \mathcal{J}_x \Phi_{I^+}(x_0, y_0)h - \mathcal{J}_y \Phi_{I^+}(x_0, y_0)v &= 0 \\ \min{}_c \{\mathcal{J}_x F_{I^0}(x_0, y_0)h + \mathcal{J}_y F_{I^0}(x_0, y_0)v, & \\ v_{I^0} - \mathcal{J}_x \Phi_{I^0}(x_0, y_0)h - \mathcal{J}_y \Phi_{I^0}(x_0, y_0)v\} &= 0. \end{aligned} \tag{6.14}$$

Proof. We use similar arguments as in the proof of Lemma 6.2 to show that the function

$$\mathcal{F}(w, y) := \begin{bmatrix} w \\ \min_c\{F(w, y), y - \Phi(w, y)\} \end{bmatrix} \tag{6.15}$$

is a Lipschitz homeomorphism near (w_0, y_0) with $w_0 = x_0$. To this purpose we consider the equation

$$\min{}_c \{F(w, y), y - \Phi(x, y)\} = s$$

which, after introducing the new variable $r := y - s$, leads to the GE

$$0 \in \begin{bmatrix} F(w, r + s) - s - \lambda \\ r - \Phi(w, r + s) \end{bmatrix} + N_{\mathbb{R}^m \times \mathbb{R}^m_+}(r, \lambda). \tag{6.16}$$

GE (6.16) is a counterpart to (6.8) and one can easily verify again that (6.16) generates the same partially linearized map as the original GE (5.41). Copying the last lines from the proof of Lemma 6.2, we conclude that $\mathcal{F}$ is a Lipschitz homeomorphism near (w_0, y_0).

Clearly, in terms of $\mathcal{F}$, the NSE (6.13) becomes

$$\mathcal{F}(w, y) = \begin{bmatrix} x \\ 0 \end{bmatrix}.$$

Hence, due to Lemma 6.1, σ_1 is directionally differentiable at x_0 and the directional derivative $\sigma_1'(x_0; h)$ equals the solution of the equation in v

$$\mathcal{H}'(x_0, y_0; h, v) = 0$$

with $\mathcal{H}(x, y) = \min_c\{F(x, y), y - \Phi(x, y)\}$. This immediately gives the equations (6.14). ■

As in the previous case, under (SRC) at (x_0, y_0), the mapping σ_1 is Fréchet differentiable whenever $I^0(x_0, y_0) = \emptyset$.

Example 6.2 (Example 5.2 continued) Consider the GE from Example 5.2 at $(x_0, y_0, \lambda_0) = (1, -2, -3, -3, -2, 0, 0, 0, 0)$. As in Example 6.1, we compute the directional derivatives of σ for $h = \pm 1$. Equation (6.14) attains the form

$$\min{}_c \left\{ \begin{array}{ll} -3h + 2v^1 - v^2, & -0.5h + 3v^1 - v^2 \\ -5h - v^1 + 2v^2 - v^3, & -2h - v^1 + 3v^2 - v^3 \\ -5h - v^2 + 2v^3 - v^4, & -2h - v^2 + 3v^3 - v^4 \\ -3h - v^3 + 2v^4, & -0.5h - v^3 + 3v^4 \end{array} \right\} = 0.$$

For $h = 1$ we easily get the solution $\sigma_1'(1;1) = (8, 13, 13, 8)$, by using the problem symmetry $(v_1 = v_4, v_2 = v_3)$. Analogously, for $h = -1$, we obtain $\sigma_1'(1;-1) = \left(-\frac{3}{5}, -\frac{13}{10}, -\frac{13}{10}, -\frac{3}{5}\right)$. The directional derivatives of the KKT vector can be computed by the formulae

$$(\sigma_2^i)'(x_0;h) = \mathcal{J}_x F^i(x_0, y_0)h + \mathcal{J}_y F^i(x_0, y_0)\sigma_1'(x_0;h), \quad i = 1, 2, 3, 4.$$

This yields

$$\begin{aligned}
&(\sigma_2^1)'(1;1) = (\sigma_2^2)'(1;1) = (\sigma_2^3)'(1;1) = (\sigma_2^4)'(1;1) = 0\\
&(\sigma_2^1)'(1;-1) = (\sigma_2^4)'(1;-1) = 3.1\\
&(\sigma_2^2)'(1;-1) = (\sigma_2^3)'(1;-1) = 4.3.
\end{aligned}$$

△

On the basis of Theorem 6.3, we can also develop conditions which characterize the Fréchet differentiability of the map σ defined by the GE (5.1) at x_0. We restrict our attention to the case

$$Q = \left\{z \in \mathbb{R}^k \mid {}^1Az = {}^1b,\ {}^2Az \leq {}^2b\right\}, \tag{6.17}$$

with ${}^1A$, ${}^2A$ being an $\ell \times k$ and an $s \times k$ matrix, ${}^1b \in R^\ell$ and ${}^2b \in \mathbb{R}^s$. Of course, this is a special case of the feasible set (4.3). The linear independence constraint qualification at $z_0 \in Q$ attains the form

(LICQ): The rows $({}^1A)^i$ for $i = 1, 2, \dots, \ell$ and $({}^2A)^i$ for $i \in \{j \in \{1, 2, \dots, s\} \mid \langle({}^2A)^j, z_0\rangle = {}^2b^j\}$ are linearly independent.

If (LICQ) holds at z_0, then the KKT system for the respective GE attains the form

$$\begin{aligned}
&F(z) + {}^1A^T\mu + {}^2A^T\lambda = 0\\
&{}^1A^Tz = {}^1b, \quad {}^2Az \leq {}^2b\\
&\lambda \geq 0\\
&\langle\lambda, {}^2Az - {}^2b\rangle = 0.
\end{aligned} \tag{6.18}$$

The critical cone

$$K = \left\{\eta \in \mathbb{R}^k \mid \langle F(x_0, z_0), \eta\rangle = 0,\ {}^1A\eta = 0,\ {}^2A_I\eta \leq 0\right\}$$

becomes

$$\left\{\eta \in \mathbb{R}^k \mid {}^1A\eta = 0,\ {}^2A_{I^+}\eta = 0,\ {}^2A_{I^0}\eta \leq 0\right\}$$

due to (6.18).

In what follows, $\mathbb{L} := K \cap (-K)$ denotes the lineality space of K, i.e.,

$$\mathbb{L} = \operatorname{Ker}{}^1A \cap \operatorname{Ker}{}^2A_I.$$

Theorem 6.7 *Consider the GE (5.1) with the reference point (x_0, z_0) and assume that Q is given by (6.17). Further suppose that (LICQ) holds at z_0 and*

$$\langle d, \mathcal{J}_z F(x_0, z_0)d\rangle > 0 \tag{6.19}$$

for all nonzero $d \in \operatorname{Ker} {}^1A \cap \operatorname{Ker} {}^2A_{I+}$. *Then the statements (i)–(iv) below are equivalent:*

(i) σ *is Fréchet differentiable at* x_0.

(ii) For all vectors $h \in \mathbb{R}^n$*, the directional derivative* $\sigma'(x_0; h) = v$ *satisfies* ${}^2A_I v = 0$.

(iii) For all vectors $h \in \mathbb{R}^n$, $\sigma'(x_0; h) = v$ *is the unique solution of the GE*

$$0 \in \mathcal{J}_x F(x_0, z_0)h + \mathcal{J}_z F(x_0, z_0)v + N_{\mathbb{L}}(v). \tag{6.20}$$

(iv) For all vectors $h \in \mathbb{R}^n$, $\sigma'(x_0; h) = v$ *satisfies*

$$\langle z, \mathcal{J}_x F(x_0, z_0)h + \mathcal{J}_z F(x_0, z_0)v \rangle = 0 \quad \textit{for } z \in K. \tag{6.21}$$

Proof. (i) ⇒ (ii). The Fréchet differentiability of σ at x_0 implies that for all vectors $h \in \mathbb{R}^n$ we have

$$v = \sigma'(x_0; h) = -\sigma'(x_0; -h),$$

and consequently $-v \in K$. Thus ${}^2A_I v = 0$ as desired.

(ii) ⇒ (iii). $\sigma'(x_0; h) = v$ is a solution of (6.20), because, due to (ii), $v \in \mathbb{L}$ and $\mathbb{L} \subset K$; it is unique due to assumption (6.19) ($\mathbb{L} \subset (\operatorname{Ker} {}^1A \cap \operatorname{Ker} {}^2A_{I+})$).

(iii) ⇒ (iv). Let $h \in \mathbb{R}^n$ be given. By (iii), $\sigma(x_0; -h)$ is the solution of the GE

$$0 \in -\mathcal{J}_x F(x_0, z_0)h + \mathcal{J}_z F(x_0, z_0)v + N_{\mathbb{L}}(v).$$

However, $-\sigma(x_0; h)$ also solves this GE and thus $\sigma'(x_0; -h) = -\sigma'(x_0; h)$.

The GE (6.10) amounts to the NSE

$$\operatorname{Proj}_K(v - \mathcal{J}_x F(x_0, z_0)h - \mathcal{J}_z F(x_0, z_0)v) = v.$$

Using the fact that for each $\xi \in \mathbb{R}^m$ one has $\operatorname{Proj}_K(\xi) - \xi \in K^*$ (cf. Hiriart-Urruty and Lemaréchal, 1993, Part I,Prop. 3.2.3), we conclude that

$$\mathcal{J}_x F(x_0, z_0)h + \mathcal{J}_z F(x_0, z_0)v \in K^*.$$

Thus, for each $z \in K$

$$\langle z, \mathcal{J}_x F(x_0, z_0)h + \mathcal{J}_z F(x_0, z_0)\sigma'(x_0; h) \rangle \geq 0$$

and, similarly,

$$\langle z, -\mathcal{J}_x F(x_0, z_0)h + \mathcal{J}_z F(x_0, z_0)\sigma'(x_0; -h) \rangle \geq 0.$$

Equation (6.21) follows, since $\sigma'(x_0; -h) = -\sigma'(x_0; h)$.

(iv) ⇒ (i). As σ is Lipschitz near x_0, it suffices to show that it is Gâteaux differentiable at x_0, i.e., that $\sigma'(x_0; h)$ is linear in h. Let $v := \sigma'(x_0; h)$ and $w := \sigma'(x_0; -h)$. Then v, w and $v + w$ belong to K and relation (6.21) implies

$$\langle v + w, \mathcal{J}_z F(x_0, z_0)(v + w) \rangle = 0.$$

By assumption (6.19) this yields $v + w = 0$. To complete the proof, let h_1, h_2 be two arbitrary vectors from $\mathbb{R}^n$ and put $v_j := \sigma'(x_0; h_j)$ for $j = 1, 2$ and $w := \sigma'(x_0; h_1 + h_2)$. In the similar way as above we observe that

$$\langle v_1 + v_2 - w, \mathcal{J}_z F(x_0, z_0)(v_1 + v_2 - w) \rangle = 0,$$

and consequently $w = v_1 + v_2$.

The assertion has been proved. ∎

By a refined proof technique (with arguments from convex analysis) Kyparisis proved the equivalence of the statements (i)–(iii) in Theorem 6.7 for polyhedral Q (without any further structure) under the weaker assumption that the GE (5.1) satisfies (SRC) at (x_0, z_0). In the following we will make use of this stronger result without proof; we refer the interested reader to Kyparisis, 1990. When applied to the GE (5.24), this result implies the following statement.

Proposition 6.8 *Consider the GE (5.24) and assume that (SRC) holds at $(x_0, y_0, \mu_0, \lambda_0)$. Moreover, let the operator σ_1 be Fréchet differentiable at x_0. Then σ is also Fréchet differentiable at x_0 and the derivatives $\mathcal{J}\sigma_1(x_0), \mathcal{J}\sigma_2(x_0), \mathcal{J}\sigma_3(x_0)$ are operators which assign $h \in \mathbb{R}^n$ the (unique) solution $(v, w, u) \in \mathbb{R}^m \times \mathbb{R}^\ell \times \mathbb{R}^s$ of the linear system*

$$\begin{aligned} \begin{bmatrix} \mathcal{J}_x \mathcal{L}(x_0, y_0, \mu_0, \lambda_0) \\ \mathcal{J}_x H(x_0, y_0) \\ -\mathcal{J}_x G_{I^+}(x_0, y_0) \end{bmatrix} h + D_{(I^+)}(x_0, y_0, \mu_0, \lambda_0) \begin{bmatrix} v \\ w \\ u_{I^+} \end{bmatrix} &= 0 \\ u_{L \cup I^0} &= 0 \end{aligned} \tag{6.22}$$

Proof. We first show that the Fréchet differentiability of σ_1 at x_0 implies the Fréchet differentiability of σ at x_0. Let $h_1, h_2 \in \mathbb{R}^n$ be two directions and $(v_1, w_1, u_{I1}), (v_2, w_2, u_{I2})$ be the corresponding solutions of the GE (6.11). Assume that σ is not even Gâteaux differentiable and for $h = h_1 + h_2$ and the corresponding solution (v, w, u_I) of (6.11) one has $v = v_1 + v_2$ but $w \neq w_1 + w_2$ and/or $u_I \neq u_{I1} + u_{I2}$. Clearly,

$$\begin{aligned} 0 &= \mathcal{J}_x \mathcal{L}(x_0, y_0, \lambda_0) h + \mathcal{J}_y \mathcal{L}(x_0, y_0, \lambda_0) \\ &\quad + (\mathcal{J}_y H(x_0, y_0))^T w + (\mathcal{J}_y G_I(x_0, y_0))^T u_I \\ 0 &= \mathcal{J}_x \mathcal{L}(x_0, y_0, \lambda_0) h_1 + \mathcal{J}_y \mathcal{L}(x_0, y_0, \lambda_0) v_1 \\ &\quad + (\mathcal{J}_y H(x_0, y_0))^T w_1 + (\mathcal{J}_y G_I(x_0, y_0))^T u_{I1} \\ 0 &= \mathcal{J}_x \mathcal{L}(x_0, y_0, \lambda_0) h_2 + \mathcal{J}_y \mathcal{L}(x_0, y_0, \lambda_0) v_2 \\ &\quad + (\mathcal{J}_y H(x_0, y_0))^T w_2 + (\mathcal{J}_y G_I(x_0, y_0))^T u_{I2}. \end{aligned}$$

By subtracting the second and the third equation from the first one, we get

$$(\mathcal{J}_y H(x_0, y_0))^T (w - w_1 - w_2) + (\mathcal{J}_y G_I(x_0, y_0))^T (u_I - u_{I1} - u_{I2}) = 0,$$

which contradicts (ELICQ) at (x_0, y_0) (a consequence of (SRC)). Thus σ is Gâteaux differentiable at x_0 and, being Lipschitz near x_0, it is even Fréchet differentiable at x_0.

Therefore we may apply implication (i) $\Rightarrow$ (iii) of the mentioned result by Kyparisis. In this case, the GE (6.20) attains the form

$$0 \in \begin{bmatrix} \mathcal{J}_x \mathcal{L}(x_0, y_0, \mu_0, \lambda_0) \\ \mathcal{J}_x H(x_0, y_0) \\ -\mathcal{J}_x G(x_0, y_0) \end{bmatrix} h + D(x_0, y_0, \mu_0, \lambda_0) \begin{bmatrix} v \\ w \\ u \end{bmatrix} + N_{\mathbb{L}}(v, w, u), \quad (6.23)$$

where

$$\mathbb{L} = \{(v, w, u) \in \mathbb{R}^m \times \mathbb{R}^\ell \times \mathbb{R}^s \mid u_L = 0,\ u_{I^0} = 0\}.$$

This yields

$$N_{\mathbb{L}}(v, w, u) = \{(w^*, w^*, u^*) \in \mathbb{R}^m \times \mathbb{R}^\ell \times \mathbb{R}^s \mid v^* = 0,\ w^* = 0,\ u^*_{I^+} = 0\}$$

and the GE (6.23) is reduced to the linear equation (6.22). ■

The same technique can be used in deriving a formula for the derivative of σ_1 in the case of the GE (5.41).

Proposition 6.9 *Consider the GE (5.41) and assume that (SRC) holds at (x_0, y_0, λ_0). Moreover, let the operator σ_1 be Fréchet differentiable at x_0. Then the operator $\mathcal{J}\sigma_1(x_0)$ assigns $h \in \mathbb{R}^n$ the unique solution $v \in \mathbb{R}^m$ of the linear system*

$$\begin{aligned} \mathcal{J}_x F_{L\cup I^0}(x_0, y_0)h + \mathcal{J}_y F_{L\cup I^0}(x_0, y_0)v &= 0 \\ v_{I^+} - \mathcal{J}_x \Phi_{I^+}(x_0, y_0)h - \mathcal{J}_y \Phi_{I^+}(x_0, y_0)v &= 0. \end{aligned} \quad (6.24)$$

Proof. We note that

(i) the Fréchet differentiability of σ_1 at x_0 implies the Fréchet differentiability of σ, and

(ii) the corresponding GE (6.20) attains the form

$$0 \in \begin{bmatrix} \mathcal{J}_x F(x_0, y_0) \\ -\mathcal{J}_x \Phi(x_0, y_0) \end{bmatrix} h + \begin{bmatrix} \mathcal{J}_y F(x_0, y_0) & -E \\ E - \mathcal{J}_y \Phi(x_0, y_0) & 0 \end{bmatrix} \begin{bmatrix} v \\ u \end{bmatrix} + N_{\mathbb{L}}(v, u), \quad (6.25)$$

where

$$\mathbb{L} = \{(v, u) \in \mathbb{R}^m \times \mathbb{R}^m \mid u_L = 0,\ u_{I^0} = 0\}.$$

Therefore $N_{\mathbb{L}}(v, u) = \{(v^*, u^*) \in \mathbb{R}^m \times \mathbb{R}^m \mid v^* = 0,\ u^*_{I^+} = 0\}$ and the GE (6.25) is reduced to

$$\begin{aligned} \mathcal{J}_x F_{L\cup I^0}(x_0, y_0)h + \mathcal{J}_y F_{L\cup I^0}(x_0, y_0)v &= 0 \\ -\mathcal{J}_x \Phi_{I^+}(x_0, y_0)h - \mathcal{J}_y \Phi_{I^+}(x_0, y_0)v + v_{I^+} &= 0 \\ \mathcal{J}_x F_{I^+}(x_0, y_0)h + \mathcal{J}_y F_{I^+}(x_0, y_0)v - u_{I^+} &= 0 \\ u_{L\cup I^0} &= 0. \end{aligned} \quad (6.26)$$

The variable u does not appear in the first two equations and so they (uniquely) determine the desired derivative $v = \mathcal{J}\sigma_1(x_0)h$. The assertion has been proved. ■

Propositions 6.8 and 6.9 will be needed in the next section dealing with generalized Jacobians of σ.

6.2 GENERALIZED JACOBIANS

If strong regularity holds, then local properties of σ can also be characterized by generalized Jacobians. So we assume (SRC) to hold and, differently from the previous section, we will impose additional structural assumptions on the feasible set Q. Let us start with the GE (5.24) at the reference point $(x_0, y_0, \mu_0, \lambda_0)$.

Lemma 6.10 *Assume that (SRC) holds at $(x_0, y_0, \mu_0, \lambda_0)$. Then there exists a neighbourhood $\mathcal{O}$ of x_0 such that*

$$I^+(x_0, y_0) \subset I^+(x, \sigma_1(x)) \subset I(x, \sigma_1(x)) \subset I(x_0, y_0) \quad \textit{for all } x \in \mathcal{O}. \tag{6.27}$$

Proof. The continuity of σ_1 implies that the inequalities which are inactive at (x_0, y_0) remain inactive in a neighbourhood of (x_0, y_0). This shows the right-hand inclusion in (6.27). The left-hand inclusion follows since, for continuous σ_3, inequalities strongly active at (x_0, y_0) remain strongly active close to (x_0, y_0). ■

In the following part, we will work with the family $\mathcal{P}(I^0(x_0, y_0))$ of all subsets of $I^0(x_0, y_0)$, because each element of this family may generate a matrix from $\partial_B \sigma(x_0)$. We denote the single elements of $\mathcal{P}(I^0(x_0, y_0))$ by $M_i(x_0, y_0)$, where i runs over a suitably chosen (finite) index set $\mathbb{K}(x_0, y_0)$. Since $M_i(x_0, y_0) \subset I^0(x_0, y_0)$, we denote $|M_i(x_0, y_0)|$ by a_i^0 for $i \in \mathbb{K}(x_0, y_0)$.

Lemma 6.11 *Assume that (SRC) holds at $(x_0, y_0, \mu_0, \lambda_0)$. Then the matrices*

$$D_{(I^+ \cup M_i)}(x_0, y_0, \mu_0, \lambda_0) := \begin{bmatrix} \mathcal{J}_y \mathcal{L}(x_0, y_0, \mu_0 \lambda_0) & (\mathcal{J}_y H(x_0, y_0))^T & (\mathcal{J}_y G_{I^+ \cup M_i}(x_0, y_0)^T) \\ \mathcal{J}_y H(x_0, y_0) & 0 & 0 \\ -\mathcal{J}_y G_{I^+ \cup M_i}(x_0, y_0) & 0 & 0 \end{bmatrix}$$

are nonsingular for all $i \in \mathbb{K}(x_0, y_0)$.

Proof. By Theorem 5.7, (SRC) holds at $(x_0, y_0, \mu_0, \lambda_0)$ if and only if $D_{(I^+)}(x_0, y_0, \mu_0, \lambda_0)$ is nonsingular and its Schur complement in $D_{(I)}(x_0, y_0, \mu_0, \lambda_0)$ is a P-matrix. Let us denote this Schur complement by $\mathcal{C}$. To prove the statement, it suffices to show that the Schur complement of $D_{(I^+)}(x_0, y_0, \mu_0, \lambda_0)$ in $D_{(I^+ \cup M_i)}(x_0, y_0, \mu_0, \lambda_0)$ is nonsingular for each $i \in \mathbb{K}(x_0, y_0)$; cf. Fiedler, 1986. However, this Schur complement is a principal submatrix of $\mathcal{C}$; hence it is also a P-matrix and thus evidently nonsingular. ■

On the basis of Lemmas 6.10, 6.11, we now get an outer approximation of $\partial \sigma_1(x_0)$.

Theorem 6.12 *Consider the GE (5.24) and assume that (SRC) holds at $(x_0, y_0, \mu_0, \lambda_0)$. Then one has*

$$\partial \sigma_1(x_0) \subset \operatorname{conv} \{ [P_i(x_0, y_0)]_m \mid i \in \mathbb{K}(x_0, y_0) \}, \tag{6.28}$$

where $P_i(x_0, y_0)$ is the (unique) solution of the linear matrix equation in Π

$$D_{(I^+ \cup M_i)}(x_0, y_0, \mu_0, \lambda_0)\Pi = \begin{bmatrix} -\mathcal{J}_x \mathcal{L}(x_0, y_0, \mu_0, \lambda_0) \\ -\mathcal{J}_x H(x_0, y_0) \\ \mathcal{J}_x G_{I^+ \cup M_i}(x_0, y_0) \end{bmatrix}, \quad i \in \mathbb{K}(x_0, y_0). \tag{6.29}$$

Proof. Under our assumptions, there exists a neighbourhood $\mathcal{N}$ of x_0 such that (SRC) holds at all (x, y, μ, λ), where $x \in \mathcal{N}$, $y = \sigma_1(x)$, $\mu = \sigma_2(x)$ and $\lambda = \sigma_3(x)$. Let $x \in \mathcal{N} \cap \mathcal{O}$ with $\mathcal{O}$ as specified in Lemma 6.10. By Proposition 6.8, σ_1 is differentiable at x if and only if σ is differentiable at x. Then one has for some $i \in \mathbb{K}(x_0, y_0)$

$$\begin{aligned} D_{(I^+ \cup M_i)}(x, y, \mu, \lambda) \begin{bmatrix} \mathcal{J}\sigma_1(x) \\ \mathcal{J}\sigma_2(x) \\ \mathcal{J}(\sigma_3)_{I^+ \cup M_i}(x) \end{bmatrix} &= \begin{bmatrix} -\mathcal{J}_x \mathcal{L}(x, y, \mu, \lambda) \\ -\mathcal{J}_x H(x, y) \\ \mathcal{J}_x G_{I^+ \cup M_i}(x, y) \end{bmatrix} \\ \mathcal{J}(\sigma_3)_{I^0 \setminus M_i}(x) &= 0 \\ \mathcal{J}(\sigma_3)_L(x) &= 0. \end{aligned} \tag{6.30}$$

Consider now a sequence $\{x_j\} \subset \mathcal{N} \cap \mathcal{O}$, $x_j \to x_0$, such that σ is differentiable at x_j and $\mathcal{J}\sigma(x_j)$ is specified for all x_j by equation (6.30) with the same index $i \in \mathbb{K}(x_0, y_0)$. As all entries of the system matrices in (6.30) are continuous functions of x on $\mathcal{N} \cap \mathcal{O}$, one has (taking into account Lemma 6.12)

$$\lim_{j \to \infty} \mathcal{J}\sigma_1(x_j) = [P_i(x_0, y_0)]_m .$$

Clearly, the set of all possible limits of $\mathcal{J}\sigma_1(x_j)$ for $x_j \to x_0$ cannot contain other operators than the matrices $[P_i(x_0, y_0)]_m$ for $i \in \mathbb{K}(x_0, y_0)$ and so inclusion (6.28) holds by definition of the generalized Jacobian. ∎

With a somewhat clumsy notation one can also construct an upper approximation of $\partial\sigma(x_0)$. For this purpose, we build from the $(m + \ell + a^+ + a_i^0) \times n$ matrices $P_i(x_0, y_0)$, $i \in \mathbb{K}(x_0, y_0)$, new $s \times n$ matrices $Q_i(x_0, y_0)$ as follows:

- if $j \in \{1, 2, \ldots, s\} \setminus (I^+(x_0, y_0) \cup M_i(x_0, y_0))$, we put $Q_i^j(x_0, y_0) := 0$;
- if $j \in I^+(x_0, y_0) \cup M_i(x_0, y_0)$, we set $Q_i^j(x_0, y_0)$ equal to that row of $P_i(x_0, y_0)$ which corresponds to the row $\nabla_x g^j(x_0, y_0)$ on the right-hand side of (6.29).

Then we introduce the $(m + \ell + s) \times n$ matrices

$$R_i(x_0, y_0) = \begin{bmatrix} [P_i(x_0, y_0)]_{m+\ell} \\ Q_i(x_0, y_0) \end{bmatrix}, \quad i \in \mathbb{K}(x_0, y_0).$$

By repeating the arguments from the proof of Theorem 6.12, we infer that under (SRC) at $(x_0, y_0, \mu_0, \lambda_0)$ it holds

$$\partial\sigma(x_0) \subset \operatorname{conv}\{R_i(x_0, y_0) \mid i \in \mathbb{K}(x_0, y_0)\}. \tag{6.31}$$

In the calculus of generalized gradients and Jacobians one usually has to be content with inclusions of type (6.28), (6.31). Equalities require additional assumptions. A step in this direction is the following simple result.

Proposition 6.13 *Let the assumptions of Theorem 6.12 be fulfilled and let $i \in \mathbb{K}(x_0, y_0)$. Then*

$$R_i(x_0, y_0) \in \partial\sigma(x_0),$$

provided there exist vectors $h \in \mathbb{R}^n$, $v \in \mathbb{R}^m$, $w \in \mathbb{R}^\ell$ *and* $u_{I^+ \cup M_i} \in \mathbb{R}^{a^+ + a_i^0}$ *such that*

$$D_{(I^+ \cup M_i)}(x_0, y_0, \mu_0, \lambda_0) \begin{bmatrix} v \\ w \\ u_{I^+ \cup M_i} \end{bmatrix} = \begin{bmatrix} -\mathcal{J}_x \mathcal{L}(x_0, y_0, \mu_0, \lambda_0) \\ -\mathcal{J}_x H(x_0, y_0) \\ \mathcal{J}_x G_{I^+ \cup M_i}(x_0, y_0) \end{bmatrix} h$$
$$u_{M_i} > 0 \qquad (6.32)$$
$$\mathcal{J}_x G_{I^0 \setminus M_i}(x_0, y_0) h + \mathcal{J}_y G_{I^0 \setminus M_i}(x_0, y_0) v < 0.$$

Proof. With $u_{(I^0 \setminus M_i) \cup L} = 0$ Theorem 6.4 shows (v, w, u) to be the directional derivative of σ at x_0 in the direction h. Therefore, for sufficiently small $\vartheta > 0$,

$$I^+(x_0 + \vartheta h, \sigma_1(x_0 + \vartheta h)) = I^+(x_0, y_0) \cup M_i(x_0, y_0)$$

and

$$L(x_0 + \vartheta h, \sigma_1(x_0 + \vartheta h)) = L(x_0, y_0) \cup (I^0(x_0, y_0) \setminus M_i(x_0, y_0)).$$

Thus Corollary 6.5 implies the Fréchet differentiability of σ at $x_0 + \vartheta h$ for small $\vartheta > 0$, and one has

$$\lim_{\vartheta \downarrow 0} \mathcal{J}\sigma(x_0 + \vartheta h) = R_i(x_0, y_0).$$

■

It is usually not easy to verify the assumptions made in Proposition 6.13. The following variant of the Mangasarian–Fromowitz constraint qualification, proposed in Kuntz and Scholtes, 1994, can be helpful here. When applied to the NSE

$$\begin{bmatrix} \mathcal{L}(x, y, \mu, \lambda) \\ H(x, y) \\ \min_c\{-G(x, y), \lambda\} \end{bmatrix} = 0$$

(equivalent to (5.24)) at its solution $(x_0, y_0, \mu_0, \lambda_0)$, this condition specializes to

(MF1): Every collection of at most $n + m + \ell + a^+ + a^0$ rows of the $(m + \ell + a^+ + 2a^0) \times (n + m + \ell + a^+ + a^0)$ matrix

$$\mathbb{S}_1 := \left[\begin{array}{c|cccc|c} \mathcal{J}_x \mathcal{L}(x_0, y_0, \mu_0, \lambda_0) & & & & & (\mathcal{J}_y G_{I^0}(x_0, y_0))^T \\ \mathcal{J}_x H(x_0, y_0) & & D_{(I^+)}(x_0, y_0, \mu_0, \lambda_0) & & & 0 \\ -\mathcal{J}_x G_{I^+}(x_0, y_0) & & & & & 0 \\ \hline -\mathcal{J}_x G_{I^0}(x_0, y_0) & -\mathcal{J}_y G_{I^0}(x_0, y_0) & 0 & 0 & & 0 \\ 0 & 0 & 0 & 0 & & E \end{array}\right] \qquad (6.33)$$

is linearly independent.

Proposition 6.14 *Let the assumptions of Theorem 6.12 be fulfilled. Assume that* $n \geq a^0$ *and the condition (MF1) holds true at* $(x_0, y_0, \mu_0, \lambda_0)$. *Then*

$$\partial\sigma(x_0) = conv\{R_i(x_0, y_0) \mid i \in \mathbb{K}(x_0, y_0)\}.$$

Proof. If $n \geq a^0$, then (MF1) amounts to the requirement that $\mathbb{S}_1$ has full row rank. Hence for each $i \in \mathbb{K}(x_0, y_0)$ a quadruple of vectors $h \in \mathbb{R}^n$, $v \in \mathbb{R}^m$, $w \in \mathbb{R}^\ell$ and $u_{I^+ \cup I^0} \in \mathbb{R}^{a^+ + a^0}$ can be found such that

$$\mathbb{S}_1 \begin{bmatrix} h \\ v \\ w \\ u_{I^+ \cup I^0} \end{bmatrix} = \begin{bmatrix} 0_{\mathbb{R}^m} \\ 0_{\mathbb{R}^\ell} \\ c \\ d \end{bmatrix}$$

with arbitrary vectors $c \in \mathbb{R}^{a^+ + a^0}$ and $d \in \mathbb{R}^{a^0}$. This implies $R_i(x_0, y_0) \subset \partial\sigma(x_0)$ for all $i \in \mathbb{K}(x_0, y_0)$ and the claim follows from (6.31). ■

Remark. The assumptions of Proposition 6.14 can be replaced by requiring the matrix

$$\begin{bmatrix} \mathcal{J}_x \mathcal{L}(x_0, y_0, \mu_0, \lambda_0) & \mathcal{J}_y \mathcal{L}(x_0, y_0, \mu_0, \lambda_0) & (\mathcal{J}_y H(x_0, y_0))^T & (\mathcal{J}_y G_{I^+}(x_0, y_0))^T \\ \mathcal{J}_x H(x_0, y_0) & \mathcal{J}_y H(x_0, y_0) & 0 & 0 \\ -\mathcal{J}_x G_{I^+ \cup I^0}(x_0, y_0) & -\mathcal{J}_y G_{I^+ \cup I^0}(x_0, y_0) & 0 & 0 \end{bmatrix}$$

to have full row rank. This requirement also plays an important role in the penalty approach investigated in Scholtes and Stöhr, 1997.

We will return to these questions in Section 7.2, where we give more handy conditions.

Let us now turn our attention to the GE (5.41) and see what counterpart to Theorem 6.12 we can prove.

Lemma 6.15 *Assume that (SRC) holds at (x_0, y_0). Then there exists a neighbourhood $\mathcal{O}$ of x_0 such that*

$$I^+(x_0, y_0) \subset I^+(x, \sigma_1(x)) \quad \textit{and} \quad L(x_0, y_0) \subset L(x, \sigma_1(x)) \quad \textit{for all } x \in \mathcal{O}.$$

The proof directly follows from the continuity of σ_1.

Lemma 6.16 *Assume that (SRC) holds at (x_0, y_0). Then the matrices*

$$\mathcal{D}_i(x_0, y_0) := \begin{bmatrix} \mathcal{J}_y F_{L \cup (I^0 \setminus M_i)}(x_0, y_0) \\ E_{I^+ \cup M_i} - \mathcal{J}_y \Phi_{I^+ \cup M_i}(x_0, y_0) \end{bmatrix}, \quad i \in \mathbb{K}(x_0, y_0),$$

are nonsingular.

Proof. By Theorem 5.10 all matrices

$$\begin{bmatrix} \mathcal{J}_y F(x_0, y_0) & -E^T_{I^+ \cup M_i} \\ E_{I^+ \cup M_i} - \mathcal{J}_y \Phi_{I^+ \cup M_i}(x_0, y_0) & 0 \end{bmatrix}, \quad i \in \mathbb{K}(x_0, y_0), \tag{6.34}$$

are nonsingular. Indeed, the matrix

$$\begin{bmatrix} \mathcal{J}_y F(x_0, y_0) & -E^T_{I^+} \\ E_{I^+} - \mathcal{J}_y \Phi_{I^+}(x_0, y_0) & 0 \end{bmatrix}$$

is nonsingular and its Schur complement in each matrix (6.34) is a principal submatrix of a P-matrix (and hence nonsingular). Assume by contradiction the existence of a nonzero vector $v \in \mathbb{R}^m$ such that $\mathcal{D}_i(x_0, y_0)v = 0$ for some $i \in \mathbb{K}(x_0, y_0)$. Consider now a vector u of dimension $a^+ + a_i^0$ such that

$$u = \mathcal{J}_y F_{I^+ \cup M_i}(x_0, y_0)v.$$

Then one can easily verify that

$$\begin{bmatrix} \mathcal{J}_y F_{L \cup (I^0 \setminus M_i)}(x_0, y_0) & 0 \\ \mathcal{J}_y F_{I^+ \cup M_i}(x_0, y_0) & -E \\ E_{I^+ \cup M_i} - \mathcal{J}_y \Phi_{I^+ \cup M_i}(x_0, y_0) & 0 \end{bmatrix} \begin{bmatrix} v \\ u \end{bmatrix} = 0. \tag{6.35}$$

However, the matrix in (6.35) is exactly the ith matrix (6.34), shown to be nonsingular. This contradiction completes the proof. ■

On the basis of Proposition 6.9 and Lemmas 6.15, 6.16 we get the following outer approximation of $\partial \sigma_1(x_0)$.

Theorem 6.17 *Consider the GE (5.41) and assume that (SRC) holds at (x_0, y_0, λ_0). Then one has*

$$\partial \sigma_1(x_0) \subset \text{conv}\left\{\mathcal{B}_i(x_0, y_0) \mid i \in \mathbb{K}(x_0, y_0)\right\}, \tag{6.36}$$

where $\mathcal{B}_i(x_0, y_0)$ for $i \in \mathbb{K}(x_0, y_0)$ is the (unique) solution of the linear matrix equation in Π

$$\mathcal{D}_i(x_0, y_0)\Pi = \begin{bmatrix} -\mathcal{J}_x F_{L \cup (I^0 \setminus M_i)}(x_0, y_0) \\ \mathcal{J}_x \Phi_{I^+ \cup M_i}(x_0, y_0) \end{bmatrix}. \tag{6.37}$$

The proof is an exact copy of the proof of Theorem 6.12 and is omitted. Note that equation (6.37) possesses a unique solution due to the nonsingularity of the matrices $\mathcal{D}_i(x_0, y_0)$ for all $i \in \mathbb{K}(x_0, y_0)$.

Whoever interested in the generalized Jacobian of the map σ_2, can just apply the generalized Jacobian chain rule (Clarke, 1983, Thm. 2.6.6) to get

$$\partial \sigma_2(x_0) \subset \mathcal{J}_x F(x_0, y_0) + \mathcal{J}_y F(x_0, y_0)\text{conv}\{\mathcal{B}_i(x_0, y_0) \mid i \in \mathbb{K}(x_0, y_0)\}. \tag{6.38}$$

For the GE (5.41), the counterpart to Proposition 6.13 attains the following form.

Proposition 6.18 *Let the assumptions of Theorem 6.17 hold and consider some $i \in \mathbb{K}(x_0, y_0)$. Then $\mathcal{B}_i(x_0, y_0) \in \partial \sigma_1(x_0)$ provided there exist vectors $h \in \mathbb{R}^n$ and $v \in \mathbb{R}^m$ such that*

$$\begin{aligned} &\mathcal{J}_x F_{L \cup (I^0 \setminus M_i)}(x_0, y_0)h + \mathcal{J}_y F_{L \cup (I_0 \setminus M_i)}(x_0, y_0)v = 0 \\ &v_{I^+ \cup M_i} - \mathcal{J}_x \Phi_{I^+ \cup M_i}(x_0, y_0)h - \mathcal{J}_y \Phi_{I^+ \cup M_i}(x_0, y_0)v = 0 \\ &\mathcal{J}_x F_{M_i}(x_0, y_0)h + \mathcal{J}_y F_{M_i}(x_0, y_0)v > 0 \\ &v_{I^0 \setminus M_i} - \mathcal{J}_x \Phi_{I^0 \setminus M_i}(x_0, y_0)h - \mathcal{J}_y \Phi_{I^0 \setminus M_i}(x_0, y_0)v > 0. \end{aligned} \tag{6.39}$$

The proof proceeds along the same lines as the proof of Proposition 6.13 and is omitted. The validity of relation (6.39) can, as in Proposition 6.14, be ensured by the nonsmooth variant of the Mangasarian–Fromowitz constraint qualification (Kuntz and Scholtes, 1994), applied to the NSE (6.13) at its solution (x_0, y_0).

(MF2): Every collection of at most $n + m$ rows of the $(m + a^0) \times (n + m)$ matrix

$$\mathbb{S}_2 := \begin{bmatrix} \mathcal{J}_x F_{L \cup I^0}(x_0, y_0) & \mathcal{J}_y F_{L \cup I_0}(x_0, y_0) \\ -\mathcal{J}_x \Phi_{I^+ \cup I^0}(x_0, y_0) & E_{I^+ \cup I^0} - \mathcal{J}_y \Phi_{I^+ \cup I^0}(x_0, y_0) \end{bmatrix} \tag{6.40}$$

is linearly independent.

Proposition 6.19 *Let the assumptions of Theorem 6.17 be fulfilled. Assume that $n \geq a^0$ and that (MF2) holds true at (x_0, y_0). Then*

$$\partial \sigma_1(x_0) = conv\{\mathcal{B}_i(x_0, y_0) \mid i \in \mathbb{K}(x_0, y_0)\}.$$

Proof. If $n \geq a_0$, then (MF2) guarantees full row rank for $\mathbb{S}_2$. Hence to each $i \in \mathbb{K}(x_0, y_0)$, vectors $h \in \mathbb{R}^n$ and $v \in \mathbb{R}^m$ can be found for which (6.39) holds true. This implies $\mathcal{B}_i(x_0, y_0) \in \partial \sigma_1(x_0)$ for all $i \in \mathbb{K}(x_0, y_0)$ and the claim follows from (6.36). ■

It is interesting to note that in the case of GE (5.37) (corresponding to a parameter dependent NCP) one has $\mathcal{B}_i^j(x_0, y_0) = 0$ for each $j \in I^+(x_0, y_0) \cup M_i(x_0, y_0)$ and $i \in \mathbb{K}(x_0, y_0)$.

As an exercise let us compute the generalized Jacobians of the maps σ and σ_1 from Examples 5.1 and 5.2.

Example 6.3 (Example 5.1 continued) Consider the GE (5.45) at the same reference point as in Examples 5.1 and 6.1. We have to work with two subsets of $I^0(x_0, y_0)$, namely $M_1(x_0, y_0) = \emptyset$ and $M_2(x_0, y_0) = \{1\}$. Hence, we have to solve two equations of form (6.29):

$$\begin{bmatrix} 2 & 0 & 0 \\ 0 & 3 & -1 \\ 0 & 1 & 0 \end{bmatrix} \begin{bmatrix} \pi^1 \\ \pi^2 \\ \pi^3 \end{bmatrix} = \begin{bmatrix} 1 \\ 0 \\ 2 \end{bmatrix}$$

and

$$\begin{bmatrix} 2 & 0 & -1 & 0 \\ 0 & 3 & 0 & -1 \\ 1 & 0 & 0 & 0 \\ 0 & 1 & 0 & 0 \end{bmatrix} \begin{bmatrix} \pi^1 \\ \pi^2 \\ \pi^3 \\ \pi^4 \end{bmatrix} = \begin{bmatrix} 1 \\ 0 \\ 1 \\ 2 \end{bmatrix}.$$

From their solutions we compose the matrices $R_i(x_0, y_0)$, $i = 1, 2$, and get by (6.31) the final inclusion

$$\partial \sigma(1) \subset \text{conv} \left\{ \begin{bmatrix} \frac{1}{2} \\ 2 \\ 0 \\ 6 \end{bmatrix}, \begin{bmatrix} 1 \\ 2 \\ 1 \\ 6 \end{bmatrix} \right\} = \begin{bmatrix} [\frac{1}{2}, 1] \\ 2 \\ [0, 1] \\ 6 \end{bmatrix}. \tag{6.41}$$

One can easily check that

$$\mathbb{S}_1 = \begin{bmatrix} -1 & 2 & 0 & 0 & -1 \\ 0 & 0 & 3 & -1 & 0 \\ -2 & 0 & 1 & 0 & 0 \\ -1 & 1 & 0 & 0 & 0 \\ 0 & 0 & 0 & 0 & 1 \end{bmatrix}$$

is nonsingular. By Proposition 6.14, inclusion (6.41) becomes equality.

Example 6.4 (Example 5.2 continued) Consider the GE from Examples 5.2 and 6.2 at $(x_0, y_0, \lambda_0) = (1, -2, -3, -3, -2, 0, 0, 0, 0)$. Due to the already mentioned problem symmetry we have to do with only four subsets of $I^0(x_0, y_0)$, namely $M_1(x_0, y_0) = \emptyset$, $M_2(x_0, y_0) = \{1, 4\}$, $M_3(x_0, y_0) = \{2, 3\}$ and $M_4(x_0, y_0) = \{1, 2, 3, 4\}$. The corresponding equations (6.37) attain the form

$$\begin{bmatrix} 2 & -1 & 0 & 0 \\ -1 & 2 & -1 & 0 \\ 0 & -1 & 2 & -1 \\ 0 & 0 & -1 & 2 \end{bmatrix} \begin{bmatrix} \pi^1 \\ \pi^2 \\ \pi^3 \\ \pi^4 \end{bmatrix} = \begin{bmatrix} 3 \\ 5 \\ 5 \\ 3 \end{bmatrix}$$

$$\begin{bmatrix} 3 & -1 & 0 & 0 \\ -1 & 2 & -1 & 0 \\ 0 & -1 & 2 & -1 \\ 0 & 0 & -1 & 3 \end{bmatrix} \begin{bmatrix} \pi^1 \\ \pi^2 \\ \pi^3 \\ \pi^4 \end{bmatrix} = \begin{bmatrix} \frac{1}{2} \\ 5 \\ 5 \\ \frac{1}{2} \end{bmatrix}$$

$$\begin{bmatrix} 2 & -1 & 0 & 0 \\ -1 & 3 & -1 & 0 \\ 0 & -1 & 3 & -1 \\ 0 & 0 & -1 & 2 \end{bmatrix} \begin{bmatrix} \pi^1 \\ \pi^2 \\ \pi^3 \\ \pi^4 \end{bmatrix} = \begin{bmatrix} 3 \\ 2 \\ 2 \\ 3 \end{bmatrix}$$

$$\begin{bmatrix} 3 & -1 & 0 & 0 \\ -1 & 3 & -1 & 0 \\ 0 & -1 & 3 & -1 \\ 0 & 0 & -1 & 3 \end{bmatrix} \begin{bmatrix} \pi^1 \\ \pi^2 \\ \pi^3 \\ \pi^4 \end{bmatrix} = \begin{bmatrix} \frac{1}{2} \\ 2 \\ 2 \\ \frac{1}{2} \end{bmatrix}.$$

By (6.36) we get the final inclusion

$$\partial\sigma_1(1) \subset \text{conv} \left\{ \begin{bmatrix} 8 \\ 13 \\ 13 \\ 8 \end{bmatrix}, \begin{bmatrix} \frac{11}{4} \\ \frac{31}{4} \\ \frac{31}{4} \\ \frac{11}{4} \end{bmatrix}, \begin{bmatrix} \frac{8}{3} \\ \frac{7}{3} \\ \frac{7}{3} \\ \frac{8}{3} \end{bmatrix}, \begin{bmatrix} \frac{3}{5} \\ \frac{13}{10} \\ \frac{13}{10} \\ \frac{3}{5} \end{bmatrix} \right\}.$$

For the generalized Jacobian of the map σ_2 we obtain from (6.38) the inclusion

$$\begin{aligned}\partial\sigma_2(1) \quad &\subset \quad \begin{bmatrix} -3 \\ -5 \\ -5 \\ -3 \end{bmatrix} + \operatorname{conv}\left\{ \begin{bmatrix} 3 \\ 5 \\ 5 \\ 3 \end{bmatrix}, \begin{bmatrix} -\frac{9}{4} \\ 5 \\ 5 \\ -\frac{9}{4} \end{bmatrix}, \begin{bmatrix} 3 \\ -\frac{1}{3} \\ -\frac{1}{3} \\ 3 \end{bmatrix}, \begin{bmatrix} -\frac{1}{10} \\ \frac{7}{10} \\ \frac{7}{10} \\ -\frac{1}{10} \end{bmatrix} \right\} \\ &= \quad \operatorname{conv}\left\{ \begin{bmatrix} 0 \\ 0 \\ 0 \\ 0 \end{bmatrix}, \begin{bmatrix} -\frac{21}{4} \\ 0 \\ 0 \\ -\frac{21}{4} \end{bmatrix}, \begin{bmatrix} 0 \\ -\frac{16}{3} \\ -\frac{16}{3} \\ 0 \end{bmatrix}, \begin{bmatrix} -\frac{31}{10} \\ -\frac{43}{10} \\ -\frac{43}{10} \\ -\frac{31}{10} \end{bmatrix} \right\}.\end{aligned}$$

The outer approximation of $\partial\sigma_1(x_0)$ and $\partial\sigma(x_0)$ provided by Theorems 6.12 and 6.17 play a key role in the rest of the book. In the next section we use them to show that the selections σ_1 and σ are semismooth at x_0.

6.3 SEMISMOOTHNESS

Chapter 3 emphasized the importance of semismoothness in numerical methods of nonsmooth analysis. To be able to apply these methods to our MPECs, we now analyse the GEs (5.24) and (5.41) with respect to semismoothness. In this task we take advantage of the equivalent characterization of semismoothness given in Theorem 2.22 (Qi and Sun, 1993).

Let us start with the GE (5.24).

Theorem 6.20 *Consider the GE (5.24) and assume that (SRC) holds at $(x_0, y_0, \mu_0, \lambda_0)$. Then the operator σ is semismooth at x_0.*

Proof. By Theorem 2.22 it suffices to show that

$$\sigma'(x_0; h) = \lim_{\substack{V \in \partial\sigma(x_0 + th) \\ t \downarrow 0}} \{Vh\}$$

and that the convergence is uniform for all $h \in \partial\mathbb{B}$. Assume by contradiction that there exists an $\varepsilon > 0$, sequences $\{h_i\} \subset \partial\mathbb{B}$, $t_i \to 0$ and matrices $V_i \in \partial\sigma(x_0 + t_i h_i)$ for which

$$\|V_i h_i - \sigma'(x_0; h_i)\| > \varepsilon \quad \text{for } i = 1, 2, \ldots$$

Since the generalized Jacobian mapping is uniformly compact and $\{h^i\} \subset \partial\mathbb{B}$, we can select convergent subsequences $h_{i'} \to h$, $V_{i'} \to V$, where $h \in \partial\mathbb{B}$ and $V \in \partial\sigma(x_0)$ due to the closedness of $\partial\sigma$. The function $\sigma'(x_0; \cdot)$ is Lipschitz and so necessarily

$$\|Vh - \sigma'(x_0; h)\| \geq \varepsilon. \tag{6.42}$$

From (6.31) we know that

$$V = \sum_{\ell \in \mathbb{K}(x_0, y_0)} \alpha_\ell R_\ell(x_0, y_0),$$

where $\alpha_\ell \geq 0$ and $\sum_{\ell \in \mathbb{K}(x_0, y_0)} \alpha_\ell = 1$. Inequality (6.42) implies the existence of some $j \in \mathbb{K}(x_0, y_0)$ such that $\alpha_j > 0$ and $R_j(x_0, y_0)h \neq \sigma'(x_0; h)$. However, by (6.11), this is possible only if

$$\nabla_x g^k(x_0, y_0)h + \nabla_y g^k(x_0, y_0)\sigma_1'(x_0; h) < 0$$

for some $k \in M_j(x_0, y_0)$. Let $\mathcal{M}(x_0, y_0)$ denote the subset of indices of weakly active constraints so that

$$\nabla_x g^\kappa(x_0, y_0)h + \nabla_y g^\kappa(x_0, y_0)\sigma_1'(x_0; h) < 0 \quad \text{for } \kappa \in \mathcal{M}(x_0, y_0). \tag{6.43}$$

Furthermore, let $\tilde{\mathbb{K}}(x_0, y_0)$ be an index set specifying the elements of $\mathcal{P}(I^0(x_0, y_0) \setminus \mathcal{M}(x_0, y_0))$ in the same way as $\mathbb{K}(x_0, y_0)$ specifies the elements of $\mathcal{P}(I^0(x_0, y_0))$. Clearly, $R_\ell(x_0, y_0)h = \sigma'(x_0; h)$ for all $\ell \in \tilde{\mathbb{K}}(x_0, y_0)$.

Due to the continuity of $\sigma_1'(x_0; \cdot)$, inequalities (6.43) imply that

$$g^\kappa(x_0 + t_{i'}h_{i'}, \sigma_1(x_0 + t_{i'}h_{i'})) < 0 \quad \text{for } \kappa \in \mathcal{M}(x_0, y_0) \cup L(x_0, y_0),$$

whenever i' is sufficiently large. Therefore, by inclusion (6.31) one has for such i'

$$V_{i'} \in \operatorname{conv}\left\{R_\ell(x_0 + t_{i'}h_{i'}, \sigma_1(x_0 + t_{i'}h_{i'})) \mid \ell \in \tilde{\mathbb{K}}(x_0, y_0)\right\}. \tag{6.44}$$

By (6.42) there exists a natural number n_0 such that for all $i' > n_0$

$$\|V_{i'}h - \sigma'(x_0; h)\| > \frac{\varepsilon}{2}, \tag{6.45}$$

where $V_{i'}$ satisfies (6.44). However, this is impossible due to the continuity of the maps $x \mapsto R_\ell(x, \sigma(x))$ in a neighbourhood of x_0. This completes the proof. ∎

The same proof idea also works in the case of the GE (5.41).

Theorem 6.21 *Consider the GE (5.41) and assume that (SRC) holds at (x_0, y_0). Then the respective map σ_1 is semismooth at x_0.*

Proof. The contradiction of the claim leads just as in the proof of Theorem 6.20 to sequences $t_i \downarrow 0$, $h_{i'} \to h$, $V_{i'} \to V$ and to inequality (6.42), where now

$$V = \sum_{\ell \in \mathbb{K}(x_0, y_0)} \alpha_\ell \mathcal{B}_\ell(x_0, y_0)$$

with $\alpha_\ell \geq 0$ and $\sum_{\ell \in \mathbb{K}(x_0, y_0)} \alpha_\ell = 1$. By Theorem 6.6, inequality (6.42) implies the existence of $j \in \mathbb{K}(x_0, y_0)$ such that $\alpha_j > 0$ and

- either there exists a $k \in I^0(x_0, y_0) \setminus M_j(x_0, y_0)$ such that

$$\nabla_x F^k(x_0, y_0)h + \nabla_y F^k(x_0, y_0)\sigma_1'(x_0; h) > 0 \tag{6.46}$$

- or there exists $k' \in M_j(x_0, y_0)$ such that

$$(\sigma_1^{k'})'(x_0; h) - \nabla_x \varphi^{k'}(x_0, y_0)h - \nabla_y \varphi^{k'}(x_0, y_0)\sigma_1'(x_0; h) > 0. \tag{6.47}$$

Let $\mathcal{M}(x_0, y_0)$, $\mathcal{M}'(x_0, y_0)$ denote the subsets of indices $k, k' \in I^0(x_0, y_0)$ for which inequalities (6.46),(6.47), respectively, hold true. Due to the continuity of $\sigma_1'(x_0; \cdot)$ inequalities (6.46),(6.47) imply for sufficiently large i' that

$$F^\kappa(x_0 + t_{i'}h_{i'}, \sigma_1(x_0 + t_{i'}h_{i'})) > 0 \quad \text{for } \kappa \in \mathcal{M}(x_0, y_0)$$

and

$$\sigma_1^\kappa(x_0 + t_{i'}h_{i'}) - \varphi^\kappa(x_0 + t_{i'}h_{i'}, \sigma(x_0 + t_{i'}h_{i'})) > 0 \quad \text{for } \kappa \in \mathcal{M}'(x_0, y_0).$$

Therefore, by Theorem 6.17, for such i' the matrices $V_{i'} \in \partial\sigma_1(x_0 + t_{i'}h_{i'})$ are of form (6.36) with index sets given by

$$\begin{aligned} L(x_0 + t_{i'}h_{i'}, \sigma_1(x_0 + t_{i'}h_{i'})) &= L(x_0, y_0) \cup \mathcal{M}'(x_0, y_0) \\ I^+(x_0 + t_{i'}h_{i'}, \sigma_1(x_0 + t_{i'}h_{i'})) &= I^+(x_0, y_0) \cup \mathcal{M}(x_0, y_0) \\ I^0(x_0 + t_{i'}h_{i'}, \sigma_1(x_0 + t_{i'}h_{i'})) &= I^0(x_0, y_0) \setminus (\mathcal{M}(x_0, y_0) \cup \mathcal{M}'(x_0, y_0)). \end{aligned}$$

Since for all $\kappa \in I^0(x_0 + t_{i'}h_{i'}, \sigma(x_0 + t_{i'}h_{i'}))$

$$\begin{aligned} &\nabla_x F^\kappa(x_0, y_0)h + \nabla_y F^\kappa(x_0, y_0)\sigma_1'(x_0; h) = \\ &(\sigma^\kappa)'(x_0; h) - \nabla_x\varphi^\kappa(x_0, y_0)h - \nabla_y\varphi^\kappa(x_0, y_0)\sigma_1'(x_0; h) = 0, \end{aligned}$$

inequality (6.45) together with the same continuity argument as in the proof of Theorem 6.20 yield a contradiction. ■

Theorem 6.20 implies in particular that for fixed $\overline{x}$ the map $y \mapsto \mathrm{Proj}_{\Delta(\overline{x})}(y - F(\overline{x}, y))$ with Δ given by (5.25) is semismooth at each $y_0 \in \mathbb{R}^m$, provided the following linear independence constraint qualification holds:

$$\left.\begin{array}{l} \text{With } v_0 := \mathrm{Proj}_{\Delta(\overline{x})}(y_0 - F(\overline{x}, y_0) \text{ the partial gradients } \nabla_y h^i(\overline{x}, v_0), \\ i = 1, 2, \ldots, \ell,\ \nabla_y g^j(\overline{x}, v_0), j \in I(\overline{x}, v_0), \text{ are linearly independent.} \end{array}\right\} \quad (6.48)$$

Indeed, v_0 is the projection of $y_0 - F(\overline{x}, y_0)$ onto $\Delta(\overline{x})$ if and only if it is the unique solution of the GE

$$0 \in \begin{bmatrix} v - y_0 + F(\overline{x}, y_0) + \sum_{i=1}^{\ell} \mu^i \nabla_y h^i(\overline{x}, v) + \sum_{i=1}^{s} \lambda^i \nabla_y g^i(\overline{x}, v) \\ H(\overline{x}, y) \\ -G(\overline{x}, y) \end{bmatrix}$$

$$+ N_{\mathbb{R}^m \times \mathbb{R}^\ell \times \mathbb{R}^s_+}(v, \mu, \lambda),$$

whose strong regularity follows from (6.48) in virtue of Theorem 5.8. It is clear that (6.48) is implied by (ELICQ) at $(\overline{x}, v_0)$.

Since the sum of semismooth functions remains semismooth (Mifflin, 1977), the Newton's method from Chapter 3 can be applied to the solution of the perturbed NSE (4.16) with feasible set Ω given by (5.25) for fixed values of the perturbation parameter. For the same reason, this Newton's method can be also applied to the solution of the perturbed NSE (4.18) and the NSE (6.13), which contain projection onto (possibly translated) nonnegative orthant.

Bibliographical notes

The sensitivity questions of this chapter were initially studied in the context of perturbed nonlinear programs. From the large number of works on this topic we mention Fiacco and McCormick, 1968 and Jittorntrum, 1984. The importance of results concerning the directional derivatives of the projection operator in the sensitivity analysis was first recognized in Haraux, 1977. The corresponding results on sensitivity for variational inequalities and complementarity problems (or generalized equations) can be found in Dafermos, 1988; Qiu and Magnanti, 1989; Kyparisis, 1990; Pang, 1990b; Robinson, 1991; Qiu and Magnanti, 1992.

The presented approach to the computation of directional derivatives relies on the basic projection result (Theorem 2.31) and a useful result from Kummer, 1992 (Lemma 6.1).

Generalized Jacobians of solution maps were computed first in the context of nonlinear programs (Outrata, 1990; Outrata, 1993). In Outrata, 1994 and Outrata and Zowe, 1995b this theory was then extended to variational inequalities and in Outrata, 1995 to quasi-variational inequalities, in the framework of GEs.

Concerning the semismoothness property, the main idea of the proofs of Theorems 6.20 and 6.21 comes from Outrata and Zowe, 1995a.

As shown in Scholtes, 1994 and Pang and Ralph, 1996, the selections σ of the solution maps, generated by the investigated GEs, are under our assumptions piecewise differentiable (PC^1-functions). This implies that all results of this chapter could alternatively be obtained by using the strong properties of these functions (Chaney, 1990). However, the applied techniques and calculus substantially differ from those used in this book.

7 OPTIMALITY CONDITIONS AND A SOLUTION METHOD

In this chapter we apply the preceding theory to establish first-order necessary optimality conditions and to construct an efficient and robust numerical method for the solution of the considered MPECs. This numerical method will extensively be used in the second ("applied") part of the book.

As explained in Chapter 1, we will deal with the optimization problem

$$\begin{aligned}&\text{minimize} \quad f(x,z)\\&\text{subject to}\\&\qquad\qquad z \in S(x)\\&\qquad\qquad x \in U_{\mathrm{ad}},\end{aligned} \tag{7.1}$$

where f maps $\mathbb{R}^n \times \mathbb{R}^k$ into $\mathbb{R}$, U_{ad} is a nonempty closed subset of $\mathbb{R}^n$ and S is the solution map, given by a GE of the type (1.1). The MPEC defined in the Introduction attains the form (7.1) provided either

- the state variable z is not subject to any constraints (i.e. $Z = \mathbb{R}^k$); or
- a possible state constraint $z \in Z$ has been added to the objective by an exact penalty and does not explicitly appear in the problem formulation.

Suppose that $(\hat{x}, \hat{z})$ is a (local) solution of (7.1). To be able to apply the implicit programming approach to (7.1), assume that the GE specifying S is strongly regular at $(\hat{x}, \hat{z})$. This ensures that the assumption (A) from the Introduction is fulfilled, i.e., there exist neighbourhoods $\mathcal{U}$ of $\hat{x}$ and $\mathcal{V}$ of $\hat{z}$ and a directionally differentiable Lipschitz selection $\sigma[\mathcal{U} \to \mathbb{R}^k]$ of S such that $\sigma(\hat{x}) = \hat{z}$ and

$$\sigma(x) = S(x) \cap \mathcal{V} \quad \text{for all } x \in \mathcal{U}.$$

Then (7.1) is locally reduced to the "easier" problem

$$\begin{array}{ll}\text{minimize} & f(x,z)\\ \text{subject to} & \\ & z=\sigma(x)\\ & x\in U_{\text{ad}}\cap\mathcal{U}.\end{array}\tag{7.2}$$

In our context, the general nature of GE (1.1) resists usable optimality conditions and numerical methods. Hence we will only consider the case when S in (7.1) is successively given by the GEs (5.24),(5.41) (and (5.37) as a special case of (5.41)).

Section 7.1 is devoted to optimality conditions. We start with the GE (5.24) and admit even optimality criteria which depend on the KKT vector (μ, λ). In particular, we examine the case, where the feasible set does not depend on x, and the bilevel programs, where the GE corresponds to necessary and sufficient optimality conditions of a lower-level optimization problem. Further we derive optimality conditions for MPEC of type (7.1) with equilibria described by complementarity problems.

Section 7.2 introduces a numerical method for the solution of MPEC of type (7.1) which is based on a bundle method of nonsmooth optimization. Above all, we discuss the question how to compute "subgradients", a necessary ingredient for all bundle methods. We also discuss how to solve the equilibrium problems by a nonsmooth variant of the Newton's method, explained in Chapter 3. In our case, this Newton's method can be suitably coupled with the bundle method used.

By Θ we denote (as in Chapter 1) the composite function $f(x, \sigma(x))$, defined on $\mathcal{U}$. If S happens to be single-valued on $\mathcal{U}$, then $\Theta(x) = f(x, S(x))$. In both sections of this chapter the function Θ plays a key role, but the goals are different: in the context of optimality conditions we are interested in "tight" outer approximation of $\partial\Theta(\hat{x})$, whereas all we need in the numerical part is one arbitrary element of $\partial\Theta(x)$ for each parameter x.

7.1 OPTIMALITY CONDITIONS

In this section we assume that f is continuously differentiable on $\mathcal{U} \times \mathbb{R}^k$, where $\mathcal{U}$ is a open neighbourhood on which the existence of σ is ensured by the strong regularity assumption.

The formulation of our necessary optimality conditions is simplified, if we employ the technique of so-called *adjoint equations*, well-known from the optimal control theory (cf., e.g., Luenberger, 1979). Consider a vector $q \in \mathbb{R}^m$, an $m \times m$ matrix A and $m \times n$ matrices P and B with $AP = B$.

Lemma 7.1 *If $\hat{p}$ solves the adjoint equation $A^T p - q = 0$, then*

$$P^T q = B^T \hat{p}.\tag{7.3}$$

Proof. The claim follows from

$$B^T\hat{p} = (AP)^T\hat{p} = P^T A^T \hat{p} = P^T q.$$

■

We first study the case, where the equilibrium is described by the GE (5.24), i.e. $z = (y, \mu, \lambda)$. If this GE is strongly regular at the local minimum $(\hat{x}, \hat{y}, \hat{\mu}, \hat{\lambda})$, then (7.1) is locally reduced to the optimization problem

$$\begin{array}{ll} \text{minimize} & f(x, y, \mu, \lambda) \\ \text{subject to} & \\ & y = \sigma_1(x),\ \mu = \sigma_2(x),\ \lambda = \sigma_3(x) \\ & x \in U_{\text{ad}} \cap \mathcal{U}, \end{array} \tag{7.4}$$

where the operators $\sigma_1, \sigma_2, \sigma_3$ were defined in the beginning of Chapter 6. Throughout this section, the index sets I, I^+, I^0, M_i and $\mathbb{K}$ (introduced in Chapters 5 and 6) are considered at the respective local minimizer $(\hat{x}, \hat{y})$ instead of a reference pair (x_0, y_0). We recall that

$$M_i(\hat{x}, \hat{y}) \in \mathcal{P}(I^0(\hat{x}, \hat{y})) \quad \text{for } i \in \mathbb{K}(\hat{x}, \hat{y}).$$

As before, a, a^+, a^0 and a_i^0 are the cardinalities of $I(\hat{x}, \hat{y}), I^+(\hat{x}, \hat{y}), I^0(\hat{x}, \hat{y})$ and $M_i(\hat{x}, \hat{y})$, respectively.

We start with the case, where f does not depend on the KKT vector (μ, λ).

Theorem 7.2 *Let $(\hat{x}, \hat{z})$ with $\hat{z} = (\hat{y}, \hat{\mu}, \hat{\lambda})$ be a (local) solution of MPEC of type (7.1), where S is given by the GE (5.24) and f does not depend on μ and λ. Assume that this GE is strongly regular at $(\hat{x}, \hat{y}, \hat{\mu}, \hat{\lambda})$ and for all $i \in \mathbb{K}(\hat{x}, \hat{y})$ the vectors $\hat{p}_i, \hat{q}_i, \hat{r}_i$ are the (unique) solutions of the adjoint equations*

$$\begin{aligned} (\mathcal{J}_y \mathcal{L}(\hat{x}, \hat{y}, \hat{\mu}, \hat{\lambda}))^T p_i + (\mathcal{J}_y H(\hat{x}, \hat{y}))^T q_i - (\mathcal{J}_y G_{I^+ \cup M_i}(\hat{x}, \hat{y}))^T r_i &= \nabla_y f(\hat{x}, \hat{y}) \\ \mathcal{J}_y H(\hat{x}, \hat{y}) p_i &= 0 \\ \mathcal{J}_y G_{I^+ \cup M_i}(\hat{x}, \hat{y}) p_i &= 0. \end{aligned} \tag{7.5}$$

Then one has

$$\begin{aligned} 0_{\mathbb{R}^n} \in \nabla_x f(\hat{x}, \hat{y}) + \operatorname{conv} \Big\{ &-(\mathcal{J}_x \mathcal{L}(\hat{x}, \hat{y}, \hat{\mu}, \hat{\lambda}))^T \hat{p}_i \\ &- (\mathcal{J}_x H(\hat{x}, \hat{y}))^T \hat{q}_i + (\mathcal{J}_x G_{I^+ \cup M_i}(\hat{x}, \hat{y}))^T \hat{r}_i \mid i \in \mathbb{K}(\hat{x}, \hat{y}) \Big\} + K_{U_{\text{ad}}}(\hat{x}). \end{aligned} \tag{7.6}$$

Proof. Problem (7.4) is simplified to

$$\begin{array}{ll} \text{minimize} & \Theta(x) \\ \text{subject to} & \\ & x \in U_{\text{ad}} \cap \mathcal{U}, \end{array} \tag{7.7}$$

where $\Theta(x) = f(x, \sigma_1(x))$. Since $\hat{x}$ is a (local) solution to (7.7), we have by Theorem 2.27

$$0 \in \partial \Theta(\hat{x}) + K_{U_{\text{ad}} \cap \mathcal{U}}(\hat{x}) = \partial \Theta(\hat{x}) + K_{U_{\text{ad}}}(\hat{x}).$$

From the generalized Jacobian chain rule (Theorem 2.20) we obtain

$$\partial \Theta(\hat{x}) = \nabla_x f(\hat{x}, \hat{y}) + \operatorname{conv} \{ \Xi^T \nabla_y f(\hat{x}, \hat{y}) \mid \Xi \in \partial \sigma_1(\hat{x}) \}$$

with (cf. (6.28)),

$$\partial\sigma_1(\hat{x}) \subset \operatorname{conv}\{[P_i(\hat{x},\hat{y})]_m \mid i \in \mathbb{K}(\hat{x},\hat{y})\}.$$

The matrix $P_i(\hat{x},\hat{y})$ is the unique solution of the linear matrix equation in Π

$$D_{(I^+\cup M_i)}(\hat{x},\hat{y},\hat{\mu},\hat{\lambda})\Pi = \begin{bmatrix} -\mathcal{J}_x\mathcal{L}(\hat{x},\hat{y},\hat{\mu},\hat{\lambda}) \\ -\mathcal{J}_x H(\hat{x},\hat{y}) \\ \mathcal{J}_x G_{I^+\cup M_i}(\hat{x},\hat{y}) \end{bmatrix}.$$

Therefore,

$$\partial\Theta(\hat{x}) \subset \nabla_x f(\hat{x},\hat{y}) + \operatorname{conv}\left\{ [P_i(\hat{x},\hat{y})]^T \begin{bmatrix} \nabla_y f(\hat{x},\hat{y}) \\ 0_{\mathbb{R}^t} \end{bmatrix} \middle| i \in \mathbb{K}(\hat{x},\hat{y}) \right\}, \tag{7.8}$$

where $t = \ell + a^+ + a_i^0$. The system matrix in (7.5) is nothing else than $(D_{I^+\cup M_i}(\hat{x},\hat{y},\hat{\mu},\hat{\lambda}))^T$ and so, by Lemma 7.1,

$$\begin{aligned} [P_i(\hat{x},\hat{y})]^T \begin{bmatrix} \nabla_y f(\hat{x},\hat{y}) \\ 0 \end{bmatrix} &= \begin{bmatrix} -\mathcal{J}_x\mathcal{L}(\hat{x},\hat{y},\hat{\mu},\hat{\lambda}) \\ -\mathcal{J}_x H(\hat{x},\hat{y}) \\ \mathcal{J}_x G_{I^+\cup M_i}(\hat{x},\hat{y}) \end{bmatrix}^T \begin{bmatrix} \hat{p}_i \\ \hat{q}_i \\ \hat{r}_i \end{bmatrix} \\ &= -(\mathcal{J}_x\mathcal{L}(\hat{x},\hat{y},\hat{\mu},\hat{\lambda}))^T\hat{p}_i - (\mathcal{J}_x H(\hat{x},\hat{y}))^T\hat{q}_i + (\mathcal{J}_x G_{I^+\cup M_i}(\hat{x},\hat{y}))^T\hat{r}_i, \end{aligned}$$

which completes the proof. ■

Consider now the general case, where the objective f depends on (x, y, μ, λ).

Theorem 7.3 *Let $(\hat{x},\hat{z})$ with $\hat{z} = (\hat{y},\hat{\mu},\hat{\lambda})$ be a (local) solution of MPEC of type (7.1), where S is given by the GE (5.24) and this GE is strongly regular at $(\hat{x},\hat{y},\hat{\mu},\hat{\lambda})$. Assume that for all $i \in \mathbb{K}(\hat{x},\hat{y})$ the vectors $\hat{p}_i, \hat{q}_i, \hat{r}_i$ are the (unique) solutions of the adjoint equations*

$$\begin{aligned} (\mathcal{J}_y\mathcal{L}(\hat{x},\hat{y},\hat{\mu},\hat{\lambda}))^T p_i + (\mathcal{J}_y H(\hat{x},\hat{y}))^T q_i - (\mathcal{J}_y G_{I^+\cup M_i}(\hat{x},\hat{y}))^T r_i &= \nabla_y f(\hat{x},\hat{y},\hat{\mu},\hat{\lambda}) \\ \mathcal{J}_y H(\hat{x},\hat{y}) p_i &= \nabla_\mu f(\hat{x},\hat{y},\hat{\mu},\hat{\lambda}) \\ \langle \nabla_y g^j(\hat{x},\hat{y}), p_i\rangle = \frac{\partial f(\hat{x},\hat{y},\hat{\mu},\hat{\lambda})}{\partial \lambda^j} \quad \text{for } j \in I^+(\hat{x},\hat{y}) \cup M_i(\hat{x},\hat{y}). & \end{aligned} \tag{7.9}$$

Then one has

$$\begin{aligned} 0_{\mathbb{R}^n} \in\ & \nabla_x f(\hat{x},\hat{y},\hat{\mu},\hat{\lambda}) + \operatorname{conv}\Big\{-(\mathcal{J}_x\mathcal{L}(\hat{x},\hat{y},\hat{\mu},\hat{\lambda}))^T\hat{p}_i - (\mathcal{J}_x H(\hat{x},\hat{y}))^T\hat{q}_i \\ & + (\mathcal{J}_x G_{I^+\cup M_i}(\hat{x},\hat{y}))^T\hat{r}_i \mid i \in \mathbb{K}(\hat{x},\hat{y})\Big\} + K_{U_{\text{ad}}}(\hat{x}). \end{aligned} \tag{7.10}$$

Proof. We proceed as in the previous proof and sketch only the main points. Problem (7.4) now becomes

$$\begin{aligned} &\text{minimize} && \Theta(x) \\ &\text{subject to} && x \in U_{\text{ad}} \cap \mathcal{U}, \end{aligned}$$

where $\Theta(x) = f(x, \sigma_1(x), \sigma_2(x), \sigma_3(x))$. From (6.31) we get

$$\partial\Theta(\hat{x}) \subset \nabla_x f(\hat{x},\hat{y},\hat{\mu},\hat{\lambda}) + \text{conv}\left\{ [R_i(\hat{x},\hat{y})]^T \begin{bmatrix} \nabla_y f(\hat{x},\hat{y},\hat{\mu},\hat{\lambda}) \\ \nabla_\mu f(\hat{x},\hat{y},\hat{\mu},\hat{\lambda}) \\ \nabla_\lambda f(\hat{x},\hat{y},\hat{\mu},\hat{\lambda}) \end{bmatrix} \middle| \, i \in \mathbb{K}(\hat{x},\hat{y}) \right\}. \tag{7.11}$$

It remains to observe that for $i \in \mathbb{K}(\hat{x},\hat{y})$

$$[R_i(\hat{x},\hat{y})]^T \begin{bmatrix} \nabla_y f(\hat{x},\hat{y},\hat{\mu},\hat{\lambda}) \\ \nabla_\mu f(\hat{x},\hat{y},\hat{\mu},\hat{\lambda}) \\ \nabla_\lambda f(\hat{x},\hat{y},\hat{\mu},\hat{\lambda}) \end{bmatrix} = [P_i(\hat{x},\hat{y})]^T \begin{bmatrix} \nabla_y f(\hat{x},\hat{y},\hat{\mu},\hat{\lambda}) \\ \nabla_\mu f(\hat{x},\hat{y},\hat{\mu},\hat{\lambda}) \\ \nabla_{\lambda_{I^+ \cup M_i}} f(\hat{x},\hat{y},\hat{\mu},\hat{\lambda}) \end{bmatrix},$$

since for $j \notin I^+(\hat{x},\hat{y}) \cup M_i(\hat{x},\hat{y})$ the columns of $[R_i(\hat{x},\hat{y})]^T$ corresponding to $\frac{\partial f(\hat{x},\hat{y},\hat{\mu},\hat{\lambda})}{\partial \lambda^j}$ are zero vectors. Application of Lemma 7.1 completes the proof. ■

If $n \geq a^0$ and (MF1) holds at $(\hat{x},\hat{y},\hat{\mu},\hat{\lambda})$, then the inclusions (7.8) and (7.11) become equalities. In this case the optimality conditions of Theorems 7.2 and 7.3 are as sharp as possible within our framework.

If the functions H and G do not depend on x, then we obtain from Theorem 7.2 the following simpler optimality conditions.

Corollary 7.4 *Let the assumptions of Theorem 7.2 hold and assume, moreover, that the mappings H and G do not depend on x. Suppose that for $i \in \mathbb{K}(\hat{x},\hat{y})$ the vectors $\hat{p}_i$ are the (unique) solutions of the adjoint (generalized) equations*

$$0 \in (\mathcal{J}_y \mathcal{L}(\hat{x},\hat{y},\hat{\mu},\hat{\lambda}))^T p_i - \nabla_y f(\hat{x},\hat{y}) + N_{L_i(\hat{x},\hat{y})}(p_i), \tag{7.12}$$

where

$$L_i(\hat{x},\hat{y}) = \{h \in \text{Ker}\mathcal{J}H(\hat{y}) \mid \langle \nabla g^i(\hat{y}), h\rangle = 0,\ j \in I^+(\hat{x},\hat{y}) \cup M_i(\hat{x},\hat{y})\}.$$

Then one has

$$0 \in \nabla_x f(\hat{x},\hat{y}) - \text{conv}\{(\mathcal{J}_x F(\hat{x},\hat{y}))^T \hat{p}_i \mid i \in \mathbb{K}(\hat{x},\hat{y})\} + K_{U_{\text{ad}}}(\hat{x}). \tag{7.13}$$

Proof. Under the assumptions we need in formula (7.6) only the p_i-components of the solution to the adjoint equations (7.5) which attain the form

$$\begin{aligned} (\mathcal{J}_y \mathcal{L}(\hat{x},\hat{y},\hat{\mu},\hat{\lambda}))^T p_i + (\mathcal{J}_y H(\hat{y}))^T q_i - (\mathcal{J}_y G_{(I^+ \cup M_i)}(\hat{y}))^T r_i &= \nabla_y f(\hat{x},\hat{y}) \\ \mathcal{J}_y H(\hat{y}) p_i &= 0 \\ \mathcal{J}_y G_{I^+ \cup M_i}(\hat{y}) p_i &= 0 \end{aligned} \tag{7.14}$$

for $i \in \mathbb{K}(\hat{x},\hat{y})$. This equation system is equivalent to the GE (7.12) in variable p_i; cf. Table 4.1. Recall that $(\hat{q}_i, -\hat{r}_i)$ is the (unique) KKT vector associated with the equality constraints defining $L_i(\hat{x},\hat{y})$ at $\hat{p}_i$. As $\mathcal{J}_x \mathcal{L}(\hat{x},\hat{y},\hat{\mu},\hat{\lambda}) = \mathcal{J}_x F(\hat{x},\hat{y})$, the proof is complete. ■

The GE (5.24) often corresponds to necessary and sufficient optimality conditions of a parameter-dependent convex optimization problem. In this case, as mentioned in Chapter 1, MPEC is a bilevel program and the adjoint equations (7.5) are reduced to *"adjoint" quadratic programs*. This is due to the symmetry of $\mathcal{J}_y F(\hat{x}, \hat{y})$ which is the Hessian of the lower-level objective with respect to y at $(\hat{x}, \hat{y})$.

Corollary 7.5 *Let the assumptions of Theorem 7.2 hold and assume, moreover, that the matrix $\mathcal{J}_y F(\hat{x}, \hat{y})$ is symmetric. Then in relation (7.6) we can replace the vectors $\hat{p}_i, \hat{q}_i, -\hat{r}_i$ for $i \in \mathbb{K}(\hat{x}, \hat{y})$, by the solution and the (unique) KKT vector of the adjoint quadratic program*

$$
\begin{aligned}
&\textit{minimize} \quad \tfrac{1}{2}\left\langle p_i, \mathcal{J}_y \mathcal{L}(\hat{x}, \hat{y}, \hat{\mu}, \hat{\lambda}) p_i \right\rangle - \langle \nabla_y f(\hat{x}, \hat{y}), p_i \rangle \\
&\textit{subject to} \\
&\qquad \mathcal{J}_y H(\hat{x}, \hat{y}) p_i = 0 \\
&\qquad \mathcal{J}_y G_{I^+ \cup M_i}(\hat{x}, \hat{y}) p_i = 0.
\end{aligned}
\tag{7.15}
$$

Proof. Due to the symmetry of $\mathcal{J}_y F(\hat{x}, \hat{y})$ the matrix

$$
\mathcal{J}_y \mathcal{L}(\hat{x}, \hat{y}, \hat{\mu}, \hat{\lambda}) = \mathcal{J}_y F(\hat{x}, \hat{y}) + \sum_{i=1}^{\ell} \hat{\mu}_i \nabla^2_{yy} h^i(\hat{x}, \hat{y}) + \sum_{i=1}^{s} \hat{\lambda}_i \nabla^2_{yy} g^i(\hat{x}, \hat{y})
$$

is also symmetric. The KKT system of the quadratic program (7.15) is thus exactly the adjoint equation (7.5) (with the opposite sign at vector $\hat{r}_i$). ∎

This additional structure can be used in the numerical approach discussed in the next section.

Remark. The strong regularity assumption guarantees that the map $\sigma : x \mapsto (y, \mu, \lambda)$ is Lipschitz near $\hat{x}$. A closer look shows that in Theorem 7.2 only the Lipschitz continuity of $\sigma_1 : x \mapsto y$ is needed. In Pang and Ralph, 1996 the corresponding conditions have been derived by using the powerful tools of degree theory. It turns out that (SRC) can then be weakened and even non-unique KKT vectors (μ, λ) can be admitted. However, this makes the evaluation of a suitable approximation of $\partial \sigma_1(\hat{x})$ more complicated and the application of the resulting conditions rather clumsy.

We now treat the case of the equilibrium described by the GE (5.41), i.e. $z = (y, \lambda)$ (note that $\lambda = F(x, y)$ in this case). We assume that the objective f does not depend on λ.

Theorem 7.6 *Let $(\hat{x}, \hat{z})$ with $\hat{z} = (\hat{y}, \hat{\lambda})$ be a (local) solution of MPEC of type (7.1), where S is given by the GE (5.41) and f does not depend on λ. Assume that this GE is strongly regular at $(\hat{x}, \hat{y}, \hat{\lambda})$ and for all $i \in \mathbb{K}(\hat{x}, \hat{y})$ the vectors $\hat{p}_i$ are the (unique) solutions of the adjoint equations*

$$
\begin{bmatrix} \mathcal{J}_y F_{L \cup (I^0 \setminus M_i)}(\hat{x}, \hat{y}) \\ E_{I^+ \cup M_i} - \mathcal{J}_y \Phi_{I^+ \cup M_i}(\hat{x}, \hat{y}) \end{bmatrix}^T p_i - \nabla_y f(\hat{x}, \hat{y}) = 0. \tag{7.16}
$$

Then one has

$$0 \in \nabla_x f(\hat{x},\hat{y}) + \text{conv}\left\{ \left[\begin{array}{c} -\mathcal{J}_x F_{L\cup(I^0\setminus M_i)}(\hat{x},\hat{y}) \\ \mathcal{J}_x \Phi_{I^+\cup M_i}(\hat{x},\hat{y}) \end{array} \right]^T \hat{p}_i \middle| i \in \mathbb{K}(\hat{x},\hat{y}) \right\} + K_{U_{\text{ad}}}(\hat{x}). \tag{7.17}$$

Proof. The proof follows the same lines as the proofs of Theorems 7.2 and 7.3. The considered MPEC is locally reduced to the problem

$$\begin{array}{ll} \text{minimize} & f(x,y) \\ \text{subject to} & \\ & y = \sigma_1(x) \\ & x \in U_{\text{ad}} \cap \mathcal{U}, \end{array}$$

where σ_1 is now defined by the GE (5.41). One has (inclusion (6.36))

$$\partial\sigma_1(\hat{x}) \subset \text{conv}\,\{\mathcal{B}_i(\hat{x},\hat{y}) \mid i \in \mathbb{K}(\hat{x},\hat{y})\},$$

where $\mathcal{B}_i(\hat{x},\hat{y})$ is the unique solution of the linear matrix equation in Π

$$\mathcal{D}_i(\hat{x},\hat{y})\Pi = \left[\begin{array}{c} -\mathcal{J}_x F_{L\cup(I^0\setminus M_i)}(\hat{x},\hat{y}) \\ \mathcal{J}_x \Phi_{I^+\cup M_i}(\hat{x},\hat{y}) \end{array} \right].$$

Therefore

$$\partial\Theta(\hat{x}) \subset \nabla_x f(\hat{x},\hat{y}) + \text{conv}\,\{(\mathcal{B}_i(\hat{x},\hat{y}))^T \nabla_y f(\hat{x},\hat{y}) \mid i \in \mathbb{K}(\hat{x},\hat{y})\} \tag{7.18}$$

and application of Lemma 7.1 completes the proof. ∎

If the objective f also depends on the KKT vector λ, i.e. if $z = (y, \lambda)$, then we can use inclusion (6.38) to get the appropriate optimality conditions.

From Proposition 6.19 we infer that, if $n \geq a^0$ and if (MF2) holds at $(\hat{x},\hat{y})$, then the sum of the first two terms on the right-hand side of (7.17) is exactly the generalized gradient $\partial\Theta(\hat{x})$. Hence our optimality conditions are again as sharp as possible in the framework of the generalized differential calculus of Clarke. Section 7.2 discusses situations in which the above assumptions hold.

In the case of the GE (5.37) (equilibria described by parameter-dependent NCPs), the optimality conditions are simplified substantially. As in the above theorem, consider the case, where f does not depend on λ. We already know that (5.37) is equivalent to the GE

$$0 \in F(x,y) + N_{\Psi+\mathbb{R}^m_+}(y); \tag{7.19}$$

see Table 4.1. Therefore the strong regularity of (5.37) at $(\hat{x},\hat{y},F(\hat{x},\hat{y}))$ is equivalent to the strong regularity of the GE (7.19) at $(\hat{x},\hat{y})$.

Theorem 7.7 *Let $(\hat{x},\hat{y})$ be a (local) solution of MPEC of type (7.1), where S is given by the GE (7.19) (with $z = y$). Assume that this GE is strongly regular at $(\hat{x},\hat{y})$ and for all $i \in \mathbb{K}(\hat{x},\hat{y})$ the vectors $\hat{\pi}_i$ are the (unique) solutions of the adjoint equations*

$$\left(\mathcal{J}_y F_{L\cup(I^0\setminus M_i),L\cup(I^0\setminus M_i)}\right)^T \pi_i - (\nabla_y f(\hat{x},\hat{y}))_{L\cup(I^0\setminus M_i)} = 0. \tag{7.20}$$

Then one has

$$0 \in \nabla_x f(\hat{x},\hat{y}) - \text{conv}\left\{(\mathcal{J}_x F_{L\cup(I^0\setminus M_i)}(\hat{x},\hat{y}))^T \hat{\pi}_i \mid i \in \mathbb{K}(\hat{x},\hat{y})\right\} + K_{U_{\text{ad}}}(\hat{x}). \quad (7.21)$$

Proof. In the conditions of Theorem 7.6 replace the function $\Phi(x,y)$ by the vector Ψ from (5.37). From relation (7.17) it is clear that we actually need only those components of the adjoint vectors $\hat{p}_i$ which belong to $L(\hat{x},\hat{y}) \cup (I^0(\hat{x},\hat{y}) \setminus M_i(\hat{x},\hat{y}))$, $i \in \mathbb{K}(\hat{x},\hat{y})$. Due to the structure of the matrix in system (7.16), this part of $\hat{p}_i$ is exactly the vector $\hat{\pi}_i$ which solves the ith adjoint equation (7.20). ■

Alternatively, we can formulate the adjoint equations (7.20) for each $i \in \mathbb{K}(\hat{x},\hat{y})$ component-wise:

$$\sum_{j\in L\cup(I^0\setminus M_i)} \frac{\partial F^j(\hat{x},\hat{y})}{\partial y^k}\pi_i^j - \frac{\partial f(\hat{x},\hat{y})}{\partial y^k} = 0, \quad k \in L(\hat{x},\hat{y}) \cup (I^0(\hat{x},\hat{y}) \setminus M_i(\hat{x},\hat{y})).$$

Then the necessary condition (7.21) becomes

$$0 \in \nabla_x f(\hat{x},\hat{y}) - \text{conv}\left\{ \sum_{j\in L\cup(I^0\setminus M_i)} \hat{\pi}_i^j \nabla_x F^j(\hat{x},\hat{y}) \mid i \in \mathbb{K}(\hat{x},\hat{y}) \right\} + K_{U_{\text{ad}}}(\hat{x}).$$

The optimality conditions from Theorems 7.2–7.7 can be used to test stationarity (which is necessary for optimality) of points computed by various numerical methods. Moreover, they are closely related to the numerical technique presented in the next section.

We close this part with two illustrative examples of MPECs in which we use the perturbed GEs from Chapter 5.

Example 7.1 (Example 6.3 continued) Consider the MPEC

$$\begin{aligned} &\text{minimize} \quad y^1 + y^2 \\ &\text{subject to} \\ &\qquad (x,y) \text{ satisfies the GE (5.45)} \\ &\qquad x \in [1,2]. \end{aligned} \quad (7.22)$$

We verify that $(\hat{x},\hat{y}) = \left(1, \frac{1}{2}, \frac{1}{2}\right)$ satisfies the necessary conditions from Theorem 7.2. The corresponding KKT vector is $\left(0, \frac{1}{2}\right)$ and all assumptions of Theorem 7.2 are fulfilled; cf. Example 5.1. We have to solve two adjoint equations (7.5):

$$\begin{bmatrix} 2 & 0 & 0 \\ 0 & 3 & 1 \\ 0 & -1 & 0 \end{bmatrix} \begin{bmatrix} p_1^1 \\ p_1^2 \\ r_1^2 \end{bmatrix} - \begin{bmatrix} 1 \\ 1 \\ 0 \end{bmatrix} = 0 \quad \text{for } M_1(\hat{x},\hat{y}) = \emptyset,$$

and

$$\begin{bmatrix} 2 & 0 & 1 & 0 \\ 0 & 3 & 0 & 1 \\ -1 & 0 & 0 & 0 \\ 0 & -1 & 0 & 0 \end{bmatrix} \begin{bmatrix} p_2^1 \\ p_2^2 \\ r_2^1 \\ r_2^2 \end{bmatrix} - \begin{bmatrix} 1 \\ 1 \\ 0 \\ 0 \end{bmatrix} = 0 \quad \text{for } M_2(\hat{x},\hat{y}) = \{1\}.$$

They provide the "adjoint vectors"

$$(\hat{p}_1^1, \hat{p}_1^2, \hat{r}_1^2) = \left(\frac{1}{2}, 0, 1\right) \quad \text{and} \quad (\hat{p}_2^1, \hat{p}_2^2, \hat{r}_2^1, \hat{r}_2^2) = (0, 0, 1, 1).$$

Since $(\mathcal{J}_x \mathcal{L}(\hat{x}, \hat{y}, \hat{\lambda}))^T = (-1, 0)$, $\mathcal{J}_x G_{\{2\}}(\hat{x}, \hat{y}) = 2$, $(\mathcal{J}_x G_{\{1,2\}}(\hat{x}, \hat{y}))^T = (1, 2)$ and $K_{U_{\rm ad}}(\hat{x}) = \mathbb{R}_-$, relation (7.6) attains the form

$$0 \in \operatorname{conv}\left\{-(-1) \cdot \frac{1}{2} + 2, 0 + 1 + 2\right\} + \mathbb{R}_- = \left[\frac{5}{2}, 3\right] + \mathbb{R}_- = (-\infty, 3].$$

Hence, $(\hat{x}, \hat{y}) = \left(1, \frac{1}{2}, \frac{1}{2}\right)$ is indeed a stationary point of MPEC (7.22). △

Example 7.2 (Example 6.4 continued) Consider the MPEC

$$\begin{aligned} &\text{minimize} \quad \tfrac{1}{2}(y^2 + 2)^2 + \tfrac{1}{2}(y^3 + 2)^2 \\ &\text{subject to} \\ &\qquad (x, y) \text{ satisfies the GE (5.41) with } F \text{ and } \Phi \text{ from Example 5.2} \\ &\qquad x \in \left[\tfrac{1}{2}, 1\right]. \end{aligned} \tag{7.23}$$

Let us show that $(\hat{x}, \hat{y}) = (1, -2, -3, -3, -2)$ satisfies the necessary conditions of Theorem 7.6. All assumptions of this theorem hold and we observe that $I^0(\hat{x}, \hat{y}) = \{1, 2, 3, 4\}$. Due to the problem symmetry, it suffices to solve the following four adjoint equations

$$\begin{bmatrix} 2 & -1 & 0 & 0 \\ -1 & 2 & -1 & 0 \\ 0 & -1 & 2 & -1 \\ 0 & 0 & -1 & 2 \end{bmatrix} \begin{bmatrix} p_4^1 \\ p_4^2 \\ p_4^3 \\ p_4^4 \end{bmatrix} - \begin{bmatrix} 0 \\ -1 \\ -1 \\ 0 \end{bmatrix} = 0$$

$$\begin{bmatrix} 3 & -1 & 0 & 0 \\ -1 & 2 & -1 & 0 \\ 0 & -1 & 2 & -1 \\ 0 & 0 & -1 & 3 \end{bmatrix} \begin{bmatrix} p_3^1 \\ p_3^2 \\ p_3^3 \\ p_3^4 \end{bmatrix} - \begin{bmatrix} 0 \\ -1 \\ -1 \\ 0 \end{bmatrix} = 0$$

$$\begin{bmatrix} 2 & -1 & 0 & 0 \\ -1 & 3 & -1 & 0 \\ 0 & -1 & 3 & -1 \\ 0 & 0 & -1 & 2 \end{bmatrix} \begin{bmatrix} p_2^1 \\ p_2^2 \\ p_2^3 \\ p_2^4 \end{bmatrix} - \begin{bmatrix} 0 \\ -1 \\ -1 \\ 0 \end{bmatrix} = 0$$

$$\begin{bmatrix} 3 & -1 & 0 & 0 \\ -1 & 3 & -1 & 0 \\ 0 & -1 & 3 & -1 \\ 0 & 0 & -1 & 3 \end{bmatrix} \begin{bmatrix} p_1^1 \\ p_1^2 \\ p_1^3 \\ p_1^4 \end{bmatrix} - \begin{bmatrix} 0 \\ -1 \\ -1 \\ 0 \end{bmatrix} = 0,$$

corresponding to $M_1(\hat{x},\hat{y}) = \emptyset, M_2(\hat{x},\hat{y}) = \{1,4\}, M_3(\hat{x},\hat{y}) = \{2,3\}$ and $M_4(\hat{x},\hat{y}) = \{1,2,3,4\}$, respectively. The solutions are

$$\hat{p}_1 = -\begin{bmatrix} 1 \\ 2 \\ 2 \\ 1 \end{bmatrix}, \quad \hat{p}_2 = -\begin{bmatrix} \frac{1}{2} \\ \frac{3}{2} \\ \frac{3}{2} \\ \frac{1}{2} \end{bmatrix}, \quad \hat{p}_3 = -\begin{bmatrix} \frac{1}{3} \\ \frac{2}{3} \\ \frac{2}{3} \\ \frac{1}{3} \end{bmatrix}, \quad \hat{p}_4 = -\begin{bmatrix} \frac{1}{5} \\ \frac{3}{5} \\ \frac{3}{5} \\ \frac{1}{5} \end{bmatrix}.$$

As matrices $\begin{bmatrix} -\mathcal{J}_x F_{I^0 \setminus M_i}(\hat{x},\hat{y}) \\ \mathcal{J}_x \Phi_{M_i}(\hat{x},\hat{y}) \end{bmatrix}$ we get for $i = 1,2,3,4$

$$\begin{bmatrix} 3 \\ 5 \\ 5 \\ 3 \end{bmatrix}, \begin{bmatrix} \frac{1}{2} \\ 5 \\ 5 \\ \frac{1}{2} \end{bmatrix}, \begin{bmatrix} 3 \\ 2 \\ 2 \\ 3 \end{bmatrix}, \begin{bmatrix} \frac{1}{2} \\ 2 \\ 2 \\ \frac{1}{2} \end{bmatrix}.$$

Since $K_{U_{\text{ad}}}(\hat{x}) = \mathbb{R}_+$, relation (7.17) attains the form

$$\begin{aligned} 0 &\in \operatorname{conv}\left\{-26, -\frac{31}{2}, -\frac{14}{3}, -\frac{13}{5}\right\} + \mathbb{R}_+ \\ &= [-26, -\frac{13}{5}] + \mathbb{R}_+ \\ &= [-26, \infty). \end{aligned}$$

Thus, $(\hat{x},\hat{y}) = (1,-2,-3,-3,-2)$ is a stationary point of MPEC (7.23). △

7.2 THE SOLUTION METHOD

Assume that U_{ad} is compact and $\tilde{A}$ is an open set containing U_{ad}. Further suppose that

(A1) f is continuously differentiable on $\tilde{A} \times \mathbb{R}^k$;

(A2) S is single-valued on $\tilde{A}$;

(A3) The considered GE is strongly regular at all points (x,z) with $x \in \tilde{A}$ and $z = S(x)$.

It is clear that (A3) ensures the local Lipschitz continuity of S on $\tilde{A}$. Then, in virtue of Proposition 1.2, (7.1) has a solution. The proposed solution method consists in the application of a bundle method to the (nonsmooth) program

$$\begin{aligned} &\text{minimize} \quad \Theta(x) \\ &\text{subject to} \\ &\qquad\qquad x \in U_{\text{ad}}, \end{aligned}$$

where $\Theta(x) := f(x, S(x))$. To this purpose, we use the bundle implementation BT discussed in Chapter 3. For this method to work, we have to require that

(i) Θ is weakly semismooth on U_{ad};

(ii) we are able to compute the state variable $z = S(x)$ and one arbitrary subgradient $\xi \in \partial\Theta(x)$ for each $x \in U_{\text{ad}}$.

Theoretically, one could use the proposed method whenever the GE in our MPEC is strongly regular at a local minimum. Then, however, we have to dispose of a good starting point sufficiently close to the local minimum, and we have to ensure that we work during the whole iteration process with the right (unique) selection of S, passing the local minimum.

In the study of the requirement (i) we make use of the following simple auxiliary statement.

Lemma 7.8 *Let $x_0 \in \mathbb{R}^n$, $\alpha[\mathbb{R}^n \to \mathbb{R}^m]$ be semismooth at x_0 and $\beta[\mathbb{R}^m \to \mathbb{R}]$ be continuously differentiable on a neighbourhood of $\alpha(x_0)$. Then $\vartheta = \beta \circ \alpha$ is weakly semismooth at x_0.*

Proof. Consider a direction $h \in \mathbb{R}^n$. Let $\xi(x_0 + t_i h) \in \partial\vartheta(x_0 + t_i h)$, where t_i is a sequence of positive reals converging to 0. Due to Theorem 2.20 one has

$$\lim_{i\to\infty} \langle \xi(x_0 + t_i h), h\rangle = \lim_{i\to\infty} \langle \nabla\beta(\alpha(x_0 + t_i h)), V(x_0 + t_i h)h\rangle,$$

where $V(x_0 + t_i h) \in \partial\alpha(x_0 + t_i h)$. The semismoothness of α at x_0 implies $\lim_{i\to\infty} V(x_0 + t_i h)h = \alpha'(x_0; h)$. Together with the continuous differentiability of β, we conclude

$$\lim_{i\to\infty} \langle \xi(x_0 + t_i h), h\rangle = \langle \nabla\beta(\alpha(x_0)), \alpha'(x_0; h)\rangle = \vartheta'(x_0; h).$$

■

Actually, even the composition of semismooth functions is again semismooth; cf. Mifflin, 1977. Lemma 7.8 will do for our purpose, though.

As in the previous section we will now consider MPECs of type (7.1) with S given successively by the GEs (5.24),(5.41) and (5.37).

Proposition 7.9 *Consider the MPEC of type (7.1) and assume that (A1)–(A3) are fulfilled. Then the function Θ is weakly semismooth on $\tilde{A}$, provided S is given by one of the GEs (5.24),(5.41) and (5.37).*

Proof. All we have to do is to combine Theorems 6.20 and 6.21 with Lemma 7.8. ■

Having shown how to guarantee (i), we now turn to the computation of subgradients. By the generalized Jacobian chain rule this task is reduced to the computation of a matrix from the generalized Jacobian of S and so we can use some of the results from the previous chapter.

We first consider the GE (5.24) and observe that under (A3) the map S splits into three Lipschitz maps S_1, S_2, S_3, which assign x the y-, the μ- and the λ-component of $S(x)$. Let $x_0 \in \tilde{A}$, $y_0 = S_1(x_0)$, $\mu_0 = S_2(x_0)$ and $z_0 = S_3(x_0)$. Proposition 6.13 suggests the following procedure:

- Select $i \in \mathbb{K}(x_0, y_0)$ and construct a feasible minimizing sequence $\{\alpha_k, h_k, v_k, w_k, (u_{I^+ \cup M_i})_k\}$ for the linear program in variables $(\alpha, h, v, w, u_{I^+ \cup M_i})$

$$\begin{aligned}
&\text{minimize } \alpha \\
&\text{subject to} \\
&D_{(I^+ \cup M_i)}(x_0, y_0, \mu_0, \lambda_0) \begin{bmatrix} v \\ w \\ u_{I^+ \cup M_i} \end{bmatrix} = \begin{bmatrix} -\mathcal{J}_x \mathcal{L}(x_0, y_0, \mu_0, \lambda_0) \\ -\mathcal{J}_x H(x_0, y_0) \\ \mathcal{J}_x G_{I^+ \cup M_i}(x_0, y_0) \end{bmatrix} h \\
&u^j \geq -\alpha \text{ for } j \in M_i(x_0, y_0) \\
&\langle \nabla_x g^k(x_0, y_0), h \rangle + \langle \nabla_y g^k(x_0, y_0), v \rangle \leq \alpha \\
&\qquad \text{for } k \in I^0(x_0, y_0) \setminus M_i(x_0, y_0).
\end{aligned} \tag{7.24}$$

- If $\alpha_k < 0$ stop; then one has

$$R_i(x_0, y_0) \in \partial S(x_0), \quad [P_i(x_0, y_0)]_m \in \partial S_1(x_0). \tag{7.25}$$

Unfortunately, even this clumsy technique is not always successful as shown in the next example.

Example 7.3 Consider the parameter-dependent optimization problem

$$\begin{aligned} &\text{minimize} \quad \tfrac{1}{2}(y - x)^2 \\ &\text{subject to} \quad y - \sin x \leq 0. \end{aligned}$$

The corresponding GE (5.24) attains the form

$$0 \in \begin{bmatrix} y - x + \lambda \\ \sin x - y \end{bmatrix} + \begin{bmatrix} 0 \\ N_{\mathbb{R}_+}(\lambda) \end{bmatrix}$$

which for $x_0 = 0$ has the unique solution $S(0) = (0, 0)$. One can easily verify that $S'(0; 1) = (1, 0)$ and $S'(0; -1) = (-1, 0)$ so that S is differentiable at 0. However, the optimal objective values in both problems (7.24) are equal to zero and so their minimization does not lead to a matrix from the generalized Jacobian of S. △

The above example reveals why the procedure may fail. This happens if for each direction $h \in \mathbb{R}^n$ there exists an index $k \in I^0(x_0, y_0)$ such that $(S_3^k)'(x_0; h) = 0$ and

$$\langle \nabla_x g^k(x_0, y_0), h \rangle + \langle \nabla_y g^k(x_0, y_0), S_1'(x_0; h) \rangle = 0.$$

In such a peculiar situation our first-order analysis is not sufficient.

To verify that index $i \in \mathbb{K}(x_0, y_0)$ generates a matrix from $\partial S(x_0)$ we can also use the following statement.

Theorem 7.10 *Consider the GE (5.24) at the reference point $(x_0, y_0, \mu_0, \lambda_0)$. Suppose that the assumptions (A2),(A3) hold and that for given $i \in \mathbb{K}(x_0, y_0)$ the following system in $(y_1^*, y_2^*, y_3^*, y_4^*, y_5^*) \in \mathbb{R}^{a^0 - a_i^0} \times \mathbb{R}^{a_i^0} \times \mathbb{R}^m \times \mathbb{R}^\ell \times \mathbb{R}^{a^+ + a_i^0}$ is inconsistent:*

$$\begin{aligned} &-\left(\mathcal{J}_y G_{I^0\setminus M_i}(x_0, y_0)\right)^T y_1^* + \left(\mathcal{J}_y \mathcal{L}(x_0, y_0, \mu_0, \lambda_0)\right)^T y_3^* \\ &\quad + \left(\mathcal{J}_y H(x_0, y_0)\right)^T y_4^* - \left(\mathcal{J}_y G_{I^+\cup M_i}(x_0, y_0)\right)^T y_5^* = 0 \end{aligned} \tag{7.26a}$$

$$y_2^* + \mathcal{J}_y G_{M_i}(x_0, y_0) y_3^* = 0 \tag{7.26b}$$

$$\begin{aligned} &-\left(\mathcal{J}_x G_{I^0\setminus M_i}(x_0, y_0)\right)^T y_1^* + \left(\mathcal{J}_x \mathcal{L}(x_0, y_0, \mu_0, \lambda_0)\right)^T y_3^* \\ &\quad + \left(\mathcal{J}_x H(x_0, y_0)\right)^T y_4^* - \left(\mathcal{J}_x G_{I^+\cup M_i}(x_0, y_0)\right)^T y_5^* = 0 \end{aligned} \tag{7.26c}$$

$$(y_1^*, y_2^*) \geq 0, \quad (y_1^*, y_2^*) \neq 0, \tag{7.26d}$$

$$y_3^* \in \operatorname{Ker}(\mathcal{J}_y H(x_0, y_0)) \cap \operatorname{Ker}(\mathcal{J}_y G_{I^+}(x_0, y_0)) \tag{7.26e}$$

Then (7.25) holds true.

Proof. By Proposition 6.13 it suffices to show the existence of vectors $h, v, w, u_{I^+\cup M_i}$ such that the linear system (6.32) of equalities and strict inequalities is consistent. The claim follows since, according to Motzkin Theorem of the Alternative (cf., e.g., Mangasarian, 1994), the consistency of (6.32) is equivalent to the inconsistency of system (7.26). ■

In the special cases $M_i(x_0, y_0) = I^0(x_0, y_0)$ or $M_i(x_0, y_0) = \emptyset$, the system (7.26) attains a simpler form. If $M_i(x_0, y_0) = I^0(x_0, y_0)$, then the variable y_1^* disappears. Consequently, (7.26a)–(7.26d) take the form

$$\begin{gathered} \left(\mathcal{J}_y \mathcal{L}(x_0, y_0, \mu_0, \lambda_0)\right)^T y_3^* + \left(\mathcal{J}_y H(x_0, y_0)\right)^T y_4^* - \left(\mathcal{J}_y G_I(x_0, y_0)\right)^T y_5^* = 0 \\ y_2^* + \mathcal{J}_y G_{I^0}(x_0, y_0) y_3^* = 0 \\ \left(\mathcal{J}_x \mathcal{L}(x_0, y_0, \mu_0, \lambda_0)\right)^T y_3^* + \left(\mathcal{J}_x H(x_0, y_0)\right)^T y_4^* - \left(\mathcal{J}_x G_I(x_0, y_0)\right)^T y_5^* = 0 \\ y_2^* \geq 0, \quad y_2^* \neq 0. \end{gathered} \tag{7.27}$$

In the case $M_i(x_0, y_0) = \emptyset$ the variable y_2^* together with equation (7.26b) disappears and (7.26d) is simplified to $y_1^* \geq 0,\ y_1^* \neq 0$.

Example 7.4 Consider the GE of the type (5.24)

$$0 \in \begin{bmatrix} -\frac{100}{3} + 2y^1 + \frac{8}{3}y^2 + \lambda^1 \\ -\frac{97}{4} + \frac{5}{4}y^1 + 2y^2 + \lambda^2 \\ 15 - y^1 - x^2 \\ 15 - y^2 - x^1 \end{bmatrix} + N_{\mathbb{R}^2 \times \mathbb{R}^2_+}(y, \lambda)$$

at the reference point $(x_0, y_0, \lambda_0) = \left(10, 5, 10, 5, 0, \frac{7}{4}\right)$. By Theorem 4.8 this GE possesses a unique solution $S(x)$ for each $x \in \mathbb{R}^2$ and Theorem 5.8 implies its strong regularity at each pair $(x, S(x))$, in particular at (x_0, y_0, λ_0). Clearly, $I^+(x_0, y_0) = \{2\}$ and $I^0(x_0, y_0) =$

$\{1\}$. We choose $M_i(x_0, y_0) = I^0(x_0, y_0)$. The first three equations of system (7.27) attain the form

$$\begin{aligned} \begin{bmatrix} 2 & \frac{5}{4} \\ \frac{8}{3} & 2 \end{bmatrix} y_3^* - \begin{bmatrix} 1 & 0 \\ 0 & 1 \end{bmatrix} y_5^* &= 0 \\ y_2^* + \begin{bmatrix} 1 & 0 \end{bmatrix} y_3^* &= 0 \\ - \begin{bmatrix} 0 & 1 \\ 1 & 0 \end{bmatrix} y_5^* &= 0, \end{aligned}$$

so that $y_2^* = 0,\ y_3^* = 0,\ y_5^* = 0$. Therefore, the conditions of Theorem 7.10 are satisfied and the solution of the matrix equation in Π

$$\begin{bmatrix} 2 & \frac{8}{3} & 1 & 0 \\ \frac{5}{4} & 2 & 0 & 1 \\ -1 & 0 & 0 & 0 \\ 0 & -1 & 0 & 0 \end{bmatrix} \Pi = \begin{bmatrix} 0 & 0 \\ 0 & 0 \\ 0 & 1 \\ 1 & 0 \end{bmatrix}$$

belongs to $\partial S(x_0)$. △

Obviously, in larger problems it is a tedious (or even impossible) job to check whether the system (7.26) is inconsistent or not. The following corollaries of Theorem 7.10 discuss some special cases, when this inconsistency can be ensured by assumptions which can be more easily verified.

Corollary 7.11 *Let the assumptions of Theorem 7.10 be fulfilled. Suppose that $\mathcal{J}_x \mathcal{L}(x_0, y_0, \mu_0, \lambda_0)$ is the null matrix and the gradients $\nabla_x h^j(x_0, y_0)$, $j = 1, 2, \ldots, \ell$, and $\nabla_x g^j(x_0, y_0)$, $j \in I(x_0, y_0)$, are linearly independent. Then (7.25) hold for each $i \in \mathbb{K}(x_0, y_0)$.*

Proof. We verify the inconsistency of (7.26) for each $i \in \mathbb{K}(x_0, y_0)$. Consider first the case $M_i(x_0, y_0) \neq \emptyset$. Under the made assumptions, we get from (7.26c) that $y_1^* = 0,\ y_4^* = 0$ and $y_5^* = 0$. Hence, (7.26a) and (7.26e) imply

$$y_3^* \in \operatorname{Ker}\left((\mathcal{J}_y \mathcal{L}(x_0, y_0, \mu_0, \lambda_0))^T\right) \cap \operatorname{Ker}\left(\mathcal{J}_y H(x_0, y_0)\right) \cap \operatorname{Ker}\left(\mathcal{J}_y G_{I^+}(x_0, y_0)\right). \quad (7.28)$$

Assume that there exists a nonzero vector y_3^* satisfying (7.28). Then the homogeneous linear equation

$$\begin{bmatrix} (\mathcal{J}_y \mathcal{L}(x_0, y_0, \mu_0, \lambda_0))^T & (\mathcal{J}_y H(x_0, y_0))^T & -(\mathcal{J}_y G_{I^+}(x_0, y_0))^T \\ \mathcal{J}_y H(x_0, y_0) & 0 & 0 \\ \mathcal{J}_y G_{I^+}(x_0, y_0) & 0 & 0 \end{bmatrix} \begin{bmatrix} v \\ w \\ u \end{bmatrix} = 0$$

possesses a nonzero solution $(y_3^*, 0, 0)$. This is a contradiction to Theorem 5.7 and so, necessarily, $y_3^* = 0$. But then (7.26b) implies that also $y_2^* = 0$ and we are done.

In the case $M_i(x_0, y_0) = \emptyset$ one can argue in a similar way. ■

Corollary 7.12 *Let the assumptions of Theorem 7.10 be fulfilled. Suppose that the functions h^i, $i = 1, 2, \ldots, \ell$, and g^j, $j = 1, 2, \ldots, s$, do not depend on x and that*

$$\operatorname{Ker}\left((\mathcal{J}_x F(x_0, y_0))^T\right) \cap \operatorname{Ker}(\mathcal{J}_y H(x_0, y_0)) \cap \operatorname{Ker}(\mathcal{J}_y G_{I^+}(x_0, y_0)) = \{0\}. \quad (7.29)$$

Then (7.25) holds for each $i \in \mathbb{K}(x_0, y_0)$.

Proof. As $\mathcal{J}_x \mathcal{L}(x_0, y_0, \mu_0, \lambda_0) = \mathcal{J}_x F(x_0, y_0)$, assumption (7.29) implies $y_3^* = 0$ and thus also $y_2^* = 0$. Furthermore, if $M_i(x_0, y_0) \neq I^0(x_0, y_0)$, then $y_1^* = 0$ due to equation (7.26a) and the (ELICQ) at (x_0, y_0). ■

Remark. Condition (7.29) evidently holds if the operator $\mathcal{J}_x F(x_0, y_0)$ $[\mathbb{R}^n \to \mathbb{R}^m]$ is surjective, since then

$$\operatorname{Ker}\left((\mathcal{J}_x F(x_0, y_0))^T\right) = \{0\}.$$

The last corollary concerns a frequently arising case, where the feasible set is given by box constraints, i.e.

$$\Omega = \left\{ y \in \mathbb{R}^m \mid a^i \le y^i \le b^i, \; i = 1, 2, \ldots, m \right\} \quad (7.30)$$

with $a, b \in \mathbb{R}^m$. Then there is no μ in the D-Lagrangian and

$$\mathcal{J}_y \mathcal{L}(x_0, y_0, \lambda_0) = \mathcal{J}_y F(x_0, y_0), \quad \mathcal{J}_x \mathcal{L}(x_0, y_0, \lambda_0) = \mathcal{J}_x F(x_0, y_0).$$

Corollary 7.13 *Let the assumptions of Theorem 7.10 be fulfilled. Assume that the feasible set is given by (7.30), $n \ge a^0$ and that the $(m - a^+) \times (n + m - a^+ - a^0)$ matrix*

$$\left[\mathcal{J}_x F_{L \cup I^0}(x_0, y_0), \mathcal{J}_y F_{L \cup I^0, L}(x_0, y_0)\right] \quad (7.31)$$

has full row rank. Then (7.25) holds for each $i \in \mathbb{K}(x_0, y_0)$.

Proof. Assume that $(y_1^*, y_2^*, y_3^*, y_5^*)$ is a solution of (7.26) and observe that for each subset K of $\{1, 2, \ldots, m\}$ the rows of the matrices $\mathcal{J}_y G_K(x_0, y_0)$ can have non-zero elements only on positions which correspond to the original row indices; further, these elements are either -1 or 1. Since $y_3^* \in \operatorname{Ker}(\mathcal{J}_y G_{I^+}(x_0, y_0))$, one has $(y_3^*)^i = 0$ for $i \in I^+(x_0, y_0)$, which enables to simplify system (7.26) by considering the remaining components only. Denote by r the subvector $(y_3^*)_{L \cup I^0}$. Then the condition $y_3^* \in \operatorname{Ker}((\mathcal{J}_x F(x_0, y_0))^T)$ is reduced to $r \in \operatorname{Ker}((\mathcal{J}_x F_{L \cup I^0}(x_0, y_0))^T)$ and equation (7.26a) becomes

$$\left(\mathcal{J}_y F_{L \cup I^0, L}(x_0, y_0)\right)^T r = 0.$$

(Note that $(-\left(\mathcal{J}_y G_{I^0 \setminus M_i}(x_0, y_0)\right)^T y_1^* - \left(\mathcal{J}_y G_{I^+ \cup M_i}(x_0, y_0)\right)^T y_5^*)^i = 0$ for all $i \in L(x_0, y_0)$.) Then, however, $r = 0$ by our assumptions so that the whole vector $y_3^* = 0$. The further reasoning is the same as in the proof of the preceding corollary. ■

If a suitable index i has been found for which relations (7.25) hold, then the corresponding subgradient $\xi \in \partial\Theta(x_0)$ can be computed by the adjoint equations technique, as explained in the previous section.

Proposition 7.14 *Let the relations (7.25) be fulfilled for $i \in \mathbb{K}(x_0, y_0)$. Then the following statements hold true:*

(i) *Assume that f does not depend on μ and λ, and the vectors $\hat{p}_i, \hat{q}_i, \hat{r}_i$ are the (unique) solutions of the adjoint equation (7.5) with $(\hat{x}, \hat{y}, \hat{\mu}, \hat{\lambda})$ replaced by $(x_0, y_0, \mu_0, \lambda_0)$. Then, with $\Theta(x) = f(x, S_1(x))$, one has*

$$\begin{aligned} \xi \;:=\; & \nabla_x f(x_0, y_0) - (\mathcal{J}_x \mathcal{L}(x_0, y_0, \mu_0, \lambda_0))^T \hat{p}_i \\ & - (\mathcal{J}_x H(x_0, y_0))^T \hat{q}_i + (\mathcal{J}_x G_{I+\cup M_i}(x_0, y_0))^T \hat{r}_i \in \partial\Theta(\hat{x}). \end{aligned} \tag{7.32}$$

(ii) *Assume that the $\hat{p}_i, \hat{q}_i, \hat{r}_i$ are the (unique) solutions of the adjoint equation (7.9) with $(\hat{x}, \hat{y}, \hat{\mu}, \hat{\lambda})$ replaced by $(x_0, y_0, \mu_0, \lambda_0)$. Then, with $\Theta(x) = f(x, S_1(x), S_2(x), S_3(x))$, one has*

$$\begin{aligned} \xi \;:=\; & \nabla_x f(x_0, y_0, \mu_0, \lambda_0) - (\mathcal{J}_x \mathcal{L}(x_0, y_0, \mu_0, \lambda_0))^T \hat{p}_i \\ & - (\mathcal{J}_x H(x_0, y_0))^T \hat{q}_i + (\mathcal{J}_x G_{I+\cup M_i}(x_0, y_0))^T \hat{r}_i \in \partial\Theta(\hat{x}). \end{aligned} \tag{7.33}$$

(iii) *Assume that f does not depend on μ and λ and H, G do not depend on x. Let $\hat{p}_i$ be the (unique) solution of the GE (7.12) with $(\hat{x}, \hat{y}, \hat{\mu}, \hat{\lambda})$ replaced by $(x_0, y_0, \mu_0, \lambda_0)$. Then, with $\Theta(x) = f(x, S_1(x))$, one has*

$$\xi := \nabla_x f(x_0, y_0) - (\mathcal{J}_x F(x_0, y_0))^T \hat{p}_i \in \partial\Theta(\hat{x}). \tag{7.34}$$

(iv) *Assume that f does not depend on μ and λ and $\mathcal{J}_y F(x_0, y_0)$ is symmetric. Let the vectors $\hat{p}_i, \hat{q}_i, -\hat{r}_i$ be the solution and the (unique) KKT vector of the adjoint quadratic program (7.15) with $(\hat{x}, \hat{y}, \hat{\mu}, \hat{\lambda})$ replaced by $(x_0, y_0, \mu_0, \lambda_0)$. Them (7.32) is fulfilled.*

Proof. All statements directly follow from (7.25) and Theorem 2.20 with Lemma 7.1 used in the same way as in Theorems 7.2,7.3. ∎

We now turn our attention to the GEs (5.41),(5.37), and denote by S_1 the Lipschitz map assigning x the y-component of $S(x)$. Let $x_0 \in \tilde{A}$, $y_0 = S_1(x_0)$ and $\lambda_0 = F(x_0, y_0)$. As counterpart of the program (7.24) in the case of the GE (5.41), we fix again $i \in \mathbb{K}(x_0, y_0)$ and work now with the linear program in $(\alpha, h, v) \in \mathbb{R} \times \mathbb{R}^n \times \mathbb{R}^m$:

$$\begin{aligned} & \text{minimize} \quad \alpha \\ & \text{subject to} \\ & \qquad \mathcal{D}_i(x_0, y_0) v = \begin{bmatrix} -\mathcal{J}_x F_{L\cup(I^0\setminus M_i)}(x_0, y_0) \\ \mathcal{J}_x \Phi_{I+\cup M_i}(x_0, y_0) \end{bmatrix} h \\ & \qquad \langle \nabla_x F^j(x_0, y_0), h \rangle + \langle \nabla_y F^j(x_0, y_0), v \rangle + \alpha \geq 0 \text{ for } j \in M_i(x_0, y_0) \\ & \qquad v^k - \langle \nabla_x \varphi^k(x_0, y_0), h \rangle - \langle \nabla_y \varphi^k(x_0, y_0), v \rangle + \alpha \geq 0 \\ & \qquad\qquad \text{for } k \in I^0(x_0, y_0) \setminus M_i(x_0, y_0). \end{aligned} \tag{7.35}$$

As soon as $\alpha < 0$ for some feasible (α, h, v), we stop the minimization and conclude that

$$\mathcal{B}_i(x_0, y_0) \in \partial S_1(x_0). \tag{7.36}$$

In the case of the GE (5.37) the constraints in (7.35) are simplified to

$$\begin{aligned} \mathcal{J}_x F_{L\cup(I^0\setminus M_i)}(x_0,y_0)h + \mathcal{J}_y F_{L\cup(I^0\setminus M_i),L\cup(I^0\setminus M_i)}(x_0,y_0)v &= 0 \\ \mathcal{J}_x F_{M_i}(x_0,y_0)h + \mathcal{J}_y F_{M_i,L\cup(I^0\setminus M_i)}(x_0,y_0)v + \alpha\mathbf{1} &\geq 0 \\ v^j + \alpha \geq 0 \text{ for } j \in I^0(x_0,y_0) \setminus M_i(x_0,y_0)&, \end{aligned} \tag{7.37}$$

where the dimension of v is reduced to $|L(x_0,y_0) \cup (I^0(x_0,y_0) \setminus M_i(x_0,y_0))|$.

If we succeed to compute a triple (α, h, v) with $\alpha < 0$, then we know that

$$\mathcal{A}_i(x_0,y_0) \in \partial S_1(x_0), \tag{7.38}$$

where $\mathcal{A}_i(x_0,y_0)$ is the unique solution of the linear matrix equation in Π

$$\begin{bmatrix} \mathcal{J}_y F_{L\cup(I^0\setminus M_i)}(x_0,y_0) \\ E_{I^+\cup M_i} \end{bmatrix} \Pi = \begin{bmatrix} -\mathcal{J}_x F_{L\cup(I^0\setminus M_i)}(x_0,y_0) \\ 0 \end{bmatrix}. \tag{7.39}$$

We leave it to the reader to derive a counterpart of Theorem 7.10 and proceed directly to the following statements which correspond to Corollaries 7.11–7.13.

Proposition 7.15 *Consider the GE (5.41) at the reference point (x_0, y_0, λ_0). Suppose that the assumptions (A2),(A3) and one of the following conditions are satisfied:*

(i) For $i \in L(x_0,y_0) \cup I^0(x_0,y_0)$ the functions F^i do not depend on x and $\mathcal{J}_x \Phi_{I^+\cup I^0}(x_0,y_0)$ has full row rank.

(ii) For $i \in I^+(x_0,y_0) \cup I^0(x_0,y_0)$ the functions φ^i do not depend on x and the matrices $\mathcal{J}_x F_{L\cup I^0}(x_0,y_0)$, $E_{I^+\cup I^0} - \mathcal{J}_y \Phi_{I^+\cup I^0}(x_0,y_0)$ have full row rank.

(iii) For $j \in I^+(x_0,y_0) \cup I^0(x_0,y_0)$ the functions φ^i are constant and the matrix (7.31) has full row rank.

Then (7.36) holds for each $i \in \mathbb{K}(x_0,y_0)$.

Proof. In the proof it suffices to show that under each of the conditions (i), (ii), (iii), the matrix $\mathbb{S}_2$ from (6.40) has full row rank and thus Proposition 6.19 applies. In the case of (i) one has to take into account that, by the strong regularity, the matrix $\mathcal{J}_y F_{L\cup I^0, L\cup I^0}(x_0,y_0)$ is nonsingular; cf. Theorem 5.9. ∎

It is apparent how the above conditions (i)–(iii) are simplified if we consider instead of GE (5.41) the simpler GE (5.37), i.e. $\varphi^i(x,y) = \psi^i$ for $i = 1, 2, \ldots, m$.

If a suitable index i has been found for which relations (7.36),(7.38) hold, then the corresponding subgradient $\xi \in \partial\Theta(x_0)$ can be computed as follows.

Proposition 7.16 *Consider the GE (5.41) and assume that (7.36) holds and f does not depend on λ. Then*

$$\xi := \nabla_x f(x_0,y_0) + \begin{bmatrix} -\mathcal{J}_x F_{L\cup(I^0\setminus M_i)}(x_0,y_0) \\ \mathcal{J}_x \Phi_{I^+\cup M_i}(x_0,y_0) \end{bmatrix}^T \hat{p}_i \in \partial\Theta(x_0), \tag{7.40}$$

where $\hat{p}_i$ is the (unique) solution of the adjoint equation (7.16) with $(\hat{x}, \hat{y})$ replaced by (x_0, y_0) (and $\Theta(x) = f(x, S_1(x))$).

Proposition 7.17 *Consider the GE (5.37) and assume that (7.38) holds and f does not depend on λ. Then*

$$\xi := \nabla_x f(x_0, y_0) - [\mathcal{J}_x F_{L\cup(I^0\setminus M_i)}(x_0, y_0)]^T \hat{\pi}_i \in \partial\Theta(x_0), \tag{7.41}$$

where $\hat{\pi}_i$ is the (unique) solution of the adjoint equation (7.20) with $(\hat{x}, \hat{y})$ replaced by (x_0, y_0) (and $\Theta(x) = f(x, S_1(x))$).

The above statements directly follow from Theorem 2.20 and relations (7.36),(7.38).

In some MPECs of small dimension or special structure, it is often possible to check the nonsingularity of matrices $\mathbb{S}_1, \mathbb{S}_2$ from (6.33),(6.40) or to verify the conditions of Corollaries 7.11–7.13 or Proposition 7.15. This holds, e.g., for the generalized Nash equilibrium (Chapter 12) or if S is a projection of x onto a convex set given by equations and inequalities (under (ELICQ)).

Unfortunately, for a general MPEC of type (7.1) of a medium or large dimension, such an analysis would be usually too time-consuming. One can think of a simple strategy to choose always $M_i = \emptyset$, because it will provide us with the correct derivatives whenever S is differentiable (Propositions 6.8, 6.9). This strategy has been applied to the mechanical problems of Chapters 9–11 which are of a considerable dimension. In this way all of them have been solved without any problems.

The rest of this chapter is devoted to the auxiliary problem of solving the GEs (5.24), (5.41) (and (5.37)) for fixed values of $x \in U_{\text{ad}}$. To this purpose we can choose from a large variety of available methods; cf. the survey by Harker and Pang, 1990. However, it is the nonsmooth variant of the Newton's method due to Kummer, 1992; Qi and Sun, 1993; Qi, 1993 and Facchinei et al., 1995 (described in Chapter 3) which is intimately connected with our method for the solution of (7.1).

Let $\overline{x} \in U_{\text{ad}}$ be fixed and consider the GE (5.24) written as the NSE (cf. (4.18))

$$\mathcal{G}_{\overline{x}}(y, \mu, \lambda) := \begin{bmatrix} \mathcal{L}(\overline{x}, y, \mu, \lambda) \\ H(\overline{x}, y) \\ \min\{-G(\overline{x}, y), \lambda\} \end{bmatrix} = 0. \tag{7.42}$$

The function $\mathcal{G}_{\overline{x}}[\mathbb{R}^m \times \mathbb{R}^\ell \times \mathbb{R}^s \to \mathbb{R}^m \times \mathbb{R}^\ell \times \mathbb{R}^s]$ is locally Lipschitz and, as explained at the end of Section 6.3, semismooth at each triple (y, μ, λ).

At the iterate (y_k, μ_k, λ_k), Newton step (3.69) attains the form

$$\begin{bmatrix} y_{k+1} \\ \mu_{k+1} \\ \lambda_{k+1} \end{bmatrix} = \begin{bmatrix} y_k \\ \mu_k \\ \lambda_k \end{bmatrix} - (\beta(y_k, \mu_k, \lambda_k))^{-1} \mathcal{G}_{\overline{x}}(y_k, \mu_k, \lambda_k), \tag{7.43}$$

where $\beta(y_k, \mu_k, \lambda_k) \in \partial_B \mathcal{G}_{\overline{x}}(y_k, \mu_k, \lambda_k)$. This Newton's method is locally superlinearly convergent to a solution $(\overline{y}, \overline{\mu}, \overline{\lambda})$, provided all matrices from $\partial_B \mathcal{G}_{\overline{x}}(\overline{y}, \overline{\mu}, \overline{\lambda})$ are nonsingular.

Proposition 7.18 *Let the GE (5.24) be strongly regular at $(\overline{x}, \overline{y}, \overline{\mu}, \overline{\lambda})$. Then all matrices from $\partial_B \mathcal{G}_{\overline{x}}(\overline{y}, \overline{\mu}, \overline{\lambda})$ are nonsingular.*

Proof. For $i \in \mathbb{K}(\overline{x}, \overline{y})$ let Ξ_{1i} be an $s \times m$ matrix given by

$$(\Xi_{1i})^j = \begin{cases} -\nabla_y g^j(\overline{x}, \overline{y}) & \text{for } j \in I^+(\overline{x}, \overline{y}) \cup M_i(\overline{x}, \overline{y}) \\ 0 & \text{otherwise} \end{cases}$$

and Ξ_{2i} be an $s \times s$ matrix given by

$$(\Xi_{2i})^j = \begin{cases} 0 & \text{for } j \in I^+(\overline{x},\overline{y}) \cup M_i(\overline{x},\overline{y}) \\ e_j & \text{otherwise.} \end{cases}$$

By Definition 2.12 each matrix from $\partial_B \mathcal{G}_{\overline{x}}(\overline{y},\overline{\mu},\overline{\lambda})$ has the form

$$\begin{bmatrix} \mathcal{J}_y \mathcal{L}(\overline{x},\overline{y},\overline{\mu},\overline{\lambda}) & (\mathcal{J}_y H(\overline{x},\overline{y}))^T & (\mathcal{J}_y G(\overline{x},\overline{y}))^T \\ \mathcal{J}_y H(\overline{x},\overline{y}) & 0 & 0 \\ \Xi_{1i} & 0 & \Xi_{2i} \end{bmatrix} \tag{7.44}$$

for some $i \in \mathbb{K}(\overline{x},\overline{y})$. By interchanging rows and columns in (7.44) we get the matrix

$$\left[\begin{array}{c:c} & (\mathcal{J}_y G_{L\cup(I^0\setminus M_i)}(\overline{x},\overline{y}))^T \\ D_{(I^+\cup M_i)}(\overline{x},\overline{y},\overline{\mu},\overline{\lambda}) & 0 \\ & 0 \\ \hdashline 0 \quad 0 \quad 0 & E \end{array}\right]. \tag{7.45}$$

Lemma 6.11 implies that each matrix $D_{(I^+\cup M_i)}(\overline{x},\overline{y},\overline{\mu},\overline{\lambda})$, $i \in \mathbb{K}(\overline{x},\overline{y})$, is nonsingular. Thus the whole matrix (7.45) is nonsingular and we are done. ∎

Actually, each matrix of the form (7.44) with $i \in \mathbb{K}(\overline{x},\overline{y})$ belongs to $\partial_B \mathcal{G}_{\overline{x}}(\overline{y},\overline{\mu},\overline{\lambda})$. Indeed, to each $i \in \mathbb{K}(\overline{x},\overline{y})$ there exists a direction $(v, w, u) \in \mathbb{R}^m \times \mathbb{R}^\ell \times \mathbb{R}^s$ such that

$$-g^j(\overline{x},\overline{y}+tv) < \lambda^j + tu^j \quad \text{for } j \in M_i(\overline{x},\overline{y})$$

and

$$-g^j(\overline{x},\overline{y}+tv) > \lambda^j + tu^j \quad \text{for } j \in I^0(\overline{x},\overline{y}) \setminus M_i(\overline{x},\overline{y})$$

for $t > 0$ sufficiently small. For this reason it is not difficult to compute the matrices from $\partial_B \mathcal{G}_{\overline{x}}(y_k, \mu_k, \lambda_k)$ at the current iterate, needed in (7.43).

Further, it is important to note that from (7.44) one can easily extract the respective matrix $D_{(I^+\cup M_i)}(\overline{x},\overline{y},\overline{\mu},\overline{\lambda})$ which, after transposition, generates the ith adjoint equation (7.5) or (7.9). In this way the Newton's method, applied to the solution of single equilibrium problems for fixed values of the parameter, cooperates with the bundle method minimizing the corresponding function Θ.

Proposition 7.18 implies that under assumption (A3) the Newton's method (7.43) effectively solves the GE (5.24) for all $x \in U_{\text{ad}}$, provided we dispose of sufficiently good starting points. If this is not so, various globally convergent modifications of the basic method (7.43) are available (Qi, 1993; Facchinei et al., 1995).

If the GE (5.24) amounts to necessary and sufficient optimality conditions of a parameter-dependent optimization problem, then it is reasonable to use to an efficient sequential quadratic programming method (Schittkowski, 1986) to its numerical solution. In this way one simultaneously computes its solution and the (unique) KKT vector, needed for the further treatment. The method is globally convergent and the convergence is also superlinear.

For $\overline{x} \in U_{\text{ad}}$ consider now the GE (5.41) written as the NSE (6.13), i.e.

$$\mathcal{H}_{\overline{x}}(y) := \min_c\{F(\overline{x},y), y - \Phi(\overline{x},y)\} = 0. \tag{7.46}$$

The function $\mathcal{H}_{\overline{x}}[\mathbb{R}^m \to \mathbb{R}^m]$ is of course also locally Lipschitz and semismooth at each y. At a current iteration y_k, the Newton's method (3.69) attains the form

$$y_{k+1} = y_k - (\delta(y_k))^{-1}\mathcal{H}_{\overline{x}}(y_k), \tag{7.47}$$

where $\delta(y_k) \in \partial_B \mathcal{H}_{\overline{x}}(y_k)$. To ensure the local superlinear convergence to a solution $\overline{y}$, we need that all matrices from $\partial_B \mathcal{H}_{\overline{x}}(\overline{y})$ are nonsingular. By Definition 2.12 each matrix from $\partial_B \mathcal{H}_{\overline{x}}(\overline{y})$ has the form

$$\begin{bmatrix} \mathcal{J}_y F_{L\cup(I^0\setminus M_i)}(\overline{x},\overline{y}) \\ E_{I^+\cup M_i} - \mathcal{J}_y F_{I^+\cup M_i}(\overline{x},\overline{y}) \end{bmatrix} \tag{7.48}$$

for some $i \in \mathbb{K}(\overline{x},\overline{y})$, and thus, due to Lemma 6.16, it is nonsingular whenever the GE (5.41) is strongly regular at $(\overline{x},\overline{y})$. Unfortunately, not each matrix (7.48) for $i \in \mathbb{K}(\overline{x},\overline{y})$ belongs to $\partial_B \mathcal{H}_{\overline{x}}(\overline{y})$. Therefore, we may theoretically have difficulties with computation of matrices $\delta(y_k)$ during the iteration process. The situation is analogous to the computation of subgradients of the composite function Θ, discussed in the first part of this section: one could give criteria ensuring that a matrix (7.48) with $\overline{y}$ replaced by y_k does belong to $\partial_B \mathcal{H}_{\overline{x}}(y_k)$, but their verification may be rather time-consuming. Our experience with the solved examples is that we never run into troubles with the choice $M_i = \emptyset$, recommended already for the computation of a subgradient of Θ.

Similarly as in the case of the GE (5.24) we note that the matrix (7.48), after transposition, directly generates the ith adjoint equation (7.16). Besides its effectiveness, this is one more argument to use the Newton's method (7.47) provided that assumption (A3) is fulfilled. Finally let us remark that, if we do not dispose of sufficiently good starting points, globally convergent modifications of (7.47) can be applied (DeLuca et al., 1996).

We close this section with three academic examples which can be successfully solved by the suggested numerical method. In all three problems we have applied the bundle-trust algorithm from Chapter 3.

Example 7.5 Consider the bilevel program with the upper-level objective

$$f(x,y) = \frac{1}{2}(y^1 - 3)^2 + \frac{1}{2}(y^2 - 4)^2$$

and the lower-level optimization problem (in variable y)

$$\begin{array}{ll} \text{minimize} & \frac{1}{2}\left[(1+0.2x)(y^1)^2 + (1+0.1x)(y^2)^2\right] - \left(3 + 1\frac{1}{3}x\right)y^1 - xy^2 \\ \text{subject to} & \end{array}$$

$$\begin{aligned} -\tfrac{1}{3}y^1 + y^2 - 1 + 0.1x &\leq 0 \\ (y^1)^2 + (y^2)^2 - 9 - 0.1x &\leq 0 \\ -y^1 &\leq 0 \\ -y^2 &\leq 0, \end{aligned} \tag{7.49}$$

which is dependent on the parameter $x \in \mathbb{R}$. Assume that $U_{\text{ad}} = [0, 10]$. For $x \in U_{\text{ad}}$ problem (7.49) may be converted to the form of the GE (5.24)

$$0 \in \begin{bmatrix} (1+0.2x)y^1 - \left(3 + 1\frac{1}{3}x\right) - \frac{1}{3}\lambda^1 + 2y^1\lambda^2 - \lambda^3 \\ (1+0.1x)y^2 - x + \lambda^1 + 2y^2\lambda^2 - \lambda^4 \\ \frac{1}{3}y^1 - y^2 + 1 - 0.1x \\ -(y^1)^2 - (y^2)^2 + 9 + 0.1x \\ y^1 \\ y^2 \end{bmatrix} + N_{\mathbb{R}^2 + \mathbb{R}^4_+}(y, \lambda). \qquad (7.50)$$

Clearly, there exists an open interval $\tilde{A} \supset U_{\text{ad}}$ such that for all $x \in \tilde{A}$ problem (7.49) possesses a unique solution y. A simple analysis of the constraints in (7.49) shows that (ELICQ) holds at each pair (x, y), where $x \in \tilde{A}$ and y is the corresponding (unique) solution of (7.49). Thus, since for all $x \in \tilde{A}$ the matrix

$$\begin{bmatrix} 1+0.2x & 0 \\ 0 & 1+0.1x \end{bmatrix}$$

is positive definite, Theorem 5.8 yields the strong regularity of the GE (7.50) for all (x, y, λ), where $x \in A$ and (y, λ) is the corresponding (unique) solution of (7.50). We observe that assumptions (A1)–(A3) from the beginning of this section are satisfied. The bundle-trust algorithm minimized the corresponding function Θ on U_{ad} from various starting points with less then 10 evaluations of Θ for the accuracy $\varepsilon = 0.0001$. The result is given in Table 7.1.

Table 7.1.

"Optimal" objective value	x	y^1	y^2	λ^1	λ^2	λ^3	λ^4
3.2077	4.0604	2.6822	1.4871	0	0.6621	0	0

At the computed approximate solution the first inequality constraint in (7.49) is weakly active. By Theorem 6.4 we get

$$\Theta'(4.0604; 1) = 0.1791$$
$$\Theta'(4.0604; 1) = 0.2855.$$

Hence, the minimized function Θ is nondifferentiable at this point. △

The next example, formulated in DeSilva, 1978, is also a bilevel program.

Example 7.6

$$\begin{aligned} &\text{minimize} \quad (x^1)^2 - 2x^1 + (x^2)^2 - 2x^2 + (y^1)^2 + (y^2)^2 \\ &\text{subject to} \\ &\qquad y \in \operatorname{argmin}\left\{(y^1 - x^1)^2 + (y^2 - x^2)^2 \mid (y^i - 1)^2 \le 0.25,\ i = 1, 2\right\} \\ &\qquad x \in U_{\text{ad}} := [0, 2] \times [0, 2]. \end{aligned}$$

By the standard tools we can verify that assumptions (A1)–(A3) hold true again. To satisfy the stopping criterion with $\varepsilon = 0.0001$, starting from zero, 7 function evaluations of Θ were needed. Table 7.2 shows the results.

Table 7.2.

"Optimal" objective value	x^1	x^2	y^1	y^2	λ^1	λ^2
−1.0	0.5	0.5	0.5	0.5	0	0

Again, at $(0.5, 0.5)$ the corresponding function Θ is nondifferentiable. In this example the matrix

$$\mathcal{J}_x F(x, y) = \begin{bmatrix} -2 & 0 \\ 0 & -2 \end{bmatrix}$$

is regular at any pair (x, y) and so the assumptions of Corollary 7.12 are fulfilled. Consequently, we cannot have any difficulties with the computation of subgradients of the corresponding function Θ. △

The last example comes from Ishizuka and Aiyoshi, 1992, where it was solved by a completely different approach. The use of a quadratic exterior penalty leads to a bilevel program, which belongs to the investigated class of MPECs.

Example 7.7

$$\begin{aligned}
&\text{minimize} \quad 2x^1 + 2x^2 - 3y^1 - 3y^2 - 60 + r\left[\max\{0, x^1 + x^2 + y^1 - 2y^2 - 40\}\right]^2 \\
&\text{subject to} \\
&\qquad y \in \operatorname{argmin}\left\{(y^1 - x^1)^2 + (y^2 - x^2)^2 + 40(y^1 + y^2) \mid -10 \le y^i,\right. \\
&\qquad\qquad\qquad\qquad\qquad\qquad \left. 2y^i - x^i + 10 \le 0,\ i = 1, 2\right\} \\
&\qquad x \in U_{\text{ad}} := [0, 50] \times [0, 50],
\end{aligned}$$

where $r = 100$ is the penalty parameter. This problem also satisfies assumptions (A1)–(A3). Starting from $(50, 50)$ and for $\varepsilon = 0.001$ we needed 8 evaluations of Θ to get the result, presented in Table 7.3.

Table 7.3.

"Optimal" objective value	x^1	x^2	y^1	y^2	λ^1	λ^2	λ^3	λ^4
0.0	0.0	0.0	-10.0	-10.0	20.0	20.0	0	0

△

Bibliographical notes

Optimality conditions of the type developed in Section 7.1 appeared, to our knowledge, first in Outrata, 1993 in the context of bilevel programming. In Outrata, 1994 they were extended to MPECs with variational inequalities and in Kočvara and Outrata, 1997 to MPECs with implicit complementarity problems. On the basis of the results from Outrata, 1995 one could extend them also to equilibria described by quasi-variational inequalities, but the necessary requirements are then in most cases too severe.

The numerical method investigated in Section 7.2 was proposed in Outrata, 1990 in the context of simple bilevel programs, where the lower-level constraints do not depend on the upper-level variable. In the subsequent papers Outrata, 1993; Outrata, 1994; Kočvara and Outrata, 1994b and Outrata and Zowe, 1995b this technique was developed to the level presented here, and applied to various types of equilibria (lower-level convex programs, variational inequalities and implicit complementarity problems).

In the papers Dempe, 1995; Dempe, 1998 this method is further extended to bilevel programs with nonunique KKT vectors in the lower-level program. Particularly, (ELICQ) is replaced by some other conditions guaranteeing that S_1 is locally Lipschitz and directionally differentiable (cf. also Ralph and Dempe, 1995; Pang and Ralph, 1996). Then, however, the computation of subgradients of Θ becomes more complicated. In Dempe and Schmidt, 1996; Outrata, 1997 and Dempe, 1998 this method is adapted even to the solution of bilevel programs with nonunique lower-level solutions. Whereas in Outrata, 1997 only a special case motivated by network design problems is considered, the main idea of Dempe and Schmidt, 1996 consists in a suitable regularization of lower-level programs so that the "regularized" problems fulfil the standard assumptions. By using the results from Kalashnikov and Kalashnikova, 1996 this approach could be used also in MPECs with monotone (but not strongly monotone) variational inequalities. Both these generalizations deserve a thorough numerical testing.

II Applications

8 INTRODUCTION

8.1 OPTIMUM SHAPE DESIGN

The idea to use the implicit programming approach in optimum shape design problems is in fact quite old. It was extensively used for solving problems with equilibrium constraints given by *variational equations*. In this case it is straightforward to plug the unique solution of the equation into the objective function and compute derivatives of such a composite function by means of the implicit-function theorem. If the equilibrium problem is more complicated, e.g., it is a *variational inequality*, this technique becomes less straightforward. In classic shape optimization, one circumvents the difficulty (switch from equality to inequality) by means of the *regularization* technique. Here the original variational inequality, e.g.,

$$\begin{aligned}&\text{Find } u \in K := \{w \in H_0^1(\Omega) \mid w \geq 0 \text{ a.e. in } \Omega\} \text{ such that}\\ &\qquad (F(u), v-u)_{L_2(\Omega)} \geq 0 \quad \text{for all } v \in K,\end{aligned}$$

is replaced by an equation with a smooth penalty term standing for the inequality constraint $u \geq 0$:

$$\begin{aligned}&\text{Find } u \in H_0^1(\Omega) \text{ such that}\\ &\qquad (F(u), v)_{L_2(\Omega)} - \tfrac{1}{\varepsilon}((u^-)^2, v)_{L_2(\Omega)} = 0 \quad \text{for all } v \in H_0^1(\Omega),\end{aligned}$$

with $u^- := \min\{0, u\}$ and $\varepsilon > 0$ sufficiently small. So the equilibrium problem is an equation again and the standard implicit approach can be used. We refer the reader to Haslinger and Neittaanmäki, 1996 for numerous examples based on this technique. In this book we approach the switch from equality to inequality in a more direct way and solve the variational inequality (or generalized equation) exactly, thus getting potentially better solution of the MPEC (e.g., if the objective functional is "flat" with respect to the state variable). This technique requires to work in a nonsmooth environment.

In a sense, the following chapters complement the book Haslinger and Neittaanmäki, 1996. While we focus on the numerical solution of discrete optimum design problems (formulated as MPECs), we neglect certain parts of the solution process analyzed in Haslinger and Neittaanmäki, 1996; in particular, the finite element analysis. Some of the examples in Chapters 9 and 10 are taken from Haslinger and Neittaanmäki, 1996; the reader has an opportunity to compare "regularized" solutions with the "nonsmooth" ones.

To simplify the parametrization of domain boundaries, we only consider rectangle-like domains with variable right "vertical" part. A step to more general domains is usually straightforward but technical.

All our optimum design problems are of a "classic" type: the control variable describes the design of either the elastic body or certain material property. For a given control, we solve a state problem which characterizes equilibrium of a mechanical structure. We also used the implicit programming technique to solve other problems in structural optimization where the MPEC structure of these problems comes either from separating design variables (Kočvara and Zowe, 1996) or from introducing an artificial design variable in order to treat complicated state constraints (Kočvara, 1997).

Let us mention the numerical tools needed in the solution of optimum design problems. The state problems are typically large-scale mathematical programs with sparse data structure. For the solution of linear systems we used a preconditioned conjugate gradient method (Axelsson, 1996), LCPs with symmetric positive definite matrix were solved by the two-step iterative method introduced in Kočvara and Zowe, 1994. General complementarity problems and variational inequalities can be efficiently solved by the nonsmooth Newton's method described in Chapter 3; see also Facchinei and Kanzow, 1997; Kanzow and Qi, 1997. For the solution of large-scale nonlinear programs one can use one of the up-to-date codes based on primal-dual interior-point methods and sequential quadratic programming; see, e.g., Byrd et al., 1997; Conn et al., 1992; Gill et al., 1997.

Another technically difficult task is the computation of the derivative of the stiffness matrix with respect to the design variable. Here one has to use the concept of material derivative, developed in Sokołowski and Zolésio, 1992; cf. also Haslinger and Neittaanmäki, 1996.

In all chapters dealing with optimum design (Chapters 9, 10 and 11), we first introduce the state problem in an infinite-dimensional setting and its discretization by the finite element method. Then we define the design problem which, after discretization, is an MPEC of the type analyzed in the theoretical part of the book. Finally, we show how to solve this MPEC by the method introduced in Chapter 7 and present results of numerical examples. As discussed in Section 7.2, the user of the solution method has to choose a suitable subset M_i of the set I^0 of weakly active constraints. In most examples, our choice was $M_i = \emptyset$.

We use notation common in partial differential equations and optimum shape design for the benefit of readers with background in these areas. All the problems are originally set in an infinite dimensional space (usually some Sobolev space). For variables in this space, we use italic letters $u, v, z, \ldots$ The letter u is usually reserved for the solution of an equilibrium problem. The problems will be discretized by the finite element method. For the discretized functions we write as $u_h, v_h, \ldots$, where h is a discretization parameter. These functions are then associated with vectors in $\mathbb{R}^m$ for which we use boldface italic letters $\boldsymbol{u}, \boldsymbol{v}, \ldots$ The discrete counterparts of linear differential operators—matrices—are denoted by boldface capitals $\boldsymbol{A}, \boldsymbol{M}, \ldots$ The subsets of the infinite dimensional space are denoted by italic capitals $K, U_{ad}, \ldots$ and their finite-dimensional counterparts by boldface capitals $\boldsymbol{K}, \mathbf{U_{ad}}, \ldots$

8.2 ECONOMIC MODELLING

Many problems in economic modelling are described by convex programs, variational inequalities or complementarity problems, and their data often depend on a number of parameters. Hence, already the study of stability and sensitivity of their solutions with respect to these parameters plays an important role. In some cases these parameters can be controlled: this generates various MPECs. Among them we can find relatively simple problems of type (7.1), but also extremely difficult ones, where the images of the solution maps are even disconnected. Whereas it is not yet clear, how these "hard" MPECs could be attacked, the solution method of Section 7.2 represents a reliable tool for the numerical solution of economic MPECs of type (7.1). The results can provide useful informations for decision-makers, and could also help to recognize certain qualitative properties of the optimal solutions, in some cases.

9 MEMBRANE WITH OBSTACLE

We start the application part of the book with a classic example of an elliptic variational inequality—an elastic membrane with an obstacle. First, we describe the state problem and then, in the subsequent sections, three shape optimization problems of graded complexity.

9.1 STATE PROBLEM

9.1.1 Problem formulation

First, we have to define the domain in which we want to solve the problem. As mentioned in the Introduction, we will consider a model problem: a rectangle with a parametrized right "vertical" part of the boundary, as shown in Figure 9.1. In order for the solutions to exist, we need to know that the domain has a Lipschitz boundary (cf. Definition B.1 and Theorem B.2). Hence the function which parametrizes the variable part of the boundary should be Lipschitz continuous. But we will be considering sequences of such functions (and boundaries), and the limits of these sequences should still be Lipschitz continuous. Thus we define the admissible[1] set U_{ad} of parametric functions (our design variables in the next sections) as a set of *uniformly bounded* and *uniformly Lipschitz continuous* functions on $[0, 1]$:

$$U_{ad} = \left\{ \alpha \in C^{0,1}([0,1]) \mid 0 < c_1 \le \alpha(\xi_2) \le c_2, \left| \frac{\mathrm{d}}{\mathrm{d}\xi_2} \alpha(\xi_2) \right| < c_3 \text{ a.e. in } (0,1) \right\}, \tag{9.1}$$

[1] We use the word "admissible", instead of "feasible", as it is common in the field of optimum shape design.

where c_1, c_2, c_3 are given positive constants such that $U_{ad} \neq \emptyset$. We further define a family $\mathcal{O}$ of domains $\Omega(\alpha)$ with variable right "vertical" part of the boundary:

$$\begin{aligned} \mathcal{O} &= \{\Omega(\alpha) \mid \alpha \in U_{ad}\}, \\ \Omega(\alpha) &= \left\{(\xi_1, \xi_2) \in \mathbb{R}^2 \mid 0 < \xi_1 < \alpha(\xi_2) \text{ for all } \xi_2 \in (0,1)\right\}; \end{aligned} \tag{9.2}$$

cf. Figure 9.1. Denote by $\widehat{\Omega} = (0, c_2) \times (0,1)$ the largest domain from this family. In

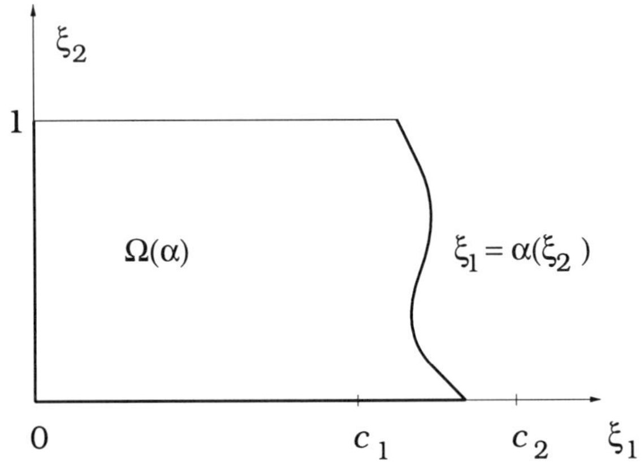

Figure 9.1. Domain $\Omega(\alpha)$

this section we only need one fixed domain from $\mathcal{O}$, say $\Omega_\alpha := \Omega(\alpha)$ associated with an arbitrary fixed $\alpha \in U_{ad}$.

Now let $f \in L_2(\Omega_\alpha)$ be a force that is perpendicularly applied to the membrane Ω_α. The deflection of the membrane is characterized by the solution of the following second-order boundary-value problem:

Find $u \in H^1(\Omega_\alpha)$ such that

$$\begin{aligned} -\Delta u &= f && \text{in } \Omega_\alpha \\ u &= 0 && \text{on } \partial\Omega_\alpha, \end{aligned} \tag{9.3}$$

where the derivatives should be understood in the distributional sense (cf. Appendix B).

We now want the membrane to lie over an obstacle given by a function $\chi : \text{cl}\widehat{\Omega} \to \mathbb{R}$, i.e. $u \geq \chi$, as illustrated in Figure 9.2. The following assumptions on the function χ are made:

$$\chi \in H^2(\text{cl}\widehat{\Omega}), \quad \chi \leq 0 \text{ on } \partial\widehat{\Omega} \cup ((c_1, c_2) \times (0,1)). \tag{9.4}$$

Hence the obstacle function is nonpositive on the boundary of all domains from $\mathcal{O}$.

Problem with rigid obstacle. First, we assume that the obstacle is rigid, i.e., its shape does not give in to the contact pressure of the membrane. The boundary-value problem for the membrane with a rigid obstacle reads:

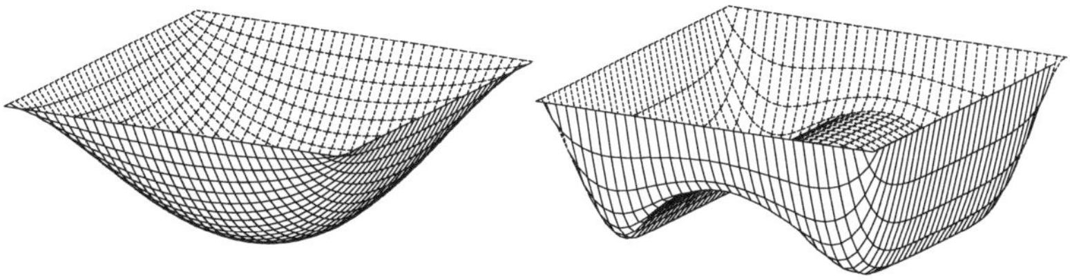

Figure 9.2. Deflection of an elastic membrane without obstacle (left) and with a rigid obstacle (right)

Find $u \in H^1(\Omega_\alpha)$ such that

$$\left.\begin{aligned} -\Delta u \geq f, \quad u \geq \chi \\ (\Delta u + f)(u - \chi) = 0 \end{aligned}\right\} \quad \text{in } \Omega_\alpha \qquad\qquad (9.5)$$
$$u = 0 \qquad \text{on } \partial\Omega_\alpha.$$

Problem (9.5) resembles an infinite-dimensional counterpart to the linear complementarity problem of form (4.6); only the boundary condition does not fit into the standard structure. However, with

$$K := \{v \in H_0^1(\Omega_\alpha) \mid v \geq \chi \text{ a.e. in } \Omega_\alpha\} \qquad (9.6)$$

it can be equivalently formulated as a variational inequality of type (4.2) (but in infinite dimension):

$$\left.\begin{aligned} &\text{Find } u \in K \text{ such that} \\ &\quad \langle F(u), v - u \rangle \geq 0 \quad \text{for all } v \in K, \end{aligned}\right\} \qquad (9.7)$$

where

$$F(u) := -\Delta u - \mathcal{E} f$$

is a (formal) differential operator on $H_0^1(\Omega_\alpha)$, $\mathcal{E}$ stands for the canonical embedding of $L_2(\Omega_\alpha)$ into $H^{-1}(\Omega_\alpha)$ and $\langle \cdot, \cdot \rangle$ denotes the duality between $H_0^1(\Omega_\alpha)$ and $H^{-1}(\Omega_\alpha)$ (for details, see Appendix B).

Variational inequality (9.7) is the state problem which will be used in Sections 9.2 and 9.3.

Remark. As explained in Appendix B, the VI (9.7) can be written in a more familiar form

$$\left.\begin{aligned} &\text{Find } u \in K \text{ such that} \\ &\quad (\nabla u, \nabla(v - u))_{0,\Omega_\alpha} \geq (f, (v - u))_{0,\Omega_\alpha} \quad \text{for all } v \in K, \end{aligned}\right\} \qquad (9.8)$$

where $(\cdot, \cdot)_{0,\Omega_\alpha}$ stands for the inner product in $L_2(\Omega_\alpha)$.

Remark. In the literature, (9.5) is known as the *obstacle problem* and is usually interpreted in the same way as above (membrane with a rigid obstacle). However, one meets the same formulation in mathematical models of other problems, such as:

- filtration of liquids in porous media (cf. Baiocchi and Capelo, 1984);

- lubrication problem (cf. Cryer, 1971).

A very similar formulation can be found in

- elastic-plastic torsion problem (cf. Duvaut and Lions, 1972).

Remark. Instead of (9.5), we can also consider a problem with a more general operator

$$A(u) := -\sum_{i,j=1}^{2} a_{ij}(\xi)\frac{\partial^2 u}{\partial \xi_i \partial \xi_j}. \tag{9.9}$$

The general problem reads

$$\left.\begin{aligned} A(u) \geq f, \quad u \geq \chi \\ (-A(u)+f)(u-\chi) = 0 \end{aligned}\right\} \quad \text{in } \Omega_\alpha \\ u = 0 \qquad \text{on } \partial\Omega_\alpha. \tag{9.10}$$

or, in the VI formulation,

$$\left.\begin{aligned} &\text{Find } u \in K \text{ such that} \\ &\langle A(u) - f, v - u\rangle \geq 0 \quad \text{for all } v \in K, \end{aligned}\right\} \tag{9.11}$$

with K from (9.6). The coefficients a_{ij} represent various physical properties. Obviously, (9.5) is a special case of (9.10) with

$$a_{ii} := 1 \quad \text{and} \quad a_{ij} := 0 \quad \text{for } i \neq j.$$

Let us assume that A is strongly elliptic, i.e.,

$$\sum_{i,j=1}^{2} a_{ij}(\xi)\eta_i\eta_j \geq c\sum_{i=1}^{2}|\eta_i|^2, \qquad \text{a.e. in } \Omega_\alpha, \text{ for all } (\eta_1,\eta_2) \neq 0 \tag{9.12}$$

with some $c > 0$; cf. Appendix B. Existence and uniqueness of the solution to problem (9.11) results from Theorem B.4, cf. also Example B.3.

Problem with compliant obstacle. We now assume that the obstacle is not rigid but *compliant* to the pressure of the membrane, as drafted in Figure 9.3. Let $k \in \mathbb{R}_+$, and set formally for $u \in H_0^1(\Omega_\alpha)$

$$G(u) = k(\Delta u + f) + \chi. \tag{9.13}$$

Here χ (defined in (9.4)) describes the original shape of the obstacle and $1/k$ characterizes the compliance of the obstacle material (so-called coefficient of compliance, e.g., $1/k \approx 0.4$ for sand). The term $\Delta u + f$ represents the force acting on the obstacle.

The boundary-value problem for the membrane with a compliant obstacle becomes: Find $u \in H^1(\Omega_\alpha)$ such that

$$\left.\begin{aligned} -\Delta u \geq f, \quad u \geq G(u) \\ (\Delta u + f)(u - G(u)) = 0 \end{aligned}\right\} \quad \text{in } \Omega_\alpha \\ u = 0 \qquad \text{on } \partial\Omega_\alpha. \tag{9.14}$$

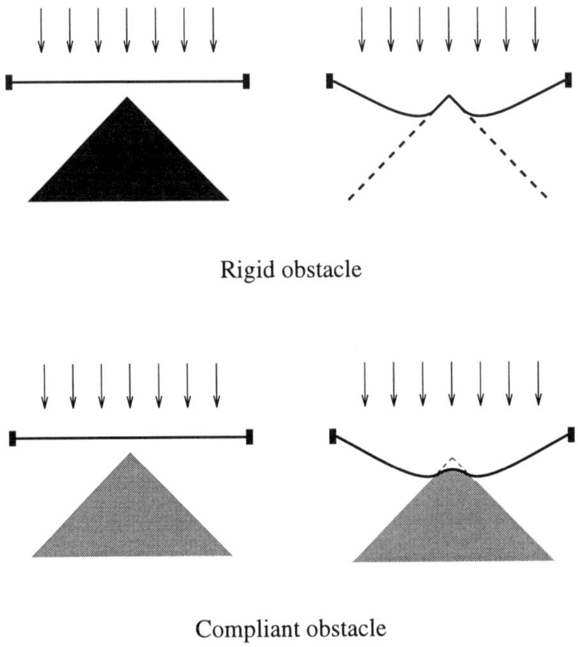

Figure 9.3. Membrane with rigid and compliant obstacle

This problem resembles an infinite-dimensional version of the implicit complementarity problem (4.9) (again, the boundary condition does not fit into the standard structure). After the discretization, the problem will be reduced and we will get a standard ICP. We note that, when $k = 0$ in (9.13), problem (9.14) becomes a standard contact problem with a rigid obstacle given by the function χ.

Let us show that there exists a unique solution to (9.14). Putting $\lambda := \Delta u + f$, we can formally rewrite (9.14) as

$$\left.\begin{aligned} -\Delta u + \lambda &= f \\ u &\geq k\lambda + \chi \\ \lambda &\leq 0 \\ \lambda(u - k\lambda - \chi) &= 0 \end{aligned}\right\} \quad \text{in } \Omega_\alpha$$
$$u = 0 \qquad \text{on } \partial\Omega_\alpha,$$

which can further be written as a nonlinear (and nonsmooth) equation

$$\begin{aligned} -\Delta u + \frac{1}{k}\min\{0, u - \chi\} &= f \quad \text{in } \Omega_\alpha \\ u &= 0 \quad \text{on } \partial\Omega_\alpha. \end{aligned} \tag{9.15}$$

The existence and uniqueness of a solution to (9.15) is guaranteed by a corollary of Theorem 26.14 from Fučík and Kufner, 1980 which, in our notation, can be stated as follows.

Theorem 9.1 *Let h be a continuous nondecreasing function on $\mathbb{R}$ and $f \in H^{-1}(\Omega_\alpha)$. Assume that there exists a real number $c > 0$ such that*

$$|h(\xi)| \le c(1 + |\xi|^\tau) \qquad \text{for all } \xi \in \mathbb{R}$$

with an arbitrary real positive τ. Then there exists a unique solution (in the weak sense) to the equation

$$\begin{aligned} -\Delta u + h(u) &= f \quad \text{in } \Omega_\alpha \\ u &= 0 \quad \text{on } \partial\Omega_\alpha. \end{aligned}$$

Setting $h(\xi) := \frac{1}{k}\min\{0, \xi - d\}$ with $k > 0$ and d an arbitrary constant, we can immediately apply Theorem 9.1 to problem (9.15) and thus (9.14). Recall that the case $k = 0$ leads to a linear complementarity problem which can be treated in the standard way.

Remark. When setting $\chi = 0$, problem (9.15) will be a generalization of the standard problem of a membrane with a compliant support

$$\begin{aligned} -\Delta u + \frac{1}{k}u &= f \quad \text{in } \Omega_\alpha \\ u &= 0 \quad \text{on } \partial\Omega_\alpha. \end{aligned}$$

This equation describes a *bilateral* support, while in many practical applications we only want to work with a *unilateral* support, described by equation (9.15); see Fig. 9.4.

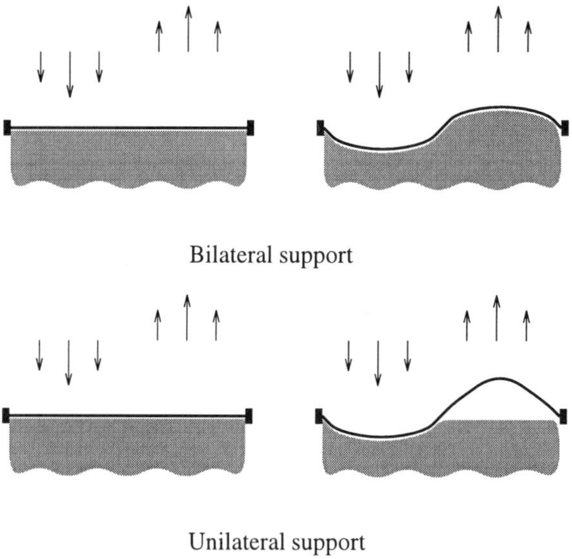

Figure 9.4. Membrane with bilateral and unilateral compliant support

Remark. Analogously to the case of a rigid obstacle, the (generalized) ICP (9.14) can be formulated as a quasi-variational inequality of form (4.7):

$$\begin{aligned} &\text{Find } u \in K(u) := \{v \in H_0^1(\Omega_\alpha) \mid v \ge G(u) \text{ a.e. in } \Omega_\alpha\} \text{ such that} \\ &\qquad \langle F(u), v - u\rangle \ge 0 \quad \text{for all } v \in K(u), \end{aligned}$$

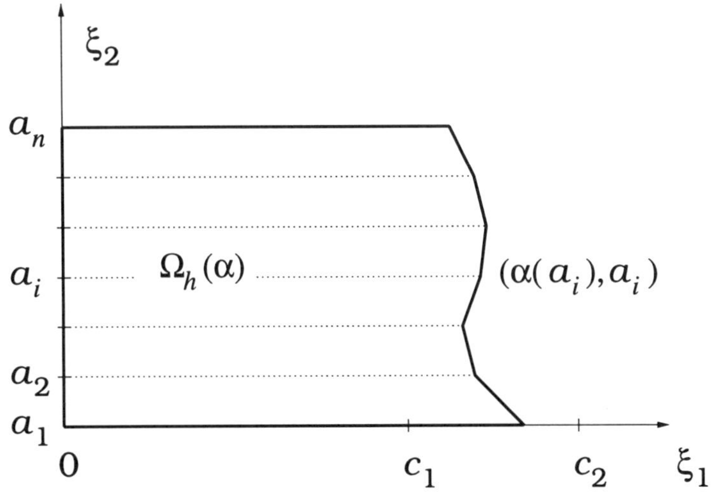

Figure 9.5. Computational domain $\Omega_h(\boldsymbol{\alpha})$

with $F(u) := -\Delta u - \mathcal{E} f$; cf. (9.7).

9.1.2 Discretization

The discretization technique is simple and standard. We use triangular finite elements with linear basis functions.

Recall that our domain Ω_α is defined as

$$\Omega_\alpha = \{(\xi_1, \xi_2) \in \mathbb{R}^2 \mid 0 < \xi_1 < \alpha(\xi_2) \text{ for all } \xi_2 \in (0,1)\}$$

with an arbitrary fixed

$$\alpha \in U_{ad} := \left\{\beta \in C^{0,1}([0,1]) \mid 0 < c_1 \leq \beta(\xi_2) \leq c_2, \left|\frac{\mathrm{d}}{\mathrm{d}\xi_2}\beta(\xi_2)\right| < c_3 \text{ a.e. in } (0,1)\right\}.$$

Let $a_1 < a_2 < \ldots < a_n$ be a uniform partition of $[0,1]$ with each segment $[a_{i-1}, a_i]$ of length $h = 1/(n-1)$.

The function α (one-dimensional real function) is approximated by a continuous piecewise linear function α_h such that $\alpha_h(a_i) = \alpha(a_i)$. The discrete counterpart of α_h is a vector $\boldsymbol{\alpha} \in \mathbb{R}^n$ of values of this function at points $a_1, \ldots, a_n$, i.e., a vector of ξ_1-coordinates of boundary nodes $(\alpha_h(a_i), a_i)$. In such a way we define a polygonal *computational domain*

$$\Omega_h(\boldsymbol{\alpha}) := \{(\xi_1, \xi_2) \in \mathbb{R}^2 \mid 0 < \xi_1 < \alpha_h(\xi_2),\ 0 < \xi_2 < 1\}$$

as shown in Figure 9.5. The set U_{ad} is replaced by the finite-dimensional set

$$\mathbf{U_{ad}} := \{\boldsymbol{\alpha} \in \mathbb{R}^n \mid \boldsymbol{\alpha}^i = \alpha(a_i), \alpha \in U_{ad},\ \alpha \text{ linear on } [a_{i-1}, a_i], i = 2, \ldots, n\}.$$

The domain $\Omega_h(\boldsymbol{\alpha})$ is discretized by triangular elements. Since the triangulation is different for different problems, we will specify it in the corresponding sections.

For the discretization of the state problems (9.5), (9.7) and (9.14) we use piecewise linear basis functions. Let m be the number of nodes of the triangulation and N the number of elements (triangles). We approximate each function $u \in H^1(\Omega_\alpha)$ by a continuous function u_h that is linear on every triangle. Such a function can be written as

$$u_h(\xi) = \sum_{k=1}^{m} \boldsymbol{u}_k \varphi_k(\xi),$$

where $\boldsymbol{u}_k$ is the value of u_h at the kth node and φ_k is the basis function associated with this node (for details, see Ciarlet, 1978). Put

$$\boldsymbol{b}_k := \left(\frac{\partial \varphi_k}{\partial \xi_1}, \frac{\partial \varphi_k}{\partial \xi_2} \right)^T$$

and, for an element (triangle) Ω_i, let $\mathcal{D}_i$ be the index set of nodes belonging to this element. Then

$$(\boldsymbol{A}_i)^{k\ell} = \int_{\Omega_i} \boldsymbol{b}_k^T \boldsymbol{b}_\ell \mathrm{d}\xi \qquad k, \ell \in \mathcal{D}_i$$

are the entries of the corresponding *element stiffness matrix* (considered as an $m \times m$ matrix, the non-defined entries being zeros). The assembled *stiffness matrix*

$$\boldsymbol{A} = \sum_{i=1}^{N} \boldsymbol{A}_i$$

is a discrete version of the Laplacian operator $-\Delta$. Analogously we obtain the components $\boldsymbol{f}_k$ of the right-hand side vector $\boldsymbol{f}$:

$$\boldsymbol{f}_k = \int_{\Omega_\alpha} f \varphi_k \mathrm{d}\xi, \qquad k = 1, \dots, m.$$

The discretized obstacle $\boldsymbol{\chi}$ is the vector of values of the function χ at the nodes of the triangulation. The matrix $\boldsymbol{A}$ and the vectors $\boldsymbol{f}$ and $\boldsymbol{\chi}$ depend on the parameter $\boldsymbol{\alpha}$ which defines the shape of Ω_α. To emphasize this, we will write $\boldsymbol{A}(\boldsymbol{\alpha}), \boldsymbol{f}(\boldsymbol{\alpha})$ and $\boldsymbol{\chi}(\boldsymbol{\alpha})$.

Summing up, we obtain as a discrete counterpart of the problem with rigid obstacle (9.5) the following LCP:

Find $\boldsymbol{u} \in \mathbb{R}^m$ such that

$$\begin{aligned} &\boldsymbol{A}(\boldsymbol{\alpha})\boldsymbol{u} - \boldsymbol{f}(\boldsymbol{\alpha}) \geq 0, \quad \boldsymbol{u} - \boldsymbol{\chi}(\boldsymbol{\alpha}) \geq 0 \\ &\langle \boldsymbol{A}(\boldsymbol{\alpha})\boldsymbol{u} - \boldsymbol{f}(\boldsymbol{\alpha}), \boldsymbol{u} - \boldsymbol{\chi}(\boldsymbol{\alpha}) \rangle = 0. \end{aligned} \tag{9.16}$$

Note that the basis functions take care of the boundary conditions from (9.5) which are now included in the stiffness matrix $\boldsymbol{A}(\boldsymbol{\alpha})$. The discrete version of the variational inequality (9.7) is

$$\left. \begin{array}{l} \text{Find } \boldsymbol{u} \in \boldsymbol{K}(\boldsymbol{\alpha}) \text{ such that} \\ \quad \langle \boldsymbol{A}(\boldsymbol{\alpha})\boldsymbol{u} - \boldsymbol{f}(\boldsymbol{\alpha}), \boldsymbol{v} - \boldsymbol{u} \rangle \geq 0 \quad \text{for all } \boldsymbol{v} \in \boldsymbol{K}(\boldsymbol{\alpha}), \end{array} \right\} \tag{9.17}$$

with

$$\boldsymbol{K}(\boldsymbol{\alpha}) := \{ \boldsymbol{y} \in \mathbb{R}^m \mid \boldsymbol{y} \geq \boldsymbol{\chi}(\boldsymbol{\alpha}) \}.$$

Note the dependence of the obstacle vector χ and thus the set $\boldsymbol{K}$ on $\boldsymbol{\alpha}$; indeed, the position of the triangulation nodes depends on $\boldsymbol{\alpha}$ and so do χ and $\boldsymbol{K}$. We can avoid this dependence and thus simplify the problem by means of a simple transformation

$$\boldsymbol{v} = \boldsymbol{u} - \chi(\boldsymbol{\alpha}).$$

Then the LCP (9.16) reads
Find $\boldsymbol{v} \in \mathbb{R}^m$ such that

$$\begin{aligned} &\boldsymbol{A}(\boldsymbol{\alpha})\boldsymbol{v} + \boldsymbol{A}(\boldsymbol{\alpha})\chi(\boldsymbol{\alpha}) - \boldsymbol{f}(\boldsymbol{\alpha}) \geq 0, \quad \boldsymbol{v} \geq 0 \\ &\langle \boldsymbol{A}(\boldsymbol{\alpha})\boldsymbol{v} + \boldsymbol{A}(\boldsymbol{\alpha})\chi(\boldsymbol{\alpha}) - \boldsymbol{f}(\boldsymbol{\alpha}), \boldsymbol{v}\rangle = 0 \end{aligned} \tag{9.18}$$

and the VI (9.17) becomes

$$\left.\begin{aligned} &\text{Find } \boldsymbol{v} \in \boldsymbol{K} := \mathbb{R}^m_+ \text{ such that} \\ &\langle \boldsymbol{A}(\boldsymbol{\alpha})\boldsymbol{v} + \boldsymbol{A}(\boldsymbol{\alpha})\chi(\boldsymbol{\alpha}) - \boldsymbol{f}(\boldsymbol{\alpha}), \boldsymbol{y} - \boldsymbol{v}\rangle \geq 0 \quad \text{for all } \boldsymbol{y} \in \boldsymbol{K}. \end{aligned}\right\} \tag{9.19}$$

The discrete counterpart of the membrane problem with compliant obstacle (9.14) is the following ICP:
Find $\boldsymbol{u} \in \mathbb{R}^m$ such that

$$\begin{aligned} &\boldsymbol{A}(\boldsymbol{\alpha})\boldsymbol{u} - \boldsymbol{f}(\boldsymbol{\alpha}) \geq 0, \quad \boldsymbol{u} - \boldsymbol{G}(\boldsymbol{\alpha}, \boldsymbol{u}) \geq 0, \\ &\langle \boldsymbol{A}(\boldsymbol{\alpha})\boldsymbol{u} - \boldsymbol{f}(\boldsymbol{\alpha}), \boldsymbol{u} - \boldsymbol{G}(\boldsymbol{\alpha}, \boldsymbol{u})\rangle = 0 \end{aligned} \tag{9.20}$$

with

$$\boldsymbol{G}(\boldsymbol{\alpha}, \boldsymbol{u}) := k(\boldsymbol{f}(\boldsymbol{\alpha}) - \boldsymbol{A}(\boldsymbol{\alpha})\boldsymbol{u}) + \chi(\boldsymbol{\alpha})$$

being the discrete version of $G(u)$ from (9.13).

9.2 PACKAGING PROBLEM WITH RIGID OBSTACLE

In the packaging problem we try to minimize the domain Ω_α (the membrane surface) under the condition that a given part of the membrane comes into contact with the obstacle. The problem was introduced by Benedict et al., 1984; for further development (existence, finite element approach), see Haslinger and Neittaanmäki, 1996.

9.2.1 Problem formulation

Consider the family of domains $\mathcal{O}$ defined in (9.2). Let Ω_0 be a given closed connected subset of $[0, c_1] \times [0, 1]$. For $\alpha \in U_{ad}$ denote by $Z(\alpha)$ the contact region $\{\xi \in \Omega_\alpha \mid u(\xi) = \chi(\xi)\}$, where u is a solution of the membrane problem with rigid obstacle (9.5). In the packaging problem we try to minimize the area of $\Omega_\alpha, \alpha \in U_{ad}$, under the condition that the contact region $Z(\alpha)$ contains the set Ω_0 (cf. one-dimensional sketch in Figure 9.6). In mathematical terms, we obtain:

$$\begin{aligned} &\text{minimize} \quad \mathcal{J}(\alpha) := \text{meas}\, \Omega_\alpha \\ &\text{subject to} \\ &\qquad\qquad u \text{ solves the VI (9.7) for } \Omega_\alpha \\ &\qquad\qquad Z(\alpha) \supset \Omega_0 \\ &\qquad\qquad \alpha \in U_{ad}. \end{aligned} \tag{9.21}$$

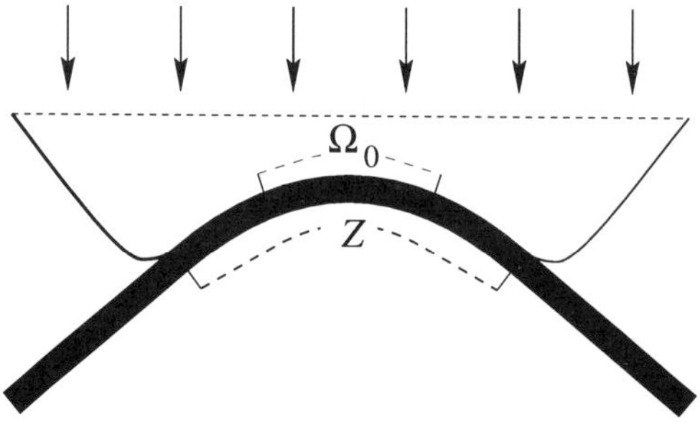

Figure 9.6. Packaging problem

If the set $\{\alpha \in U_{ad} \mid Z(\alpha) \supset \Omega_0\}$ is nonempty, then (9.21) has at least one solution; cf. Haslinger and Neittaanmäki, 1996, Thm. 9.1.

For the treatment of the state constraint $Z(\alpha) \supset \Omega_0$, a quadratic penalty technique was proposed and analyzed in Haslinger and Neittaanmäki, 1996. However, as $u \geq \chi$, the numerical method from Chapter 7 allows us to add this state constraint to the objective as a nondifferentiable *exact* penalty. Note that this is impossible in the regularization technique used in Haslinger and Neittaanmäki, 1996, where the relationship $u \geq \chi$ does not hold. We then obtain an augmented objective functional

$$\mathcal{J}_r(\alpha, u) := \text{meas}\, \Omega_\alpha + r \int_{\Omega_0} (u - \chi)\mathrm{d}\xi, \tag{9.22}$$

where $r > 0$ is a penalty parameter. Hence, instead of (9.21), we will solve the problem

$$\begin{array}{ll} \text{minimize} & \mathcal{J}_r(\alpha, u) \\ \text{subject to} & \\ & u \text{ solves the VI (9.7) for } \Omega_\alpha \\ & \alpha \in U_{ad} \end{array} \tag{9.23}$$

In Haslinger and Neittaanmäki, 1996, Thms. 9.3, 9.4 it is shown that for a quadratic penalty of type $r \int_{\Omega_0} (u - \chi)^2 \mathrm{d}\xi$ the penalized problem has at least one solution for each $r > 0$, and if $r \to \infty$, then u_r, the solutions of (9.23), converge uniformly to u^*, the solution of (9.21). Analogously, the same can be proved for $\mathcal{J}_r$ from (9.22) with the linear penalty.

9.2.2 Discretization

In order to discretize the problem, we have to specify the triangulation of the computational domain $\Omega_h(\boldsymbol{\alpha})$. The triangulation will be regular and kept fixed on a rectangle $[0, c_0] \times [0, 1]$ with $c_0 < c_1$. The nodes $(\alpha_h(a_i), a_i)$ on the moving boundary are called *principal moving nodes*. Each segment $[(c_0, a_i), (\alpha_h(a_i), a_i)], i = 1, \ldots, n$, is partitioned equidistantly to get the *associated moving nodes*; cf. Figure 9.7. To make things simple, we assume that the set Ω_0 is included in the rectangle $[0, c_0] \times [0, 1]$, and so its triangulation does not depend on $\boldsymbol{\alpha}$.

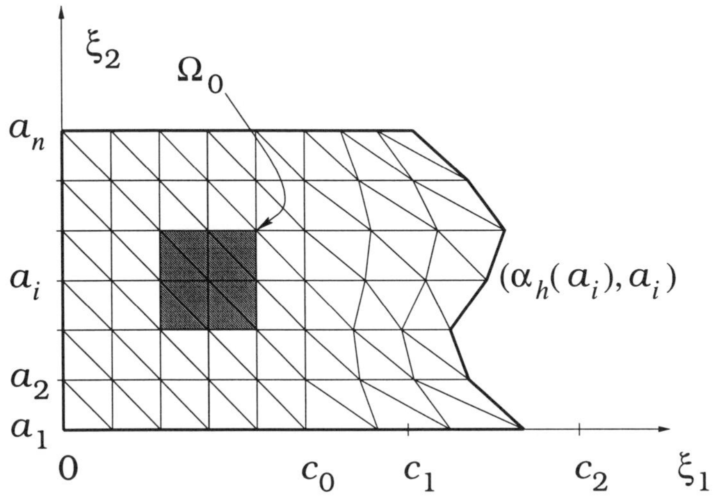

Figure 9.7. Triangulation of $\Omega_h(\boldsymbol{\alpha})$

The discretization of (9.23) is now straightforward. Let $\mathcal{D}_0$ be the set of indices of nodes lying in Ω_0. Then the discretized problem reads as

$$\begin{aligned} &\text{minimize} \quad \boldsymbol{J}_r(\boldsymbol{\alpha}, \boldsymbol{v}) := \text{meas}\, \Omega_h(\boldsymbol{\alpha}) + rh^2 \sum_{i \in \mathcal{D}_0} \boldsymbol{v}^i \\ &\text{subject to} \\ &\qquad \boldsymbol{v} \text{ solves the LCP (9.18)} \\ &\qquad \boldsymbol{\alpha} \in \mathbf{U_{ad}} \end{aligned} \tag{9.24}$$

with $r > 0$. The factor h^2 in the penalty part of the objective function comes from the approximate integration over the cells of size h^2. It is known (Haslinger and Neittaanmäki, 1996, Thm. 9.5) that if $h \to 0+$ then α_{rh}, the solutions of (9.24), converge uniformly to α_r, the solution of (9.23).

9.2.3 Numerical method

Problem (9.24) is an MPEC of type (7.1). We now want to apply the numerical method from Section 7.2. Hence, we have to verify the assumptions of this method and show how to compute a subgradient, in order to apply a bundle code. The equilibrium problem is an LCP, hence it can also be written as a generalized equation of type (5.24):

$$0 \in \begin{bmatrix} \boldsymbol{A}(\boldsymbol{\alpha})\boldsymbol{v} + \boldsymbol{A}(\boldsymbol{\alpha})\boldsymbol{\chi}(\boldsymbol{\alpha}) - \boldsymbol{f}(\boldsymbol{\alpha}) - \boldsymbol{\lambda} \\ \boldsymbol{v} \end{bmatrix} + N_{\mathbb{R}^m \times \mathbb{R}^m_+}(\boldsymbol{v}, \boldsymbol{\lambda}). \tag{9.25}$$

We have to show that with some $\tilde{A} \supset \mathbf{U_{ad}}$

(A1) $\boldsymbol{J}_r$ is continuously differentiable on $\tilde{A} \times \mathbb{R}^m$;

(A2) for all $\boldsymbol{\alpha} \in \tilde{A}$ the GE (9.25) has a unique solution $S(\boldsymbol{\alpha})$;

(A3) the GE (9.25) is strongly regular at all points $(\boldsymbol{\alpha}, \boldsymbol{v})$ with $\boldsymbol{\alpha} \in \tilde{A}, \boldsymbol{v} = S_1(\boldsymbol{\alpha})$;

cf. assumptions (A1)–(A3) from Section 7.2. The validity of assumption (A1) directly comes from the definition of $\boldsymbol{J}_r$ and from the construction of the finite element mesh. Due to the definition of U_{ad}, the boundary $\partial\Omega_\alpha$ is uniformly Lipschitz on U_{ad}, and thus the stiffness matrix $\boldsymbol{A}(\boldsymbol{\alpha})$ is positive definite on $\mathbf{U_{ad}}$. Then, according to Theorem 4.7, LCP (9.18) has a unique solution (and thus the GE (9.25) has a unique $\boldsymbol{v}$−part of a solution) and (A2) is satisfied. The validity of assumption (A3) results from Theorem 5.8: indeed, the gradients of active constraints in LCP are trivially linearly independent and for all $\boldsymbol{\alpha} \in \mathbf{U_{ad}}$ the matrix $\boldsymbol{A}(\boldsymbol{\alpha})$ is positive definite, thus the GE (9.25) is strongly regular at all $(\boldsymbol{\alpha}, \boldsymbol{v}, \boldsymbol{\lambda})$. The above assumptions also guarantee weak semismoothness of the function $\Theta(\boldsymbol{\alpha}) := \boldsymbol{J}_r(\boldsymbol{\alpha}, S_1(\boldsymbol{\alpha}))$, where S_1 assigns $\boldsymbol{\alpha}$ the $\boldsymbol{v}$−component of the solution to (9.25) (Proposition 7.9).

The second task is to compute a subgradient from $\partial\Theta(\boldsymbol{\alpha})$. First, for given control $\boldsymbol{\alpha}$, we have to solve the adjoint problem, which in our situation amounts to solving the quadratic program (cf. Corollary 7.5)

$$\begin{array}{ll} \text{minimize} & \frac{1}{2}\langle \boldsymbol{p}, \boldsymbol{A}(\boldsymbol{\alpha})\boldsymbol{p}\rangle - \langle \nabla_{\boldsymbol{v}} \boldsymbol{J}_r(\boldsymbol{\alpha}, \boldsymbol{v}), \boldsymbol{p}\rangle \\ \text{subject to} & \\ & \boldsymbol{p}_j = 0, \quad j \in I^+(\boldsymbol{\alpha}, \boldsymbol{v}) \cup M_i(\boldsymbol{\alpha}, \boldsymbol{v}) \end{array}$$

with

$$\begin{aligned} I(\boldsymbol{\alpha}, \boldsymbol{v}) &= \{i \in \{1, 2, \ldots, m\} \mid \boldsymbol{v}^i = 0\} \\ I^+(\boldsymbol{\alpha}, \boldsymbol{v}) &= \{i \in I(\boldsymbol{\alpha}, \boldsymbol{v}) \mid \boldsymbol{\lambda}^i > 0\} \\ I^0(\boldsymbol{\alpha}, \boldsymbol{v}) &= I(\boldsymbol{\alpha}, \boldsymbol{v}) \setminus I^+(\boldsymbol{\alpha}, \boldsymbol{v}). \end{aligned}$$

Further, $M_i(\boldsymbol{\alpha}, \boldsymbol{v})$ is a suitably chosen subset of $I^0(\boldsymbol{\alpha}, \boldsymbol{v})$. The subgradient associated with M_i can be computed from

$$\nabla_{\boldsymbol{\alpha}} \boldsymbol{J}_r(\boldsymbol{\alpha}, \boldsymbol{v}) - [\mathcal{J}_{\boldsymbol{\alpha}}(\boldsymbol{A}(\boldsymbol{\alpha})\boldsymbol{v} + \boldsymbol{A}(\boldsymbol{\alpha})\chi(\boldsymbol{\alpha}) - \boldsymbol{f}(\boldsymbol{\alpha}))]^T \boldsymbol{p};$$

cf. Proposition 7.14.

The nonsmooth optimization problem is solved by the BT code introduced in Chapter 3. To solve LCP (9.18) we used a two-step algorithm introduced in Kočvara and Zowe, 1994. This algorithm combines the successive overrelaxation method with projection and the conjugate gradient method (preconditioned by incomplete factorization). The sparsity of $\boldsymbol{A}$ is preserved by this method; this significantly contributes to the efficiency of the overall algorithm.

Note that the use of nonsmooth code becomes compulsory in our approach. Usually, test runs with smooth codes (like SQP) broke down at points far from the true solution.

9.2.4 Examples

Example 9.1 (Haslinger and Neittaanmäki, 1996) Consider the packaging problem with

$$\Omega_0 = [0.25, 0.5] \times [0.25, 0.75], \quad f(\xi_1, \xi_2) = -1.0 \quad \text{and} \quad \chi(\xi_1, \xi_2) = -0.05\xi_1$$

(cf. Figure 9.8). The set U_{ad} is specified by the parameters $c_1 = 0.6$, $c_2 = 1.0$, $c_3 = 3.0$, the triangulation parameter $c_0 = 0.5$.

First, we used (as in Haslinger and Neittaanmäki, 1996) the quadratic penalty term $\frac{r}{2} h^2 \sum_{i \in \mathcal{D}_0} (v^i)^2$ with penalty parameter $r = 10^{-4}$. The results obtained for $h = \frac{1}{16}$

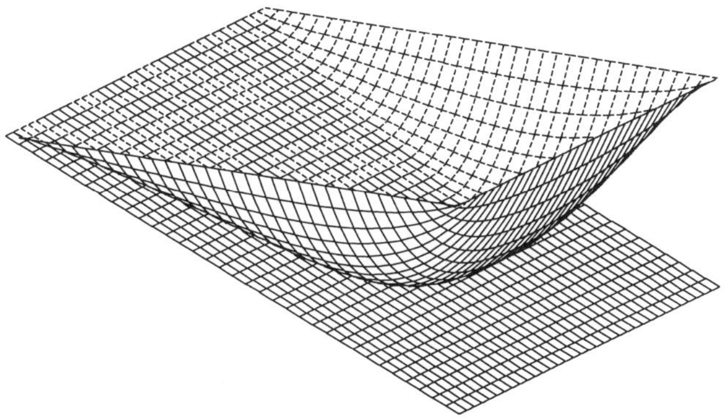

Figure 9.8. Obstacle and deflected membrane for Example 9.1

($n = 17$) are close to those of Haslinger and Neittaanmäki, 1996, at least concerning the optimal value of the objective ($\boldsymbol{J}_r^{opt} = 0.784213$). However, the used quadratic penalty led to considerable inaccuracies in satisfying the state constraint (up to 4 % of the deflection at the front corners of Ω_0). Therefore, we switched to the exact linear penalty $rh^2 \sum_{i \in \mathcal{D}_0} \boldsymbol{v}^i$ and used the four discretizations with $h = \frac{1}{8}, \frac{1}{16}, \frac{1}{32}, \frac{1}{64}$. The penalty parameters r were chosen large enough such that the state constraint was satisfied *exactly*. Their values together with the corresponding optimal objective values $\boldsymbol{J}_r^{opt}$ are given in the following table.

n	r	$\boldsymbol{J}_r^{opt}$
9	$8 \cdot 10^3$	0.787932
17	$8 \cdot 10^4$	0.826013
33	$8 \cdot 10^5$	0.850895
65	$8 \cdot 10^6$	0.866364

The final design for $n = 65$ is shown in Figure 9.9. We see that in the set Ω_0 the contour lines of the solution follow the contour lines of the obstacle (which are parallel to x_2−coordinate).

Comparing the values of $\boldsymbol{J}_r^{opt}$ for the quadratic and the linear penalty, respectively, we see a significant difference (5% for $h = \frac{1}{16}$). Also the resulting optimal design is quite sensitive to the exact satisfaction of the state constraint. This becomes even more evident in the next example. △

Example 9.2 (Haslinger and Neittaanmäki, 1996) Let $\chi(\xi_1, \xi_2) = -0.05(\xi_1^2 + (\xi_2 - 0.25)^2)$ (cf. Figure 9.10) and all other data as in Example 9.1. Again, the linear penalty was used with four discretizations given by $h = \frac{1}{8}, \frac{1}{16}, \frac{1}{32}, \frac{1}{64}$. The values of the penalty parameters r and the corresponding optimal objective values $\boldsymbol{J}_r^{opt}$ are given in the following table.

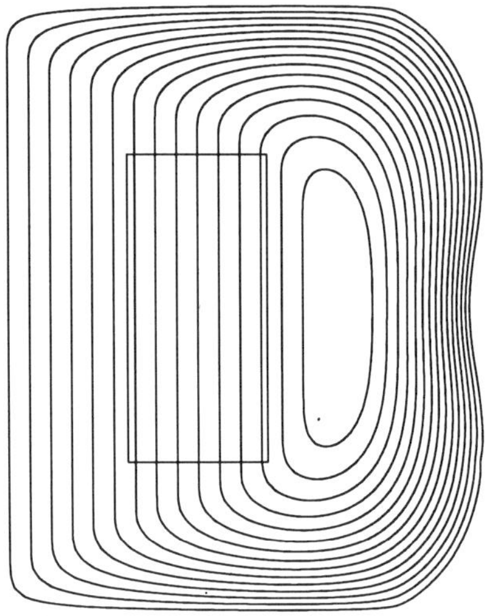

Figure 9.9. Optimal design for Example 9.1, $n = 65$.

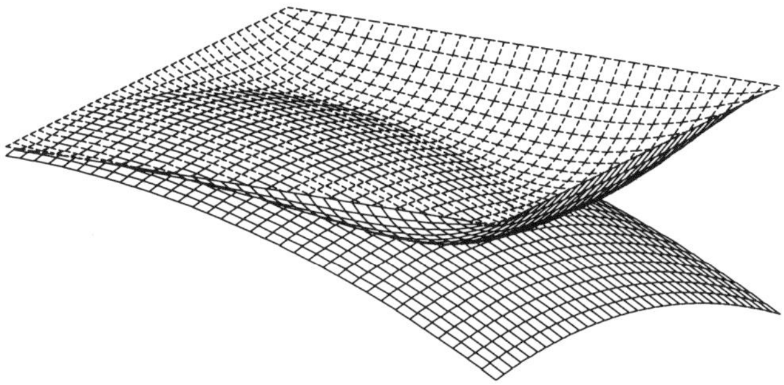

Figure 9.10. Example 9.2, obstacle and deflected membrane

n	r	$\boldsymbol{J}_r^{opt}$
9	$1,6 \cdot 10^4$	0.780361
17	$3 \cdot 10^5$	0.900842
33	$5 \cdot 10^5$	0.934860
65	$1 \cdot 10^6$	0.980475

The design for the finest mesh ($n = 65$) is shown in Figure 9.11. Again, a closer look shows that in Ω_0 the contour lines of the solution coincide with the contour lines of the obstacle, which are depicted in Figure 9.12.

It is quite interesting to compare for this example the optimal design with the one computed via regularization technique and with quadratic penalty. The maximal components of

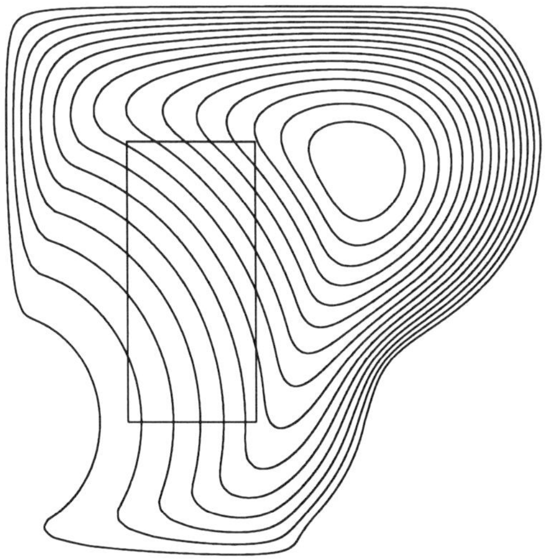

Figure 9.11. Optimal design for Example 9.2, $n = 65$.

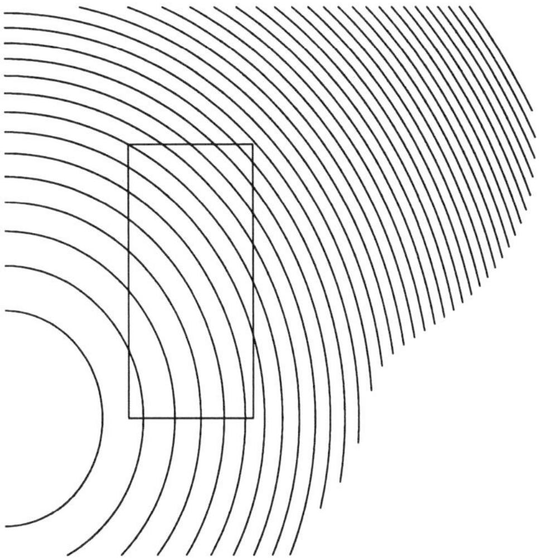

Figure 9.12. Contour lines of obstacle for Example 9.2.

the design vectors differ for the same discretization parameter $h = \frac{1}{32}$ by 11%! Obviously, the price for having to deal with a nonsmooth problem is more than worth it. $\triangle$

9.3 PACKAGING PROBLEM WITH COMPLIANT OBSTACLE

We consider the packaging problem, now with a compliant obstacle, instead of the rigid one. The state problem is the ICP (9.14) with G introduced in (9.13). Again, we try to find a function α such that the membrane comes into contact with the obstacle on a given set $\Omega_0 \subset \widehat{\Omega}$, while the surface of the membrane (i.e., the measure of Ω_α) becomes minimal.

9.3.1 Problem formulation

Let Ω_0 be a closed connected subset of $[0, c_1] \times [0, 1]$. For $\alpha \in U_{ad}$, denote by $Z(\alpha)$ the contact region $\{\xi \in \Omega_\alpha \mid u(\xi) = G(u)(\xi)\}$, where u is the solution of (9.14). The packaging problem is defined as follows:

$$\begin{aligned} &\text{minimize} \quad \mathcal{J}(\alpha) := \text{meas}\, \Omega_\alpha \\ &\text{subject to} \\ &\qquad\qquad u \text{ solves the ICP (9.14) with } \Omega_\alpha \\ &\qquad\qquad Z(\alpha) \supset \Omega_0 \\ &\qquad\qquad \alpha \in U_{ad}. \end{aligned} \tag{9.26}$$

As in the previous section, an exact penalty technique is used for the treatment of the state constraint $Z(\alpha) \supset \Omega_0$. This leads to an augmented objective functional

$$\mathcal{J}_r(\alpha, u) := \text{meas}\, \Omega_\alpha + r \int_{\Omega_0} (u - G(u)) \mathrm{d}\xi$$

with penalty parameter $r > 0$ and $G(u)$ from (9.13). Hence, the problem to solve reads

$$\begin{aligned} &\text{minimize} \quad \mathcal{J}_r(\alpha, u) \\ &\text{subject to} \\ &\qquad\qquad u \text{ solves the ICP (9.14) with } \Omega_\alpha \\ &\qquad\qquad \alpha \in U_{ad}. \end{aligned} \tag{9.27}$$

9.3.2 Discretization

The triangulation of $\Omega_h(\boldsymbol{\alpha})$ is the same as in the previous section. It is regular and fixed on a rectangle $[0, c_0] \times [0, 1]$ with $c_0 < c_1$. Each segment $[(c_0, a_i), (\boldsymbol{\alpha}(a_i), a_i)], i = 1, \ldots, n$, is partitioned equidistantly; cf. Figure 9.7. Again, we assume that the set Ω_0 is included in the rectangle $[0, c_0] \times [0, 1]$, whose triangulation does not depend on $\boldsymbol{\alpha}$.

In order to discretize the problem (9.27), we introduce the set $\mathcal{D}_0$ of indices of nodes lying in Ω_0. Given $r > 0$, the discretized problem reads as follows:

$$\begin{aligned} &\text{minimize} \quad \boldsymbol{J}_r(\boldsymbol{\alpha}, \boldsymbol{u}) := \text{meas}\, \Omega_h(\boldsymbol{\alpha}) + rh^2 \sum_{i \in \mathcal{D}_0} (\boldsymbol{u} - \boldsymbol{G}(\boldsymbol{\alpha}, \boldsymbol{u}))^i \\ &\text{subject to} \\ &\qquad\qquad \boldsymbol{u} \text{ solves the ICP (9.20)} \\ &\qquad\qquad \boldsymbol{\alpha} \in \mathbf{U_{ad}} \end{aligned} \tag{9.28}$$

with $r > 0$.

9.3.3 Numerical method

Problem (9.28) is a mathematical program with equilibrium constraint in the form of an ICP; this ICP can be written as a generalized equation of type (5.41):

$$0 \in \begin{bmatrix} \boldsymbol{A}(\boldsymbol{\alpha})\boldsymbol{u} - \boldsymbol{f}(\boldsymbol{\alpha}) - \boldsymbol{\lambda} \\ \boldsymbol{u} - \boldsymbol{G}(\boldsymbol{\alpha}, \boldsymbol{u}) \end{bmatrix} + N_{\mathbb{R}^m \times \mathbb{R}^m_+}(\boldsymbol{u}, \boldsymbol{\lambda}). \tag{9.29}$$

We have to show that with some $\widetilde{A} \supset \mathbf{U_{ad}}$

(A1) $\boldsymbol{J}_r$ is continuously differentiable on $\tilde{A} \times \mathbb{R}^m$;

(A2) for all $\boldsymbol{\alpha} \in \tilde{A}$ the GE (9.29) has a unique solution $S(\boldsymbol{\alpha})$;

(A3) the GE (9.29) is strongly regular at all points $(\boldsymbol{\alpha}, \boldsymbol{v})$ with $\boldsymbol{\alpha} \in \tilde{A}, \boldsymbol{v} = S_1(\boldsymbol{\alpha})$

cf. assumptions (A1),(A2),(A3) from Section 7.2. The continuous differentiability of $\boldsymbol{J}_r(\boldsymbol{\alpha}, \boldsymbol{u})$ results from the construction of the discretization and from the definition. Due to the definition of U_{ad}, the boundary $\partial\Omega_\alpha$ is uniformly Lipschitz on U_{ad}, hence the stiffness matrix $\boldsymbol{A}(\boldsymbol{\alpha})$ is positive definite on $\mathbf{U_{ad}}$ and the operator $\boldsymbol{A}(\boldsymbol{\alpha})\boldsymbol{u} - \boldsymbol{f}(\boldsymbol{\alpha})$ is strongly monotone with respect to $\boldsymbol{u}$ on $\mathbb{R}^m$. Since $\boldsymbol{G}(\boldsymbol{\alpha}, \boldsymbol{u})$ is a linear function of $\boldsymbol{u}$, assumption (4.34) of Theorem 4.9 simplifies to

$$\langle \boldsymbol{h}, \boldsymbol{A}^{-1}(-k\boldsymbol{A})\boldsymbol{h} \rangle = -k\langle \boldsymbol{h}, \boldsymbol{h} \rangle \le 0 \quad \text{for all } \boldsymbol{h} \in \mathbb{R}^m,$$

which holds since $k \ge 0$. Hence, according to Theorem 4.9, ICP (9.20) and thus the GE (9.29) has a unique solution and (A2) is satisfied. The strong regularity of (9.29) is due to Proposition 5.11 and the lines above. The above assumptions also guarantee weak semismoothness of $\boldsymbol{J}_r(\boldsymbol{\alpha}, S_1(\boldsymbol{\alpha}))$ (Proposition 7.9).

In order to apply a nonsmooth code, we have to compute a subgradient from $\partial\Theta(\boldsymbol{\alpha})$, $\Theta(\boldsymbol{\alpha}) := \boldsymbol{J}_r(\boldsymbol{\alpha}, S_1(\boldsymbol{\alpha}))$, for each $\boldsymbol{\alpha}$. To this we use Theorem 7.6. First, we recall the definition of the index sets

$$\begin{aligned} I^+(\boldsymbol{\alpha}, \boldsymbol{u}) &= \left\{ i \in \{1, 2, \ldots, m\} \mid [\boldsymbol{A}(\boldsymbol{\alpha})\boldsymbol{u} - \boldsymbol{f}(\boldsymbol{\alpha})]^i > 0 \right\} \\ L(\boldsymbol{\alpha}, \boldsymbol{u}) &= \{ i \in \{1, 2, \ldots, m\} \mid \boldsymbol{u}^i > \boldsymbol{G}^i(\boldsymbol{\alpha}, \boldsymbol{u}) \} \\ I^0(\boldsymbol{\alpha}, \boldsymbol{u}) &= \left\{ i \in \{1, 2, \ldots, m\} \mid [\boldsymbol{A}(\boldsymbol{\alpha})\boldsymbol{u} - \boldsymbol{f}(\boldsymbol{\alpha})]^i = 0,\ \boldsymbol{u}^i = \boldsymbol{G}^i(\boldsymbol{\alpha}, \boldsymbol{u}) \right\}. \end{aligned}$$

As before, $M_i(\boldsymbol{\alpha}, \boldsymbol{u})$ will be a suitably chosen subset of $I^0(\boldsymbol{\alpha}, \boldsymbol{u})$. The adjoint system reads as

$$\begin{bmatrix} \boldsymbol{A}(\boldsymbol{\alpha})_{L \cup (I^0 \setminus M_i)} \\ \boldsymbol{E}_{I^+ \cup M_i} - \mathcal{J}_{\boldsymbol{u}} \boldsymbol{G}_{I^+ \cup M_i}(\boldsymbol{\alpha}, \boldsymbol{u}) \end{bmatrix}^T \boldsymbol{p} - \nabla_{\boldsymbol{u}} \boldsymbol{J}_r(\boldsymbol{\alpha}, \boldsymbol{u}) = 0.$$

The element from $\partial \boldsymbol{J}_r(\boldsymbol{\alpha}, S(\boldsymbol{\alpha}))$ corresponding to $M_i(\boldsymbol{\alpha}, \boldsymbol{u})$ can be computed as

$$\nabla_{\boldsymbol{\alpha}} \boldsymbol{J}_r(\boldsymbol{\alpha}, \boldsymbol{u}) + \begin{bmatrix} -\mathcal{J}_{\boldsymbol{\alpha}}(\boldsymbol{A}(\boldsymbol{\alpha})\boldsymbol{u} - \boldsymbol{f}(\boldsymbol{\alpha}))_{L \cup (I^0 \setminus M_i)} \\ \mathcal{J}_{\boldsymbol{\alpha}} \boldsymbol{G}_{I^+ \cup M_i}(\boldsymbol{\alpha}, \boldsymbol{u}) \end{bmatrix}^T \boldsymbol{p};$$

cf. Proposition 7.16.

The ICP (9.20) will be solved (for fixed $\boldsymbol{\alpha}$) by the nonsmooth Newton's method as introduced in Chapter 3. In the notation of this section, the nonsmooth equation (in $\boldsymbol{u}$) to be solved reads as

$$\mathcal{H}(\boldsymbol{\alpha}, \boldsymbol{u}) := \min\nolimits_c \{ \boldsymbol{A}(\boldsymbol{\alpha})\boldsymbol{u} - \boldsymbol{f}(\boldsymbol{\alpha}), \boldsymbol{u} - \boldsymbol{G}(\boldsymbol{\alpha}, \boldsymbol{u}) \} = 0$$

(cf. (9.20) and (4.23)) and one step of the Newton's method (with fixed $\boldsymbol{\alpha}$) is defined by

$$\boldsymbol{u}_{k+1} = \boldsymbol{u}_k - \boldsymbol{V}_k^{-1}\mathcal{H}(\boldsymbol{\alpha}, \boldsymbol{u}_k)$$

with $\boldsymbol{V}_k := \boldsymbol{V}(\boldsymbol{\alpha}, \boldsymbol{u}_k) \in \partial_B\mathcal{H}(\boldsymbol{\alpha}, \boldsymbol{u}_k)$. The matrix $\boldsymbol{V}_k$ is composed of the rows

$$(\nabla_2 \left[\boldsymbol{A}(\boldsymbol{\alpha})\boldsymbol{u} - \boldsymbol{f}(\boldsymbol{\alpha})\right]^i (\boldsymbol{\alpha}, \boldsymbol{u}))^T = \boldsymbol{A}^i(\boldsymbol{\alpha}) \tag{9.30}$$

and

$$(e_i - \nabla_2 \boldsymbol{G}^i(\boldsymbol{\alpha}, \boldsymbol{u}))^T = e_i^T + k\boldsymbol{A}^i(\boldsymbol{\alpha}). \tag{9.31}$$

We have to verify (Lemma 3.13) the nonsingularity of $\boldsymbol{V}$ at $(\boldsymbol{\alpha}, \boldsymbol{u}) \in \mathbf{U_{ad}} \times \mathbb{R}^m$. But since $k \geq 0$ and $\boldsymbol{A}(\boldsymbol{\alpha})$ is nonsingular for all $\boldsymbol{\alpha} \in \mathbf{U_{ad}}$, then V_k is also nonsingular for all $\boldsymbol{\alpha} \in \mathbf{U_{ad}}$. (This also follows from the strong regularity of the GE (5.41) shown above.) According to the equivalence of the ICP with a nonlinear complementarity problem (cf. the proof of Theorem 4.9), the question of convergence of the discretized solutions to the "continuous" one may be treated in the same way as in Haslinger and Neittaanmäki, 1996.

Let us make some remarks on the numerical algorithms used for solving the example below. The nonsmooth problem corresponding to (9.28) was solved by the bundle algorithm BT which, in the example, needed 69 BT iterations. The ICP (9.20) was solved by the nonsmooth Newton's method. This method was surprisingly efficient—the average number of iterations was five in the first ten BT iterations and three in the rest; note that the dimension of the problem is $m = 1089$. The Newton iterations were stopped when either the norm $\|\mathcal{H}(\boldsymbol{\alpha}, \boldsymbol{u}_k)\|$ or the difference of two successive iterates $\|\boldsymbol{u}_k - \boldsymbol{u}_{k-1}\|$ was smaller then 10^{-11}. At each Newton step, one has to solve a system of linear equations with the matrix $\boldsymbol{V}_k$. In our application, this matrix is given by (9.30) and (9.31), i.e., it is sparse, symmetric and positive definite, as the stiffness matrix $\boldsymbol{A}$. Therefore, for solving this linear system we used the conjugate gradient method with preconditioning by incomplete factorization. This method retains the sparsity of V and contributes significantly to the efficiency of the overall algorithm.

9.3.4 Example

Example 9.3 Consider the packaging problem in which $\Omega_0 = [0.25, 0.5] \times [0.25, 0.75]$, $f(\xi_1, \xi_2) = -1.0$ and $\chi(\xi_1, \xi_2) = -0.04[(\xi_1)^2 + ((\xi_2)^2 - 0.25)^2]$. The set U_{ad} is specified by the parameters $c_1 = 0.6$, $c_2 = 1.0$, $c_3 = 3.0$, the parameter c_0 (determining the triangulation) was set to 0.5 and the discretization parameter was chosen as $h = 1/32$, i.e., we used $33 \times 33 = 1089$ discretization nodes and 2048 finite elements. Figure 9.13 shows the contour lines of the initial obstacle. For the compliance parameter $k = 2$ we obtained the optimal shape shown in Figure 9.14. The penalty parameter was $r = 10^6$, the resulting value of the objective functional $\boldsymbol{J}_r = 0.6918$. We compared this result with that obtained for $k = 0$, i.e. for a rigid obstacle, described in the previous section. In that case the resulting objective value was $\boldsymbol{J}_r = 0.7616$ with optimal shape depicted in Figure 9.15. One easily sees that the objective values and the shapes substantially differ from each other. △

9.4 INCIDENCE SET IDENTIFICATION PROBLEM

In this problem, we put aside the surface of the membrane as the point of our interest. Instead, our objective is to bring the membrane into contact with the obstacle *exclusively* on

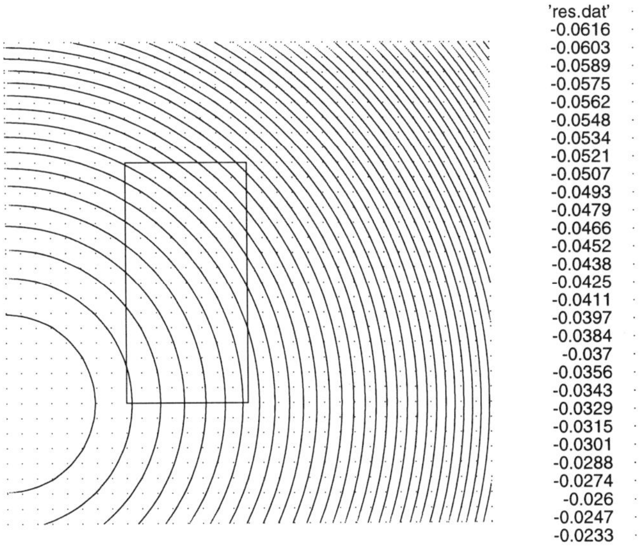

Figure 9.13. Contour lines of the initial obstacle χ

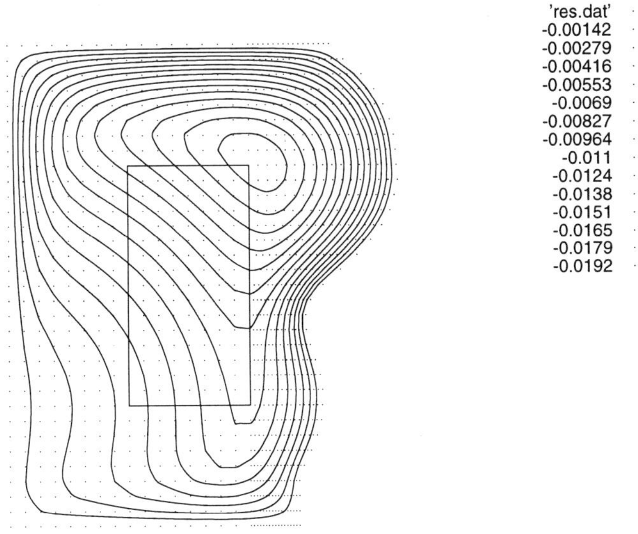

Figure 9.14. Optimal shape for a compliant obstacle

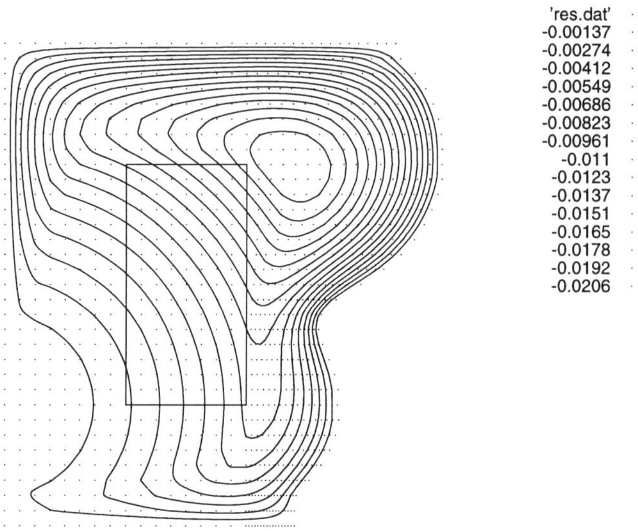

Figure 9.15. Optimal shape for a rigid obstacle

a prescribed subset $\Omega_0 \subset \widehat{\Omega}$. Moreover, contrary to the packaging problem, the shape of Ω_0 is part of the design variable. Such problems were studied, e.g., by Barbu, 1984; Zolesio, 1983. Here we follow the approach of Haslinger and Neittaanmäki, 1996.

9.4.1 Problem formulation

We consider the family of domains $\mathcal{O} = \{\Omega_\alpha \mid \alpha \in U_{ad}\}$ from (9.2) with the admissible set U_{ad} from (9.1). Again, let $Z(\alpha) := \{\xi \in \Omega_\alpha \mid u(\xi) = \chi(\xi)\}$ be the incidence set where u is the solution of the membrane problem (9.5).

We have mentioned that not only the shape of Ω_α but also the shape of Ω_0 can be controlled. The set Ω_0 is parametrized by means of a function ω. For given $\gamma, \delta \in \mathbb{R}$ with $0 < \gamma < \delta < 1$ and given positive $\varepsilon, \varpi, c_4, c_5 \in \mathbb{R}$, we define the additional admissible set for ω:

$$U'_{ad} = \{\omega = (\omega_1, \omega_2) | \quad \omega_i \in C^{0,1}([\gamma, \delta]), \quad \varepsilon \leq \omega_1, \omega_1 + \varpi \leq \omega_2 \leq c_4, \quad |\omega'_i| \leq c_5 \text{ a.e. in } [\gamma, \delta], i = 1, 2\}. \tag{9.32}$$

We assume that $U'_{ad} \neq \emptyset$. With $\omega \in U'_{ad}$ we associate the set

$$\Omega_0(\omega) = \{(\xi_1, \xi_2) \in \mathbb{R} \times [\gamma, \delta] \mid \omega_1(\xi_2) \leq \xi_1 \leq \omega_2(\xi_2)\},$$

see Figure 9.16. Assuming $c_4 < c_1$, we have $\Omega_0(\omega) \subset \Omega_\alpha$ for any $(\alpha, \omega) \in U_{ad} \times U'_{ad}$. The original formulation of the problems reads as

$$\left.\begin{array}{l} \text{Find } (\alpha^*, \omega^*) \in U_{ad} \times U'_{ad} \text{ such that} \\ \quad u(\alpha^*) \text{ solves (9.5) for } \Omega(\alpha^*) \\ \quad Z(\alpha^*) = \Omega_0(\omega^*). \end{array}\right\} \tag{9.33}$$

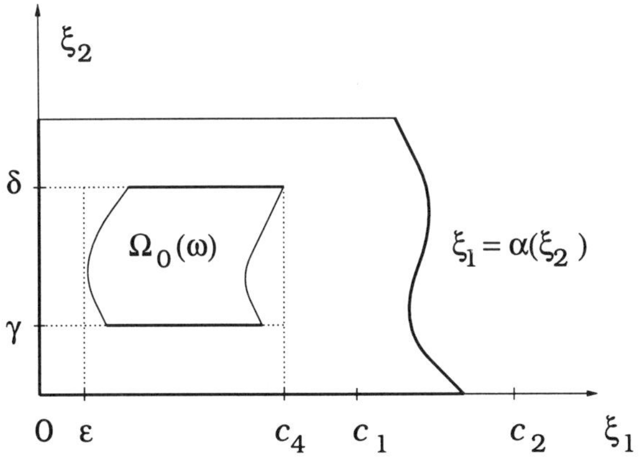

Figure 9.16. Domain $\Omega(\alpha)$

In general, we cannot expect a solution of (9.33) to exist. But we can reformulate (9.33) as a minimization problem and try to find (α^*, ω^*) such that $Z(\alpha^*)$ and $\Omega_0(\omega^*)$ are as close as possible to each other. In Haslinger and Neittaanmäki, 1996, one such objective functional, expressing the "identification" requirement, was suggested. The authors propose to identify the difference $(u - \chi)$ with a given function z_d positive outside $\Omega_0(\omega)$, i.e., they minimize

$$\mathcal{J}(\alpha, \omega, u) := \frac{1}{p}\|u - \chi - z_d\|^p_{L_p(\Omega_\alpha \setminus \Omega_0(\omega))},$$

where $p \in (1, \infty)$, $z_d \in L_p(\widehat{\Omega})$. The hope is that, for appropriately chosen z_d, the difference $(u - \chi)$ is positive on a substantial part of $\Omega_\alpha \setminus \Omega_0(\omega)$.

In Kočvara and Outrata, 1994c we proposed another approach utilizing the complementarity condition in (9.5). We define a new objective functional

$$\mathcal{J}_r(\alpha, \omega, u) := \int_{\Omega_\alpha \setminus \Omega_0(\omega)} (-\Delta u - f)\mathrm{d}\xi + r \int_{\Omega_0(\omega)} (u - \chi)\mathrm{d}\xi \tag{9.34}$$

with $r > 0$. As u is the solution of (9.5), we know that both integrals in (9.34) are nonnegative. So if $\mathcal{J}(\alpha, \omega, u) = 0$ for some (α, ω), then the set $\Omega_0(\omega)$ is covered by the membrane and outside $\Omega_0(\omega)$ we must have $(-\Delta u - f) = 0$. Still this does not mean that $u > \chi$ outside $\Omega_0(\omega)$, but the situation, when both vectors in the complementarity condition in (9.5) are equal to zero, is in practice only limited to points (typically corner points of $\Omega_0(\omega)$). Hence, instead of (9.33), we solve the following problem

$$\begin{array}{ll} \text{minimize} & \mathcal{J}_r(\alpha, \omega, u) \\ \text{subject to} & \\ & u \text{ solves the LCP (9.5) with } \Omega_\alpha \\ & (\alpha, \omega) \in U_{ad} \times U'_{ad}. \end{array} \tag{9.35}$$

9.4.2 Discretization

First, we have to discretize the new admissible set U'_{ad}. Let us consider the partition $a_1 < a_2 < \ldots < a_n$ of $[0,1]$ known from Section 9.1.2. Further, let us identify the segment $[\gamma,\delta]$ with $[a_k, a_\ell]$ for some k, ℓ such that $1 < k < \ell < n$. Denote by $\mathcal{D}_\omega := \{i \mid k \le i \le \ell\}$ the set of indices of all points within this segment; the number of these points is denoted by n'. Now the functions ω_1, ω_2 are approximated by piecewise linear functions ω_{h1}, ω_{h2} such that $\omega_{hi}(a_k) = \omega_i(a_k)$ for $i = 1,2$ and $k \in \mathcal{D}_\omega$. The discrete counterpart to $\omega_{hi}, i = 1,2$, is a vector $\boldsymbol{\omega}_i \in \mathbb{R}^{n'}$ of values of this function at points $a_k, k \in \mathcal{D}_\omega$. The set U'_{ad} is now replaced by the finite-dimensional set

$$\mathbf{U}'_{\mathbf{ad}} := \Big\{\boldsymbol{\omega} = (\boldsymbol{\omega}_1, \boldsymbol{\omega}_2) \in \mathbb{R}^{2n'} \mid \quad \boldsymbol{\omega}_i^k = \omega_i(a_k), \omega \in U'_{ad}, \\ \omega_i \text{ linear on } [\gamma,\delta], i = 1,2, \text{ and } k \in \mathcal{D}_\omega\Big\}.$$

The triangulation of the computational domain $\Omega_h(\boldsymbol{\alpha})$ is done by means of *principal moving nodes* given by couples

$$\begin{array}{ll} (\alpha_h(a_i), a_i), & i = 1,\ldots,n, \\ (\omega_{1h}(a_i), a_i),\ (\omega_{2h}(a_i), a_i), & i \in \mathcal{D}_\omega. \end{array}$$

The ξ_1–coordinates of the associated moving nodes are given by an equidistant partition of segments

$$\begin{array}{rl} [0, \alpha_h(a_i)] & \text{for } i \notin \mathcal{D}_\omega \\ [0, \omega_{1h}(a_i)],\ [\omega_{1h}(a_i), \omega_{2h}(a_i)],\ [\omega_{2h}(a_i), \alpha_h(a_i)] & \text{for } i \in \mathcal{D}_\omega, \end{array}$$

see Figure 9.17.

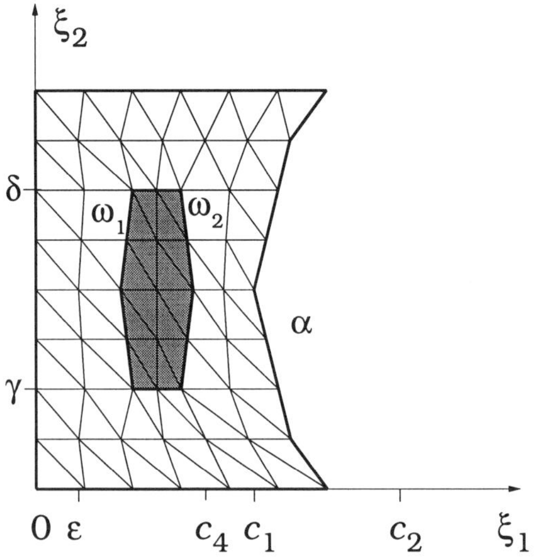

Figure 9.17. Discretization of Ω, α, ω_1 and ω_2.

In order to discretize the cost functional $\mathcal{J}(\alpha, \omega, u)$, we recall that we work with the "transformed" version (9.18) of the LCP (9.16). Let $\mathcal{D}_0$ be the set of indices of nodes lying in $\Omega_0(\boldsymbol{\omega})$; $\mathcal{D}_0$ does not depend on $\boldsymbol{\omega}$ since the nodes are moving together with it. Further, put $\mathcal{D}_1 := \{1, 2, \ldots, N\} \setminus \mathcal{D}_0$. Then the discrete form of (9.35) reads as

$$\begin{aligned}
&\text{minimize} \quad \boldsymbol{J}_r(\boldsymbol{\alpha}, \boldsymbol{\omega}, \boldsymbol{v}) := \sum_{i \in \mathcal{D}_1} (\boldsymbol{A}(\boldsymbol{\alpha}, \boldsymbol{\omega})(\boldsymbol{v} + \boldsymbol{\chi}(\boldsymbol{\alpha}, \boldsymbol{\omega})) - \boldsymbol{f}(\boldsymbol{\alpha}, \boldsymbol{\omega}))^i + rh^2 \sum_{i \in \mathcal{D}_0} \boldsymbol{v}^i \\
&\text{subject to} \\
&\qquad \boldsymbol{v} \text{ solves the LCP (9.18)} \\
&\qquad (\boldsymbol{\alpha}, \boldsymbol{\omega}) \in \mathbf{U_{ad}} \times \mathbf{U}'_{\mathbf{ad}}.
\end{aligned} \tag{9.36}$$

9.4.3 Numerical method

Problem (9.36) is MPEC of type (7.1). The assumptions which allow to use the numerical method from Chapter 7 are identical to assumptions (A1)–(A3) from Section 9.2.3 and so is their verification. Also a subgradient from $\partial\Theta(\boldsymbol{\alpha})$, $\Theta(\boldsymbol{\alpha}) := \boldsymbol{J}_r(\boldsymbol{\alpha}, \boldsymbol{\omega}, S(\boldsymbol{\alpha}, \boldsymbol{\omega}))$ can be computed in the same way as in Section 9.2.3; the actual computation is a bit more complicated because of the nontrivial dependence of the functional $\boldsymbol{J}_r$ on the control variables $\boldsymbol{\alpha}, \boldsymbol{\omega}$. We can, however, simplify the procedure. First, let us denote by $\widetilde{\boldsymbol{\alpha}} := (\boldsymbol{\alpha}, \boldsymbol{\omega})$ the joint control variable and write the LCP (9.18) as a GE of type (5.24):

$$0 \in \begin{bmatrix} \boldsymbol{A}(\widetilde{\boldsymbol{\alpha}})\boldsymbol{v} + \boldsymbol{A}(\widetilde{\boldsymbol{\alpha}})\boldsymbol{\chi}(\widetilde{\boldsymbol{\alpha}}) - \boldsymbol{f}(\widetilde{\boldsymbol{\alpha}}) - \boldsymbol{\lambda} \\ \boldsymbol{v} \end{bmatrix} + N_{\mathbb{R}^m \times \mathbb{R}^m_+}(\boldsymbol{v}, \boldsymbol{\lambda}).$$

Then the objective functional can be simplified to

$$\boldsymbol{J}_r(\widetilde{\boldsymbol{\alpha}}, \boldsymbol{v}, \boldsymbol{\lambda}) := \sum_{i \in \mathcal{D}_1} \boldsymbol{\lambda}^i + \frac{r}{h^2} \sum_{i \in \mathcal{D}_0} \boldsymbol{v}^i.$$

According to Theorem 7.3, the adjoint quadratic programming problem reads as

$$\begin{aligned}
&\text{minimize} \quad \tfrac{1}{2} \langle \boldsymbol{p}, \boldsymbol{A}(\widetilde{\boldsymbol{\alpha}})\boldsymbol{p} \rangle - \langle \nabla_{\boldsymbol{v}} \boldsymbol{J}_r(\widetilde{\boldsymbol{\alpha}}, \boldsymbol{v}, \boldsymbol{\lambda}), \boldsymbol{p} \rangle \\
&\text{subject to} \\
&\qquad \boldsymbol{p}_j = \frac{\partial \boldsymbol{J}_r(\widetilde{\boldsymbol{\alpha}}, \boldsymbol{v}, \boldsymbol{\lambda})}{\partial \boldsymbol{\lambda}^j}, \quad j \in I^+ \cup M_i
\end{aligned}$$

with I^+ and M_i defined as in Section 9.2.3. The subgradient associated with M_i can be computed as

$$\nabla_{\widetilde{\boldsymbol{\alpha}}} \boldsymbol{J}_r(\widetilde{\boldsymbol{\alpha}}, \boldsymbol{v}, \boldsymbol{\lambda}) - \left[\mathcal{J}_{\widetilde{\boldsymbol{\alpha}}}(\boldsymbol{A}(\widetilde{\boldsymbol{\alpha}})\boldsymbol{v} + \boldsymbol{A}(\widetilde{\boldsymbol{\alpha}})\boldsymbol{\chi}(\widetilde{\boldsymbol{\alpha}}) - \boldsymbol{f}(\widetilde{\boldsymbol{\alpha}}))\right]^T \boldsymbol{p};$$

cf. Proposition 7.14.

Again, BT from Chapter 3 is the code to deal with the nonsmooth problem. The LCP (9.18) was solved by a two-step algorithm introduced in Kočvara and Zowe, 1994 which combines the successive overrelaxation method with projection and the preconditioned conjugate gradient method; the preconditioning by incomplete factorization was used.

9.4.4 Example

Example 9.4 Consider the problem of the incidence set identification where $f(\xi_1, \xi_2) = -1.0$ and $\chi(\xi_1, \xi_2) = -0.03$. The sets U_{ad}, U'_{ad} are specified by parameters $c_1 = 0.7$, $c_2 =$

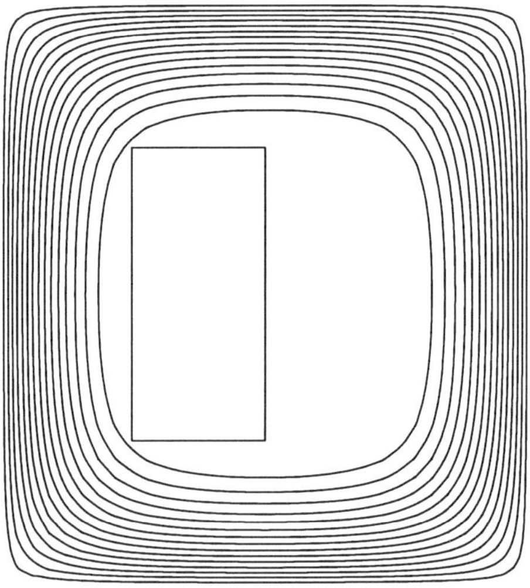

Figure 9.18. Initial design for Example 9.20, $n = 33$.

$1.2,\ c_3 = 2.5,\ \gamma = 0.25,\ \delta = 0.75,\ \varepsilon = 0.15,\ \varpi = 0.05,\ c_4 = 0.65$. Proper choice of the (linear) penalty parameter r is more difficult than in the obstacle problem, because both terms of the objective functional $\boldsymbol{J}_r$ are of the same nature and thus r determines a scaling between them. For $r = 33$, $h = \frac{1}{16}$ and $h = \frac{1}{32}$, the final values of the objective functional are $\boldsymbol{J}_r^{opt} = 0$ and $\boldsymbol{J}_r^{opt} = 0.395618 \cdot 10^{-4}$, respectively.

In Figure 9.18 we see the initial design of Ω and Ω_0 and the corresponding contour lines of the solution. Figure 9.19 shows the optimal design for $h = \frac{1}{32}$; the boundary of Ω_0 in fact coincides with the contour line -0.03. Finally, Figure 9.20 shows a 3D view of the solution and the obstacle with changed scaling in the vertical axis. We see that the problem constraints are clearly satisfied.

$\triangle$

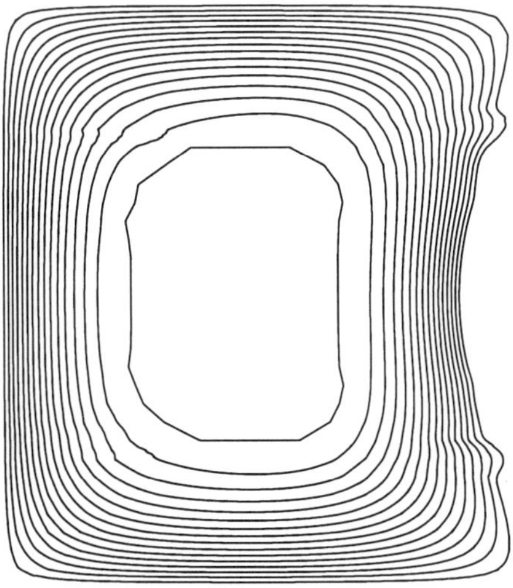

Figure 9.19. Optimal design for Example 9.20, $n = 33$.

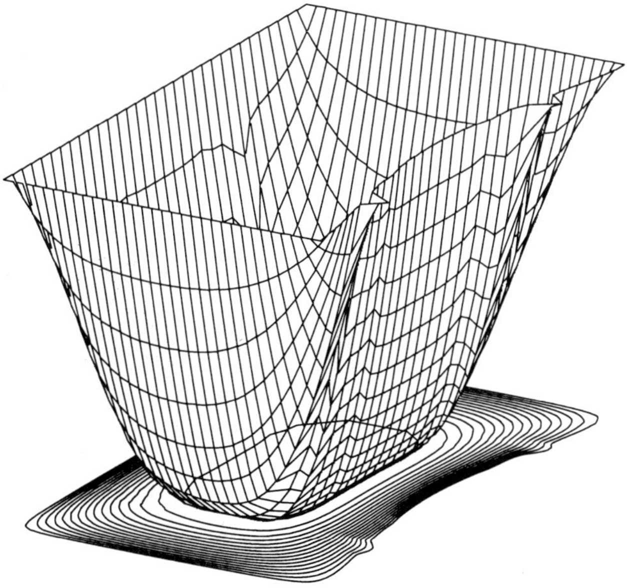

Figure 9.20. Optimal design for Example 9.20, $n = 33$.

10 ELASTICITY PROBLEMS WITH INTERNAL OBSTACLES

Although the membrane problems discussed in the previous chapter have practical applications, they are usually considered as only model problems. Truly interesting practical examples arise from modelling of two- and three-dimensional elastic bodies. Their behaviour is again described by elliptic variational equations or inequalities.

A direct generalization of the obstacle problems in the previous chapter leads to problems of contact mechanics. The next Chapter 11 deals with one such example. In this chapter, the "obstacle" constraints do not restrict the displacements of boundary nodes, but the behaviour of the elastic material at each point of the body. The problems in this chapter are therefore called problems with internal obstacles.

In Section 10.1 we introduce the problem of linear elasticity, modelled by an elliptic variational equation. Sections 10.2 and 10.3 deal with two optimum shape problems: design of elastic-perfectly plastic bodies and design of masonry structures. In the latter two sections, the state problem is an elasticity problem with internal obstacle; its behaviour is modelled by elliptic variational inequality.

10.1 LINEAR ELASTICITY PROBLEM

The purpose of this section is certainly not to develop a mathematical theory of linear elasticity; this subject has been covered in other books (see, e.g., Ciarlet, 1988; Duvaut and Lions, 1972; Nečas and Hlaváček, 1981). On the other hand, we do not want to introduce a mathematical formulation of the problem without any explanation of its physical background. In our opinion, a basic knowledge of such background substantially helps to formulate optimization problems and examples, and to understand the results. In the following section we try to explain some basic notions of the theory of elastic bodies and introduce two formulations of the linear elasticity problem. A reader familiar with this subject can directly pass on to the formulation of the optimum design problems, Sections 10.2 and 10.3.

10.1.1 Problem formulation

In this section we closely follow the book by Nečas and Hlaváček, 1981.

We start with the definition of the linear elasticity problem. We only consider problems in two-dimensional space. There is, however, no essential difficulty in generalizing our problems in the three-dimensional space. The two-dimensional model we use is the *plane-strain* model. That means, we consider the two-dimensional body to be a cross-section of a thick three-dimensional one (e.g., a long cylinder).

For the purpose of this section, let $\Omega \subset \mathbb{R}^2$ be a fixed domain with a Lipschitz boundary $\partial\Omega$. We assume that the domain Ω represents an *elastic body* (think, for instance, of metal parts in your car). The body is fixed at a part of its boundary, called Γ_u. The rest of the boundary, called Γ_g, is subject to external forces (surface tractions) $T = (T_1, T_2)$. We also assume that there are given body forces $F = (F_1, F_2)$ (like gravity, inertia, etc.) acting on the body. For some typical examples—cantilever beam, (one fourth of) pressure vessel, L-shaped body—cf. Figure 10.1.

So we consider a partitioning of the boundary $\partial\Omega$ of Ω into two parts: $\partial\Omega = \mathrm{cl}\Gamma_u \cup \mathrm{cl}\Gamma_g$, where Γ_u and Γ_g are open in $\partial\Omega$ and $\Gamma_u \cap \Gamma_g = \emptyset$.

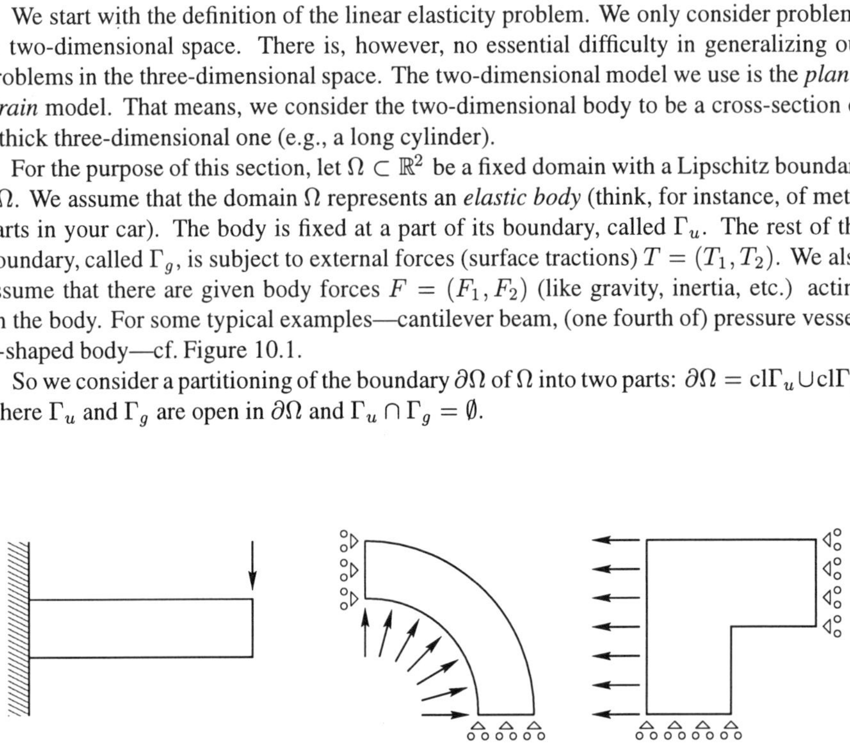

Figure 10.1. Some typical examples of elastic bodies

After loading, all the body points will be under stress and all the points (except those on Γ_u) will change their position—the body will be deformed. It is a primal subject of structural engineering to determine (measure, compute) this deformation (called strain) and, in particular, the stress within the loaded body. In the following paragraphs we will introduce the notions of stress and strain tensor, their relationship, and the primal and dual formulation of the problem of elasticity.

Let Ω_0 be an open subset of Ω with a Lipschitz boundary $\partial\Omega_0$ such that $\mathrm{cl}\Omega_0 \subset \Omega$. At $\xi \in \partial\Omega_0$ let n be the outer normal to $\partial\Omega_0$; cf. Figure 10.2. The internal forces can be schematically represented by a surface density of forces $\widetilde{T}$. The *stress vector* $\widetilde{T}(\xi, n) = (\widetilde{T}_1(\xi, n), \widetilde{T}_2(\xi, n))$ characterizes the density of surface forces acting on $\mathrm{cl}\Omega_0$ at point ξ from the rest of the body $\Omega \setminus \mathrm{cl}\Omega_0$. It depends on the point ξ and the normal n.

We can ask whether the stress vector $\widetilde{T}(\xi, n)$ can be characterized by some of its values for certain specific normals. Indeed, if we choose the normals as unit vectors ϵ_1, ϵ_2, parallel

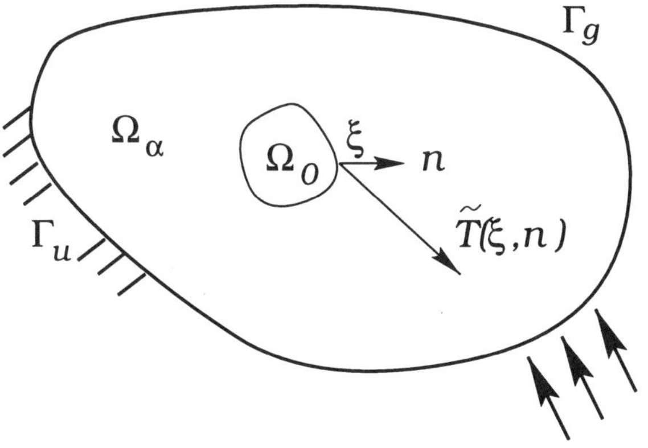

Figure 10.2. Domain $\Omega(\alpha)$, partition of the boundary, stress vector

to Cartesian axes, and define

$$\begin{aligned} \sigma_{1i}(\xi) &:= \widetilde{T}_i(\xi, \epsilon_1), \qquad \epsilon_1 = (1,0), \qquad i = 1,2 \\ \sigma_{2i}(\xi) &:= \widetilde{T}_i(\xi, \epsilon_2), \qquad \epsilon_2 = (0,1), \qquad i = 1,2 \end{aligned}$$

then the matrix $\sigma(\xi)$ fully characterizes the state of stress at point ξ. This 2×2 matrix is actually a second-order tensor[1] and is called *(Cauchy) stress tensor*. The diagonal components σ_{11} and σ_{22} are called *normal stresses* and the off-diagonal components *friction stresses* or *tangential stresses*. It is easy to see that $\sigma(x)$ is *symmetric*. The stress vector $\widetilde{T}(\xi, n)$ can be recovered from the stress tensor $\sigma(x)$ as

$$\widetilde{T}_i(\xi, n) = \sum_{j=1}^{2} n_j \sigma_{ji}(\xi) \qquad i = 1,2;$$

cf. Nečas and Hlaváček, 1981, Chap. 1.3. These relations are schematically shown in Figure 10.3.

The equilibrium of internal forces on a ball $B \subset \Omega$

$$\int_B F_i \mathrm{d}\xi + \int_{\partial B} \sum_{j=1}^{2} \sigma_{ij} n_j \mathrm{d}\partial B = 0, \qquad i = 1,2$$

gives us (after applying the Green's theorem) the *equilibrium conditions*

$$\sum_{j=1}^{2} \frac{\partial \sigma_{ij}}{\partial \xi_j}(\xi) + F_i(\xi) = 0, \qquad i = 1,2, \qquad \text{for all } \xi \in \Omega. \tag{10.1}$$

[1] Let $\xi_i' = \sum_{j=1}^{n} A_{ij}\xi_j + c_i, i = 1, \ldots, n$ be an orthogonal transformation in $\mathbb{R}^n$. An $n \times n$ matrix M is a second-order tensor if its components are transformed according to $M_{ij}' = \sum_{k,\ell=1}^{n} A_{ik} A_{j\ell} M_{k\ell}$. This yields, e.g., $Mx = y \Longleftrightarrow M'x' = y'$.

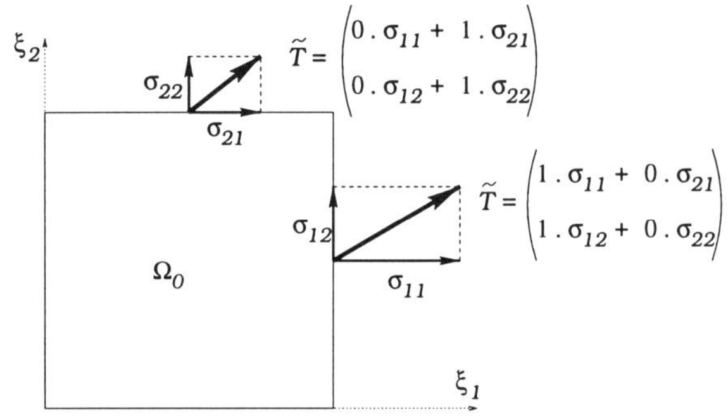

Figure 10.3. Components of the stress vector and of the representative matrix of the stress tensor

To these equations we have to add the equilibrium of forces on the boundary Γ_g:

$$T_i(\xi) = \sum_{j=1}^{2} n_j \sigma_{ji}(\xi) \qquad i = 1, 2, \qquad \text{for all } \xi \in \Gamma_g \tag{10.2}$$

(recall that $T(\xi)$ is the given external force). The component values of $\sigma(\xi)$ depend on the choice of the coordinate system. However, due to its tensor character, the eigenvalues of the matrix $\sigma(\xi)$ are independent of the orthogonal coordinate transformation. These eigenvalues are called *principal stresses* and the corresponding (orthogonal) eigenvectors *principal stress directions*.

Remark. Recall that the considered model is the plane-strain model; we are working with a cross-section of a "long" three-dimensional body. Actually, in this model we should also consider stress with respect to the "third" axis, perpendicular to the plane spanned by ξ_1, ξ_2. This is the component σ_{33} of the stress tensor of fully three-dimensional model. This component is non-zero in our two-dimensional idealization. However, we are not interested in its value and simply ignore this stress in the following text.

Now when we load our elastic body Ω, it will be deformed to a body Ω'. We assume that the deformation is described by a function

$$\eta(\xi) = \xi + u(\xi)$$

where $u(\xi)$ is the *displacement vector*. But to know the absolute displacements of the points of Ω may not be as important as to have the information about relative displacements of the nodes with respect to each other. This is measured by a *strain tensor*[2] $e(\xi)$ defined as

$$e_{ij}(\xi) = \left(\frac{\partial u_i}{\partial \xi_j}(\xi) + \frac{\partial u_j}{\partial \xi_i}(\xi) \right), \qquad i, j = 1, 2.$$

[2]We assume that the deformations are very small so that we may neglect higher-order terms. From this reason, the strain tensor introduced here is also called *small strain tensor*.

Obviously, the 2×2 matrix $e(\xi)$ is symmetric. What is the physical meaning of the components of $e(\xi)$? The diagonal components $e_{11}(\xi)$ and $e_{22}(\xi)$ characterize relative prolongation of an infinitesimally small segment located at point ξ and parallel to axis ξ_1 and ξ_2, respectively. For example, consider a wire of length ℓ, parallel to ξ_1. After loading, the wire prolongs by $\delta\ell$. Then at any point of the wire we have

$$e_{11} \approx \frac{\delta\ell}{\ell}.$$

The meaning of the off-diagonal elements of $e(\xi)$ is the following: Consider two perpendicular infinitesimally small segments located at point ξ. Then the number $2e_{12}(\xi)$ characterizes the change of the right angle between these two segments; cf. Figure 10.4 where $2e_{12}(\xi) \approx \alpha - \alpha'$.

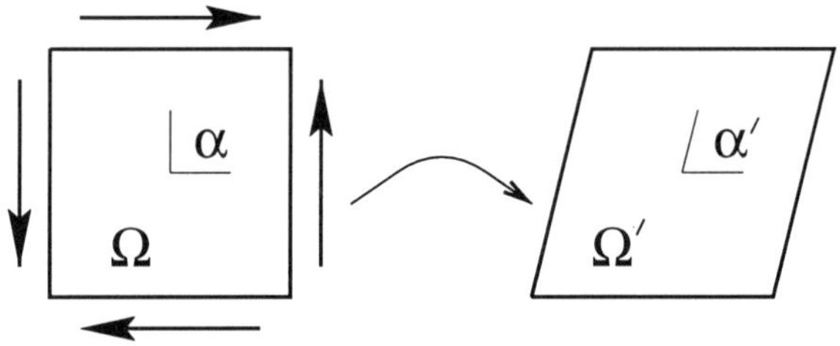

Figure 10.4. Physical meaning of e_{12}

The relation between the stress and strain tensors is given by *generalized Hooke's law.* First let us consider a one-dimensional model, a bar loaded at its two ends with forces of equal value and opposite signs. Experiments show that, for certain materials like steel, there is a linear relation between the load and the relative prolongation of the bar, i.e., between the stress and strain tensors. This holds for reasonably small values of the load, until the material starts to behave nonlinearly. Hence, in one dimension and for small strains, we have

$$\sigma_{11}(\xi) = E(\xi)e_{11}(\xi).$$

If we consider the material properties of our bar to be the same at all its points, i.e., bar made of *homogeneous material*, then is the above relation simplified to

$$\sigma_{11}(\xi) = Ee_{11}(\xi).$$

This is the one-dimensional Hooke's law for homogeneous materials; the number E is called *Young's modulus of elasticity*.

The situation in two dimensions is a bit more complicated. Denote by $\mathbb{R}_s^{2\times 2}$ the set of symmetric 2×2 matrices and introduce

$$S(\Omega) := \{\tau : \Omega \to \mathbb{R}_s^{2\times 2} \mid \tau_{ij} \in L_2(\Omega),\ i,j = 1,2\}. \tag{10.3}$$

Generalized Hooke's law is defined by means of an isomorphism

$$\mathcal{H} : S(\Omega) \to S(\Omega), \qquad \mathcal{H} : e \mapsto \sigma.$$

This mapping is determined by material properties of our elastic body and is represented by a fourth-order tensor with components $\mathcal{H}_{ijk\ell}$. We again assume that $\mathcal{H}$ is linear.

To get a physically reasonable and solvable problem, we assume that the $\mathcal{H}_{ijk\ell}(\xi)$ are bounded and measurable functions on Ω and that $\mathcal{H}$ is symmetric and elliptic, i.e., respectively,

$$\mathcal{H}_{ijk\ell} = \mathcal{H}_{jik\ell} = \mathcal{H}_{k\ell ij} \quad \text{for } i,j,k,\ell = 1,2 \quad \text{a.e. in } \Omega \tag{10.4}$$

$$\sum_{i,j,k,\ell=1}^{2} \mathcal{H}_{ijk\ell}(\xi) e_{ij} e_{k\ell} \geq c_0 \sum_{i,j=1}^{2} e_{ij} e_{ij} \quad \text{for all } e \in \mathbb{R}^{2\times 2}_s \quad \text{a.e. in } \Omega \tag{10.5}$$

with some $c_0 > 0$ (cf. (9.12)).

For simplicity, suppose that the material is homogeneous (i.e., independent of ξ) and isotropic (i.e., independent of the coordinate system). In this simple case, the generalized Hooke's law can be written as

$$\begin{aligned}
\sigma_{11} &= \frac{E}{(1+\nu)(1-2\nu)}((1-\nu)e_{11} + \nu e_{12}) \\
\sigma_{22} &= \frac{E}{(1+\nu)(1-2\nu)}((1-\nu)e_{22} + \nu e_{12}) \\
\sigma_{12} &= \frac{E}{(1+\nu)} e_{12},
\end{aligned} \tag{10.6}$$

where E is the Young's modulus of elasticity (which we know from the one-dimensional case) and ν is the *Poisson's ratio.*

Remark. Instead of E and ν, one can equivalently work with the so-called Lamé coefficients λ and μ defined by

$$\nu = \frac{E}{2(1+\nu)}, \qquad \lambda = \frac{E\nu}{(1+\nu)(1-2\nu)}.$$

The advantage (and reason why these coefficients are often used in mathematical literature) is a more compact form of the generalized Hooke's law

$$\sigma_{ij}(\xi) = \lambda(\xi)\delta_{ij}(e_{11}(\xi) + e_{22}(\xi)) + 2\mu(\xi)e_{ij}(\xi),$$

with the Kronecker symbol δ_{ij}.

The following example should help to explain the meaning of E and ν. We consider a square body loaded with a hydrostatic pressure p (Figure 10.5(A)), with pressure T on two opposite sides (Figure 10.5(B)) and with shear forces T on all the sides (Figure 10.5(C)). Also shown are the associated stress tensors. The corresponding strain tensors can be computed according to inverse Hooke's law introduced later in this section:

$$e_A = \begin{pmatrix} -\frac{(1+\nu)(1-2\nu)}{E}p & 0 \\ 0 & -\frac{(1+\nu)(1-2\nu)}{E}p \end{pmatrix}$$

$$e_B = \begin{pmatrix} \frac{(1+\nu)\nu}{E}T & 0 \\ 0 & -\frac{(1+\nu)(1-\nu)}{E}T \end{pmatrix}$$

$$e_C = \begin{pmatrix} 0 & \frac{1+\nu}{E}T \\ \frac{1+\nu}{E}T & 0 \end{pmatrix}.$$

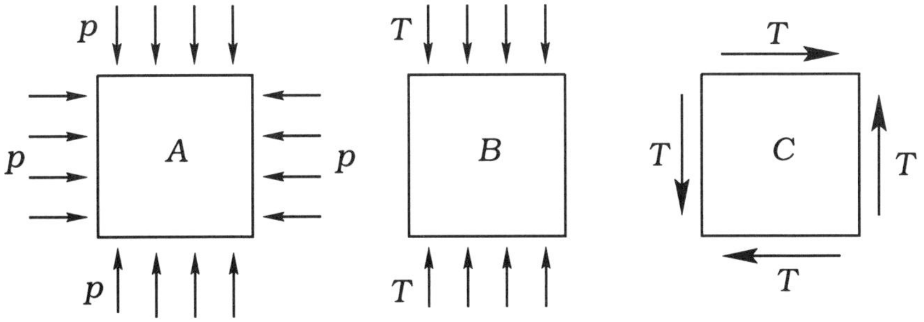

$$\sigma_A = \begin{pmatrix} -p & 0 \\ 0 & -p \end{pmatrix} \qquad \sigma_B = \begin{pmatrix} 0 & 0 \\ 0 & -T \end{pmatrix} \qquad \sigma_C = \begin{pmatrix} 0 & T \\ T & 0 \end{pmatrix}$$

Figure 10.5. Examples to interpretation of stress-strain relation

(The reader can simply check the correctness of $e_{A,B,C}$ by substituting $e_{A,B,C}$ and $\sigma_{A,B,C}$ in (10.6).) The Young's modulus E can be interpreted as the stiffness of the material. The bigger E, the higher stiffness and, accordingly, the smaller strain (deformation). Poisson's ratio ν is related to the cross contraction of the body; cf. example (B) where $\sigma_{11} = 0$ but $e_{11} > 0$. The body, though pressed, tries to keep its volume and expands in the "free" direction ξ_1. The value $\nu = 0$ corresponds to zero contraction. Physically reasonable limits for E and ν are

$$E > 0, \qquad -1 < \nu < \frac{1}{2};$$

these conditions also guarantee positive definiteness of the (fourth-order) tensor associated with the mapping $\mathcal{H}$.

Now, having the given loading and material properties at our disposal, we ask whether we can compute the displacements, strains and/or stresses in the elastic body. A traditional way is to plug Hooke's law (10.6) into the equilibrium equations (10.1), and to add the boundary conditions to obtain a system of second-order partial differential equations, the so-called Lamé equations. Here we prefer a different way based on energy principles.

For two vector-valued functions $v = (v_1, v_2)$ and $w = (w_1, w_2)$ with $v, w \in (L_2(\Omega))^2$, let $\langle \cdot, \cdot \rangle$ denote the standard inner product:

$$\langle v, w \rangle = v_1 w_1 + v_2 w_2.$$

For matrix-valued functions $\varepsilon, \tau \in S(\Omega)$ we use an inner product

$$\langle \varepsilon, \tau \rangle := \text{trace}(\varepsilon\tau) = \varepsilon_{11}\tau_{11} + \varepsilon_{22}\tau_{22} + 2\varepsilon_{12}\tau_{12}.$$

As *potential energy* of the elastic body we define

$$\Phi(u) = \frac{1}{2}\int_\Omega \langle \mathcal{H}e(u), e(u) \rangle \mathrm{d}\xi - \int_\Omega \langle F, u \rangle \mathrm{d}\xi - \int_{\Gamma_g} \langle T, u \rangle \mathrm{d}\Gamma_g. \qquad (10.7)$$

We assume that the body is fixed on a nonempty part Γ_u of the boundary, i.e.,

$$u(\xi) = 0 \qquad \text{on } \Gamma_u. \qquad (10.8)$$

A function u^* solves the problem of elasticity if it minimizes the potential energy $\Phi(u)$ over all functions satisfying (10.8). It is shown, e.g., in Nečas and Hlaváček, 1981 that if $F \in (L_2(\Omega))^2$ and $T \in (L_2(\Gamma_g))^2$, then there exists a unique minimizer

$$u^* \in \mathcal{U}(\Omega) := \{v \in (H^1(\Omega))^2 \mid v = 0 \text{ on } \Gamma_u\}.$$

The problem

$$\min_{u \in \mathcal{U}} \Phi(u) \tag{10.9}$$

is called *primal problem of elasticity* or *principle of minimum potential energy*. It can further be shown from the optimality conditions of (10.9) that if the solution u^* is sufficiently smooth, then the corresponding stress tensor (computed from Hooke's law) satisfies the equilibrium conditions (10.1)–(10.2). Hence the energy approach is a generalization of the classic elasticity problem.

Let us now derive a problem dual to (10.9). We denote by $\mathcal{E}$ the set of admissible stress tensors that satisfy the equilibrium conditions in a weak sense:

$$\mathcal{E}(\Omega) := \left\{ \tau \in S(\Omega) \mid \int_\Omega \langle \tau, e(v) \rangle \mathrm{d}\xi = \int_\Omega \langle F, v \rangle \mathrm{d}\xi + \int_{\Gamma_g} \langle T, v \rangle \mathrm{d}\Gamma_g, \ \forall v \in \mathcal{U}(\Omega) \right\}. \tag{10.10}$$

The mapping $\mathcal{H}$, defining Hooke's law for homogeneous isotropic material, can be inverted. We denote this inverse by $\mathcal{A} : \sigma \mapsto e$. It is given as follows:

$$\begin{aligned} e_{11} &= \frac{1+\nu}{E}[\sigma_{11} - \nu(\sigma_{11} + \sigma_{22})] \\ e_{22} &= \frac{1+\nu}{E}[\sigma_{22} - \nu(\sigma_{11} + \sigma_{22})] \\ e_{12} &= \frac{1+\nu}{E}\sigma_{12}. \end{aligned}$$

We will refer to $\mathcal{A}$ as to *inverse Hooke's law*.

Finally, we define the functional of *complementary energy* $\Psi(\sigma)$ by

$$\Psi(\sigma) = \frac{1}{2} \int_\Omega \langle \mathcal{A}\sigma, \sigma \rangle \mathrm{d}\xi. \tag{10.11}$$

We say that $\sigma^* \in \mathcal{E}$ solves the *dual* or *reciprocal problem of elasticity* if it minimizes the complementary energy

$$\min_{\sigma \in \mathcal{E}} \Psi(\sigma). \tag{10.12}$$

The problem is also called *Castigliano-Menabre principle of minimum complementary energy*. Again, it is shown in Nečas and Hlaváček, 1981 that if $F \in (L_2(\Omega))^2$ and $T \in (L_2(\Gamma_g))^2$, then $\sigma^* \in \mathcal{E}$ minimizes $\Psi(\sigma)$ over $\mathcal{E}$ if and only if

$$\sigma^* := \sigma^*(u^*) = \mathcal{H}e(u^*),$$

where u^* minimizes the potential energy $\Phi(u)$ over $\mathcal{U}(\Omega)$.

Why do we call the above two minimization problems (10.9) and (10.12) "primal" and "dual"? It holds that

$$-\Psi(\sigma^*(u^*)) = \Phi(u^*).$$

10.1.2 Discretization

The discretization technique is just a bit more complicated than in the previous chapter. Again, we use triangular finite elements. Both our optimum design problems work with the dual problem of elasticity (10.12), hence we concentrate on the discretization of this problem.

We first introduce a family of domains in which we will solve the problem. We consider the same type of domains as in the membrane problems, a rectangle with a parametrized right "vertical" part of the boundary; cf. Figure 9.1. For the discretization of the state problem, it suffices to work with a domain

$$\Omega(\alpha) = \{(\xi_1, \xi_2) \in \mathbb{R}^2 \mid 0 < \xi_1 < \alpha(\xi_2) \text{ for all } \xi_2 \in (0,1)\}$$

with an arbitrary fixed $\alpha \in C^{0,1}([0,1])$.

Let $a_1 < a_2 < \ldots < a_n$ be a uniform partition of $[0,1]$ with segments $[a_{i-1}, a_i]$ of length $h = 1/(n-1)$. The function α (one-dimensional real function) is approximated by a continuous piece-wise linear function α_h such that $\alpha_h(a_i) = \alpha(a_i)$. The discrete counterpart of α_h is a vector $\boldsymbol{\alpha} \in \mathbb{R}^n$ of values of this function at points $a_1, \ldots, a_n$, i.e., a vector of ξ_1-coordinates of boundary nodes $(\alpha_h(a_i), a_i)$. In this way we define a polygonal *computational domain*

$$\Omega_h(\boldsymbol{\alpha}) := \{(\xi_1, \xi_2) \in \mathbb{R}^2 \mid 0 < \xi_1 < \alpha_h(\xi_2),\ 0 < \xi_2 < 1\},$$

as shown in Figure 9.5. The parts of the boundary of $\Omega_h(\boldsymbol{\alpha})$ corresponding to Γ_u and Γ_g are denoted by $\Gamma_{uh}(\boldsymbol{\alpha})$ and $\Gamma_{gh}(\boldsymbol{\alpha})$, respectively. The domain $\Omega_h(\boldsymbol{\alpha})$ is discretized by triangular elements which we construct by means of *principal moving nodes* (design nodes)

$$(\alpha_h(a_i), a_i), \quad i = 1, \ldots, n$$

and *associated moving nodes*, whose ξ_1–coordinates are given by an equidistant partitioning of the segments $[(0, a_i), (\alpha_h(a_i), a_i)], i = 1, \ldots, n$ (cf. Figure 9.5 with $c_0 = 0$). Thus for a fixed $h > 0$, the triangulation depends continuously on α_h. The number of elements of the triangulation is denoted by M and the number of nodes by m.

Now we have two kinds of variables to discretize: the stress tensor σ and the displacement vector u (in the definition of the admissible set $\mathcal{E}$). For the purpose of discretization, we will use standard convention from structural analysis. The 2×2 symmetric strain and stress tensors are interpreted as 3–vectors:

$$\begin{pmatrix} e_{11} & e_{12} \\ e_{12} & e_{22} \end{pmatrix} \to \begin{pmatrix} e_{11} \\ e_{22} \\ \sqrt{2}e_{12} \end{pmatrix}, \quad \begin{pmatrix} \sigma_{11} & \sigma_{12} \\ \sigma_{12} & \sigma_{22} \end{pmatrix} \to \begin{pmatrix} \sigma_{11} \\ \sigma_{22} \\ \sqrt{2}\sigma_{12} \end{pmatrix}.$$

Accordingly, the Hooke's law will be represented by a 3×3 matrix

$$D = \begin{pmatrix} D_{1111} & D_{1122} & \sqrt{2}D_{1112} \\ & D_{2222} & \sqrt{2}D_{2212} \\ \text{sym.} & & 2D_{1212} \end{pmatrix}.$$

As before, the displacement functions $u = (u_1, u_2) \in (H^1(\Omega_\alpha))^2$ are approximated by continuous functions $u_h = (u_{1h}, u_{2h})$ that are linear on every triangle. Such function can be written as

$$u_{ih}(\xi) = \sum_{k=1}^{m} \boldsymbol{u}_{ik}\varphi_k(\xi), \qquad i = 1, 2,$$

where $\boldsymbol{u}_{ik}$ is the value of u_{ih} at the kth node and φ_k is the basis function associated with this node (for details, see Ciarlet, 1978). For the basis functions $\varphi_k(\xi), k = 1, \ldots, m$, we define a 3×2 matrices

$$B_k(\xi) = \begin{pmatrix} \frac{\partial \varphi(\xi)}{\partial \xi_1} & 0 \\ 0 & \frac{\partial \varphi(\xi)}{\partial \xi_2} \\ \frac{1}{2}\frac{\partial \varphi(\xi)}{\partial \xi_2} & \frac{1}{2}\frac{\partial \varphi(\xi)}{\partial \xi_1} \end{pmatrix}$$

in order to determine the approximate strain tensor as $e_h(\xi) = \sum_{k=1}^{m} B_k(\xi)\boldsymbol{u}_k$. Further, let $\boldsymbol{\varphi}_k$ denote the couple (φ_k, φ_k).

The stress tensor $\sigma = (\sigma_{11}, \sigma_{22}, \sigma_{12}) \in (L_2(\Omega_\alpha))^2$ is approximated by functions $\sigma_h = (\sigma_{h11}, \sigma_{h22}, \sigma_{h12})$ that are constant on every triangle. Such function can be written as

$$\sigma_{hij}(\xi) = \sum_{k=1}^{m} \boldsymbol{\sigma}_{ij,k}\psi_k(\xi), \qquad i = 1, 2,$$

where $\boldsymbol{\sigma}_{ij,k}$ is the value of σ_{hij} on the kth triangle and ψ_k is the basis function associated with this triangle. We denote by $\boldsymbol{\sigma}_k$ the triple $(\boldsymbol{\sigma}_{11,k}, \boldsymbol{\sigma}_{22,k}, \boldsymbol{\sigma}_{12,k})$ and by $\boldsymbol{\psi}_k$ the triple (ψ_k, ψ_k, ψ_k). Further let $\boldsymbol{\sigma}$ be the vector $(\boldsymbol{\sigma}_1, \ldots, \boldsymbol{\sigma}_M) \in \mathbb{R}^{3M}$ (recall that M is the number of elements of the mesh).

The inverse Hooke's law for homogeneous and isotropic material is defined by the *inverse elasticity matrix*

$$D^{-1} = \begin{pmatrix} \frac{(1+\nu)(1-\nu)}{E} & -\frac{(1+\nu)\nu}{E} & 0 \\ -\frac{(1+\nu)\nu}{E} & \frac{(1+\nu)(1-\nu)}{E} & 0 \\ 0 & 0 & \frac{1+\nu}{E} \end{pmatrix}.$$

The discrete version of the functional of complementary energy $\Psi(\sigma)$ is defined by means of *flexibility matrix*

$$(\boldsymbol{F}(\boldsymbol{\alpha}))_{k\ell} = \int_{\Omega_h(\boldsymbol{\alpha})} (D^{-1}\boldsymbol{\psi}_k)^T \boldsymbol{\psi}_\ell \mathrm{d}\xi, \qquad k, \ell = 1, \ldots, M$$

as

$$\boldsymbol{\Psi}(\boldsymbol{\alpha}, \boldsymbol{\sigma}) = \frac{1}{2}(\boldsymbol{F}(\boldsymbol{\alpha})\boldsymbol{\sigma})^T \boldsymbol{\sigma}.$$

The flexibility matrix $\boldsymbol{F}(\boldsymbol{\alpha})$ is a $3M \times 3M$ block diagonal symmetric positive definite matrix with blocks

$$(\boldsymbol{F}(\boldsymbol{\alpha}))_{kk} = \int_{K_k(\boldsymbol{\alpha})} (D^{-1}\boldsymbol{\psi}_k)^T \boldsymbol{\psi}_k \mathrm{d}\xi, \qquad k = 1, \ldots, M,$$

where $K_k(\boldsymbol{\alpha})$ is the kth elements of the triangulation. The size of this element and hence the value of $(\boldsymbol{F}(\boldsymbol{\alpha}))_{kk}$ depends continuously on $\boldsymbol{\alpha}$.

In the next step we derive the discrete counterpart to the admissible set $\mathcal{E}$. The equilibrium equations in the definition of $\mathcal{E}$ (cf. (10.10)) has to be satisfied for each test function from $\mathcal{U}$. First we replace $\mathcal{U}$ by a finite dimensional set with a basis $\{\varphi_i\}_{i=1}^n$. Then the equilibrium equation has only to be satisfied for every basis function φ_i. Next we define the $3M \times 2m$ *equilibrium matrix*

$$(\boldsymbol{A}(\boldsymbol{\alpha}))^{ij} = \int_{K_i(\boldsymbol{\alpha})} \psi_i(\xi)^T B_j(\xi) \mathrm{d}\xi, \qquad i = 1, \ldots, M, \ j = 1, \ldots, m,$$

where, again, $K_i(\boldsymbol{\alpha})$ is the ith element of the triangulation (and support of ψ_i). Note that $\boldsymbol{A}(\boldsymbol{\alpha})$ is a $3M \times 2m$ sparse matrix. Similarly, the discrete right-hand side vector for the equilibrium equation is defined as

$$\boldsymbol{f}_j(\boldsymbol{\alpha}) = \int_{\Omega_h(\boldsymbol{\alpha})} F\varphi_j \mathrm{d}\xi + \int_{\Gamma_{gh}(\boldsymbol{\alpha})} T\varphi_j \mathrm{d}\Gamma_{gh}, \qquad j = 1, \ldots, m.$$

The discrete version of the admissible set $\mathcal{E}$ (10.10) is

$$\mathbb{E}(\boldsymbol{\alpha}) := \{\boldsymbol{\tau} \in \mathbb{R}^{3M} \mid \boldsymbol{A}(\boldsymbol{\alpha})\boldsymbol{\tau} = \boldsymbol{f}(\boldsymbol{\alpha})\}$$

and the discrete dual problem of elasticity reads as

$$\min_{\sigma \in \mathbb{E}(\boldsymbol{\alpha})} \frac{1}{2} \Psi(\boldsymbol{\alpha}, \boldsymbol{\sigma}). \tag{10.13}$$

This is a convex quadratic program.

In the next two sections we will work with state problems of more complicated structure; the stress tensor will have to satisfy additional constraints, nonlinear in the first case and linear in the second one.

10.2 DESIGN OF ELASTIC-PERFECTLY PLASTIC STRUCTURES

Life is not all roses. If we increase the loading of our elastic body, i.e., if the stress increases, the material does not behave according to linear Hooke's law anymore. When a certain limit stress is exceeded, the stress-strain relationship becomes nonlinear and if we further increase the stress, the material starts to yield (now we speak e.g. of steel). If, after unloading, the body returns to its original shape, we still speak of elasticity (even nonlinear). If not, then we speak of *plastic* behaviour of the material. There are many materials and many models of elastic and plastic behaviour. The problem with modelling plastic behaviour is that we can make reliable experiments with only one-dimensional bodies (thin bars). The one-dimensional stress-strain relationship is, however, very difficult to "extrapolate" in higher dimensions.

In the next paragraph we will introduce a simple model which still approximates the behaviour of many materials well enough and which is extensively used in structural mechanics, the *Hencky's model of plasticity*. The idea is quite simple: either the material behaves linearly elastically or, when stress reaches certain *elastic limit* σ_E, it yields (it is *perfectly plastic*). The stress-strain relationship for one-dimensional problem is shown in Figure 10.6. In two (and three) dimensions, one has to introduce the *yield function* $\Upsilon : \mathbb{R}^{2\times 2}_s \to \mathbb{R}$ that is supposed to be convex, Lipschitz and such that

$$\Upsilon(\lambda\sigma) = |\lambda|\Upsilon(\sigma) \qquad \text{for all } \lambda \in \mathbb{R} \text{ and } \sigma \in \mathbb{R}^{2\times 2}_s.$$

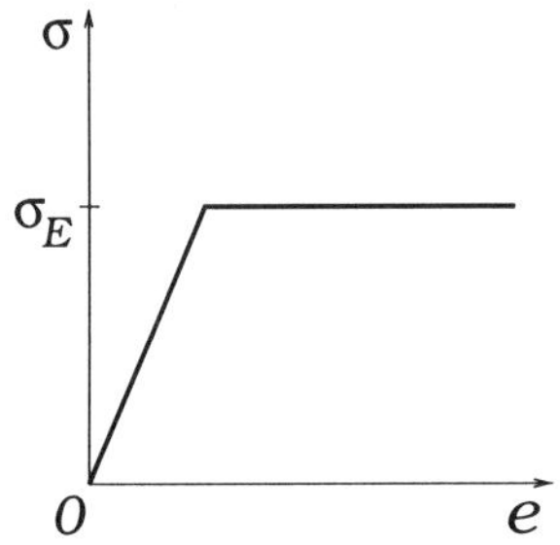

Figure 10.6. Stress-strain relationship for Hencky's model of plasticity

The mostly used yield function is the von Mises function, which for the plane-strain model takes the form

$$\Upsilon(\sigma) = \left((\sigma_{11} - \sigma_{22})^2 + ((1-\nu)\sigma_{22} - \nu\sigma_{11})^2 + ((1-\nu)\sigma_{11} - \nu\sigma_{22})^2 + 6\sigma_{12}^2\right)^{1/2}$$

It can be shown that $\Upsilon(\sigma)$ is independent of the orthogonal coordinate transformation.

Given an elastic limit $\sigma_E \in \mathbb{R}_+$, we can define the set of *plastically admissible* stresses as

$$\mathcal{P}(\Omega) = \{\tau \in S(\Omega) \mid \Upsilon(\tau) \leq \sigma_E \text{ a.e. in } \Omega\} \tag{10.14}$$

with $S(\Omega)$ defined in (10.3). Now it is advantageous to work with the dual model of elasticity (10.12); we say that σ^* solves the (dual) elastic-perfectly plastic problem if it minimizes the complementary energy Ψ (10.11) over the set $\mathcal{E}(\Omega) \cap \mathcal{P}(\Omega)$ (cf. (10.10) for the definition of $\mathcal{E}(\Omega)$), i.e., if it solves the problem

$$\begin{aligned} &\text{minimize} \quad \tfrac{1}{2}\Psi(\sigma) \\ &\text{subject to} \\ &\qquad\qquad \sigma \in \mathcal{E}(\Omega) \cap \mathcal{P}(\Omega). \end{aligned} \tag{10.15}$$

One intuitively feels that a plastically deformed body is less safe than the elastically deformed one. When the plastic region (the set where $\Upsilon(\sigma) = \sigma_E$) is not too large, the forces can still be carried by the rest of the body. Our goal in this section is to find such a shape of an elastic-perfectly plastic body, that the plastic region is minimized, while the volume of the body remains constant.

The existence of solutions to the corresponding state and shape optimization problem was shown by Hlaváček, 1986. He also proved the convergence of the approximate solutions to the exact ones.

10.2.1 Problem formulation

We introduce the set of *admissible design variables*

$$U_{ad} = \left\{ \alpha \in C^{0,1}([0,1]) \mid 0 < c_1 \leq \alpha \leq c_2, \right.$$
$$\left. \left| \frac{\mathrm{d}}{\mathrm{d}\xi_2}\alpha(\xi_2) \right| \leq c_3, \left| \frac{\mathrm{d}^2}{\mathrm{d}\xi_2^2}\alpha(\xi_2) \right| \leq c_4, \int_0^1 \alpha(\xi_2)\mathrm{d}\xi_2 = V \right\},$$

where c_1, c_2, c_3 and V are given positive numbers such that $U_{ad} \neq \emptyset$. Consider a family of admissible domains $\Omega(\alpha)$ with variable right "vertical" part of the boundary:

$$\Omega_\alpha = \{(\xi_1, \xi_2) \in \mathbb{R}^2 \mid 0 < \xi_1 < \alpha(\xi_2) \text{ for all } \xi_2 \in (0,1)\}.$$

In the shape optimization problem we allow the stress to reach the elastic limit, but we require the "overall" stress to be as low as possible. Here is the problem:

$$\begin{aligned} &\text{minimize} && \mathcal{J}(\sigma(\alpha)) := \textstyle\int_{\Omega_\alpha} \Upsilon^2(\sigma(\alpha))\mathrm{d}\xi \\ &\text{subject to} && \\ & && \sigma(\alpha) \text{ solves (10.15) with } \Omega := \Omega_\alpha \\ & && \alpha \in U_{ad}. \end{aligned} \tag{10.16}$$

10.2.2 Discretization

In addition to what we have introduced in Section 10.1.2, we only need a discrete version of the set of admissible design variables U_{ad}, the set of plastically admissible stresses $\mathcal{P}(\Omega)$ (10.14) and of the objective functional $\mathcal{J}(\sigma(\alpha))$ from (10.16).

Recall that $a_1 < a_2 < \ldots < a_n$ is a uniform partition of $[0,1]$ and that α (one-dimensional real function) is approximated by a continuous piece-wise linear function α_h such that $\alpha_h(a_i) = \alpha(a_i)$. The discrete counterpart of α_h is a vector $\boldsymbol{\alpha} \in \mathbb{R}^n$ of values of this function at points $a_1, \ldots, a_n$, i.e., a vector of ξ_1-coordinates of boundary nodes $(\alpha_h(a_i), a_i)$. The set U_{ad} is represented by the finite-dimensional set

$$\mathbf{U_{ad}} := \{\boldsymbol{\alpha} \in \mathbb{R}^n \mid \boldsymbol{\alpha}^i = \alpha(a_i), \alpha \in U_{ad},\ \alpha \text{ linear on } [a_{i-1}, a_i], i = 2, 3, \ldots, n\}.$$

The discrete set of plastically admissible stresses is defined as

$$\mathbb{P} := \{\boldsymbol{\sigma} \in \mathbb{R}^{3M} \mid \Upsilon^k(\boldsymbol{\sigma}) \le 0,\ i = 1, 2, \ldots, M\},$$

where

$$\Upsilon^k(\boldsymbol{\sigma}) := \Upsilon(\boldsymbol{\sigma}_{11,k}, \boldsymbol{\sigma}_{22,k}, \boldsymbol{\sigma}_{12,k}) - \sigma_E;$$

note that $\mathbb{P}$ does not depend on $\boldsymbol{\alpha}$. The discrete state problem of elasto-plasticity is a convex programming problem with quadratic objective and nonlinear constraints:

$$\begin{aligned} &\text{minimize} && \frac{1}{2}(\boldsymbol{F}(\boldsymbol{\alpha})\boldsymbol{\sigma})^T\boldsymbol{\sigma} \\ &\text{subject to} && \\ & && \boldsymbol{\sigma} \in \mathbb{E}(\boldsymbol{\alpha}) \cap \mathbb{P}. \end{aligned} \tag{10.17}$$

Denote again by $K_k, k = 1, 2, \ldots, M$, the elements of the triangulation. Then the discrete counterpart to the shape optimization problem (10.16) reads as

$$\begin{aligned} &\text{minimize} && \boldsymbol{J}(\boldsymbol{\alpha}, \boldsymbol{\sigma}) := \sum_{k=1}^{M} (\text{meas } (K_k)\Upsilon^k(\boldsymbol{\sigma}))^2 \\ &\text{subject to} && \\ & && \boldsymbol{\sigma} \text{ solves (10.17)} \\ & && \boldsymbol{u} \in \mathbf{U_{ad}}. \end{aligned} \tag{10.18}$$

10.2.3 Numerical method

Problem (10.18) is MPEC of type (7.1). In order to apply the numerical method from Section 7.2, we have to verify the assumptions required by this method and show how to compute a subgradient. The equilibrium problem (10.17) is a convex program that can be written as a generalized equation of type (5.24):

$$0 \in \begin{bmatrix} \boldsymbol{F}(\boldsymbol{\alpha})\boldsymbol{\sigma} + \boldsymbol{\mu}\boldsymbol{A}(\boldsymbol{\alpha}) + \sum_{i=1}^{M} \boldsymbol{\lambda}^i \nabla \Upsilon^i(\boldsymbol{\sigma}) \\ \boldsymbol{A}(\boldsymbol{\alpha})\boldsymbol{\sigma} - \boldsymbol{f}(\boldsymbol{\alpha}) \\ -\Upsilon^1(\boldsymbol{\sigma}) \\ \vdots \\ -\Upsilon^M(\boldsymbol{\sigma}) \end{bmatrix} + N_{\mathbb{R}^{3M}\times\mathbb{R}^{2m}\times\mathbb{R}^M_+}(\boldsymbol{\sigma}, \boldsymbol{\mu}, \boldsymbol{\lambda}). \tag{10.19}$$

We have to show that with some $\tilde{A} \supset \mathbf{U_{ad}}$

(A1) $\boldsymbol{J}$ is continuously differentiable on $\tilde{A} \times \mathbb{R}^{3M}$;

(A2) for all $\boldsymbol{\alpha} \in \tilde{A}$ the GE (10.19) has a unique solution $S(\boldsymbol{\alpha})$;

(A3) the GE (10.19) is strongly regular at all points $(\boldsymbol{\alpha}, \boldsymbol{\sigma})$ with $\boldsymbol{\alpha} \in \tilde{A}, \boldsymbol{\sigma} = S_1(\boldsymbol{\alpha})$;

cf. assumptions (A1)–(A3) from Section 7.2. The validity of assumption (A1) directly comes from the definition of $\boldsymbol{J}$ and from the construction of the finite element mesh. Due to the definition of U_{ad}, the boundary $\partial\Omega_\alpha$ is uniformly Lipschitz on U_{ad}, hence the flexibility matrix $\boldsymbol{F}(\boldsymbol{\alpha})$ is positive definite on $\mathbf{U_{ad}}$. The functions $\Upsilon^i(\boldsymbol{\sigma})$ are convex and their gradients are linearly independent (the support of each function is one finite element). Then, according to Theorem 4.4(ii), Proposition 4.5(ii) and Theorem 4.8, the GE (10.19) has a unique solution and (A2) is satisfied. The validity of assumption (A3) results from Theorem 5.8: indeed, for all $\boldsymbol{\alpha} \in \mathbf{U_{ad}}$ the matrix $\boldsymbol{F}(\boldsymbol{\alpha})$ is positive definite, thus the GE (10.19) is strongly regular at all $(\boldsymbol{\alpha}, \boldsymbol{\sigma}, \boldsymbol{\mu}, \boldsymbol{\lambda})$. The above assumptions also guarantee weak semismoothness of the function $\Theta(\boldsymbol{\alpha}) := \boldsymbol{J}(\boldsymbol{\alpha}, S_1(\boldsymbol{\alpha}))$, where S_1 is the first component of the solution to (10.19) (Proposition 7.9).

The second task is to compute a subgradient from $\partial\Theta(\boldsymbol{\alpha})$. For given $\boldsymbol{\alpha}$ and the associated solution $\boldsymbol{\sigma}$ of (10.17), we have to solve an adjoint quadratic program (cf. Corollary 7.5)

$$\begin{aligned} &\text{minimize} \quad \tfrac{1}{2}\langle \boldsymbol{p}, \boldsymbol{Q}(\boldsymbol{\alpha})\boldsymbol{p}\rangle - \langle \nabla_{\boldsymbol{\sigma}} \boldsymbol{J}(\boldsymbol{\alpha}, \boldsymbol{\sigma}), \boldsymbol{p}\rangle \\ &\text{subject to} \\ &\qquad \boldsymbol{A}(\boldsymbol{\alpha})\boldsymbol{p} = 0 \\ &\qquad \langle \nabla \Upsilon^j(\boldsymbol{\sigma}), \boldsymbol{p}\rangle = 0, \quad j \in I^+(\boldsymbol{\alpha}, \boldsymbol{\sigma}) \cup M_i(\boldsymbol{\alpha}, \boldsymbol{\sigma}) \end{aligned}$$

with

$$\boldsymbol{Q}(\boldsymbol{\alpha}) = \boldsymbol{F}(\boldsymbol{\alpha}) + \sum_{i=1}^{M} \boldsymbol{\lambda}^i \nabla^2 \Upsilon^i(\boldsymbol{\sigma})$$

and

$$\begin{aligned} I(\boldsymbol{\alpha}, \boldsymbol{\sigma}) &= \{i \in \{1, 2, \dots, M\} \mid \Upsilon^i(\boldsymbol{\sigma}) = 0\} \\ I^+(\boldsymbol{\alpha}, \boldsymbol{\sigma}) &= \{i \in I(\boldsymbol{\alpha}, \boldsymbol{\sigma}) \mid \boldsymbol{\lambda}^i > 0\} \\ I^0(\boldsymbol{\alpha}, \boldsymbol{\sigma}) &= I(\boldsymbol{\alpha}, \boldsymbol{\sigma}) \setminus I^+(\boldsymbol{\alpha}, \boldsymbol{\sigma}). \end{aligned}$$

$M_i(\boldsymbol{\alpha}, \boldsymbol{\sigma})$ is a suitably chosen subset of $I^0(\boldsymbol{\alpha}, \boldsymbol{\sigma})$. The solution of the adjoint problem is a triple $(\boldsymbol{p}, \boldsymbol{q}, \boldsymbol{r})$; here $\boldsymbol{q}$ and $\boldsymbol{r}$, respectively, are components of the KKT vector associated with the first and second set of the equality constraints. The subgradient associated with M_i can be computed as

$$\nabla_{\boldsymbol{\alpha}} \boldsymbol{J}(\boldsymbol{\alpha}, \boldsymbol{\sigma}) - \left[\mathcal{J}_{\boldsymbol{\alpha}}(\boldsymbol{F}(\boldsymbol{\alpha})\boldsymbol{\sigma}) + \sum_{i=1}^{2m} \mu^i \mathcal{J} \boldsymbol{A}^i(\boldsymbol{\alpha}) \right]^T \boldsymbol{p} - [\mathcal{J}_{\boldsymbol{\alpha}}(\boldsymbol{A}(\boldsymbol{\alpha})\boldsymbol{\sigma}) - \mathcal{J} \boldsymbol{f}(\boldsymbol{\alpha})]^T \boldsymbol{q};$$

cf. Proposition 7.14.

10.2.4 Examples

Example 10.1 Consider the optimum design problem of elastic-perfectly plastic body with the initial domain $\Omega(\boldsymbol{\alpha}_0)$ shown in Figure 10.7. Let $\Gamma_u = \{(\xi_1, \xi_2) \mid 0 \le \xi_1 \le 1.2, \xi_2 = 0\}$

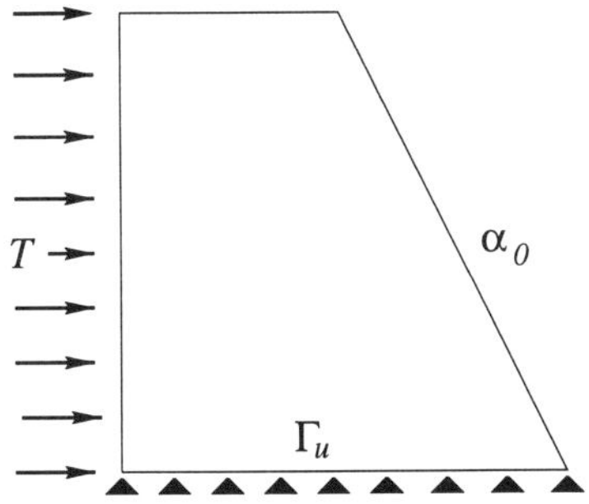

Figure 10.7. Initial data for Example 10.1

be the part of the boundary $\partial\Omega(\boldsymbol{\alpha}_0)$ with prescribed zero displacements, and let a constant surface traction $T = (1, 0)$ apply on the left vertical part of $\partial\Omega(\boldsymbol{\alpha}_0)$, that is, on the segment $\{(\xi_1, \xi_2) \mid \xi_1 = 0, 0 \le \xi_2 \le 1\}$. The surface traction on the rest of $\partial\Omega(\boldsymbol{\alpha}_0)$ is set to zero, as well as the body force F. The domain is discretized by triangles defined by 11×11 grid of nodes. The set of admissible controls U_{ad} is characterized by numbers

$$c_1 = 0.4, \ c_2 = 1.6, \ c_3 = 5.0, \ c_4 = 4.0, \ V = 1.$$

Thus we have 600 state variables (200 elements $\times$ 3 unknowns per element) and 11 design variables (ξ_1–coordinates of principal moving nodes). The discretized (nonlinear) state problems (10.17) were solved by the SQP code NLPQLD due to Schittkowski, 1986, while the nonsmooth optimization problems (10.18) by the code BT; in particular, by the nonconvex version BTNCLC which can handle linear constraints (see Schramm and Zowe, 1991). Figure 10.8(a) shows the initial design and Figure 10.9(a) the optimal design obtained after 106 iterations of BTNCLC. The initial value of the objective functional was $\boldsymbol{J}_{ini} = 2.4486$ and the optimal value was $\boldsymbol{J}_{opt} = 2.2252$ (i.e., 91% of $\boldsymbol{J}_{ini}$). In both figures, the value of the plasticity function $\Upsilon(\boldsymbol{\sigma})$ on the elements is depicted by intensity of grey. Full black means that the element lies in the plastic region. We realize that this region dramatically shrank after optimization, from 17 elements in the initial design to only 5 elements in the optimal one! For better orientation, these elements are separately plotted in Figures 10.8(b) and 10.9(b). △

(a) (b)

Figure 10.8. Initial design for Example 10.1

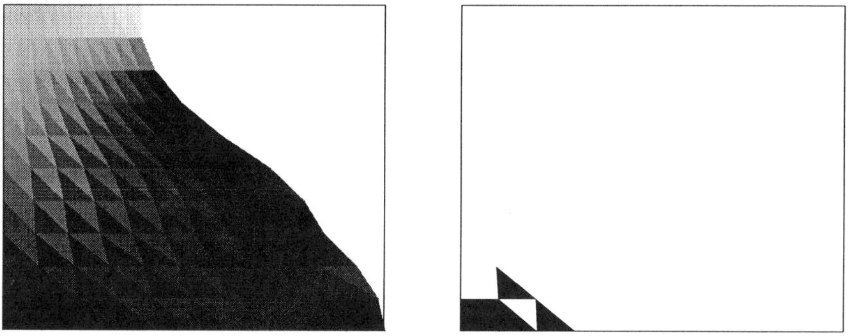

(a) (b)

Figure 10.9. Optimal design for Example 10.1

10.3 DESIGN OF MASONRY STRUCTURES

We shall investigate masonry-like material that behaves elastically under pressure forces, but is extremely weak if we try to pull it apart. In general, quite a small tension is able to produce a fracture: the material breaks up and shows lines of fracture. However, the appearance of fracture is not necessarily destructive; in most cases it is compatible with global (weak) equilibrium of the structure. The mathematical description of the equilibrium of such structures was given in Giaquinta and Giusti, 1985. Our goal in this section is to optimize the shape of these structures. Shape optimization of masonry structures was also studied in Hlaváček and Křížek, 1992; the authors, however, use a different mathematical model of describing such structures. Here we follow the "variational inequality approach" of Giaquinta and Giusti, 1985.

Let us first derive the state problem by modifying the dual problem of elasticity (10.12); for the notation, cf. Section 10.1. The masonry-like materials can be characterized in different ways. The most restrictive one is to require that the stress tensor $\sigma(\xi), \xi \in \Omega$, must not have positive eigenvalues (Giaquinta and Giusti, 1985), i.e., that the principal stresses are nonpositive. We consider, however, a simpler situation which can be used, e.g., in modelling of supporting piers in cathedrals: we require that the "vertical" stress component is nonpositive, i.e., $\sigma_{22} \leq 0$ a.e. in Ω. In piers, typically, the external force coming from the weight of the supported roof causes a bending which is compensated by the weight of the pier itself. An example of such a pier can be found in the famous cathedral in Chartres (Figure 10.10).

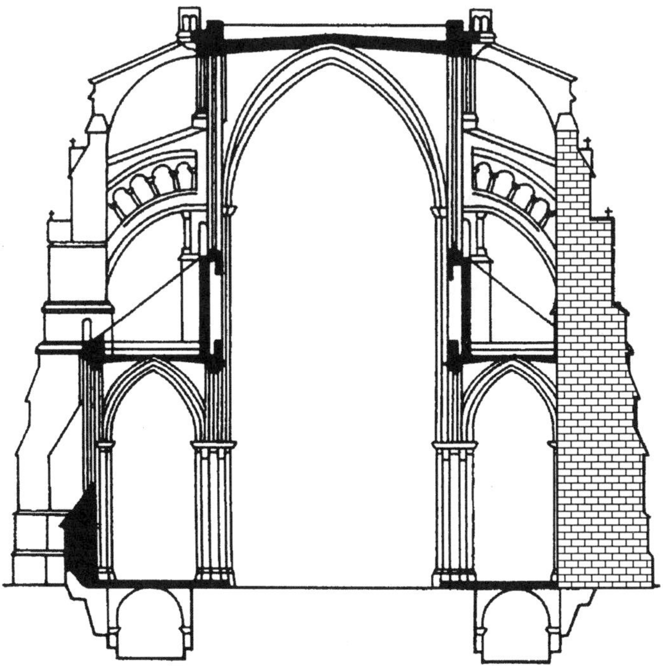

Figure 10.10. Cathedral in Chartres

Our aim is to find the lightest pier still capable to support the roof. For convenience, let us introduce the set of *admissible stresses* as

$$\mathcal{M}(\Omega) = \{\sigma \in S(\Omega) \mid \sigma_{22} \leq 0 \text{ a.e. in } \Omega\}. \qquad (10.20)$$

Then the state *problem of masonry structures* can be formulated as a minimization problem

$$\begin{array}{ll} \text{minimize} & \frac{1}{2}\Psi(\sigma) \\ \text{subject to} & \\ & \sigma \in \mathcal{E}(\Omega) \cap \mathcal{M}(\Omega); \end{array} \qquad (10.21)$$

(note the analogy with the elastic-perfectly plastic problem (10.15)). Recall that $\Psi(\sigma)$ is the complementary energy (10.11) and $\mathcal{E}(\Omega)$ the set of stresses satisfying the weak equilibrium conditions (10.10). It can be shown that, under some additional conditions on the size of external forces T, problem (10.21) has a unique solution.

10.3.1 Problem formulation

In a familiar way, we introduce the set of *admissible design variables*

$$U_{ad} = \Big\{ \alpha \in C^{0,1}([0,1]) \mid 0 < c_1 \leq \alpha \leq c_2, \\ \left| \frac{\mathrm{d}}{\mathrm{d}\xi_2} \alpha(\xi_2) \right| \leq c_3, \left| \frac{\mathrm{d}^2}{\mathrm{d}\xi_2^2} \alpha(\xi_2) \right| \leq c_4 \Big\},$$

where c_1, c_2, c_3 are given positive numbers such that $U_{ad} \neq \emptyset$. Consider a family of admissible domains $\Omega(\alpha)$ with variable right "vertical" part of the boundary:

$$\Omega_\alpha = \{(\xi_1, \xi_2) \in \mathbb{R}^2 \mid 0 < \xi_1 < \alpha(\xi_2) \text{ for all } \xi_2 \in (0,1)\}.$$

The goal of the optimum design problem is to minimize the structure volume, denoted by meas $\Omega(\alpha)$. To prevent collapse, we require that the inequality constraint in the definition of $\mathcal{M}(\Omega)$ are strongly active only in a given subset $\Omega_0(\alpha) \subset \Omega_\alpha$. In other words, denoting by $\lambda = \lambda(\alpha)$ the KKT vector corresponding to this constraint, we require $\lambda = 0$ in $\Omega_\alpha \setminus \Omega_0(\alpha)$ (cf. Figure 10.11). Of course, in this way one cannot prevent some constraints in $\Omega_\alpha \setminus \Omega_0(\alpha)$

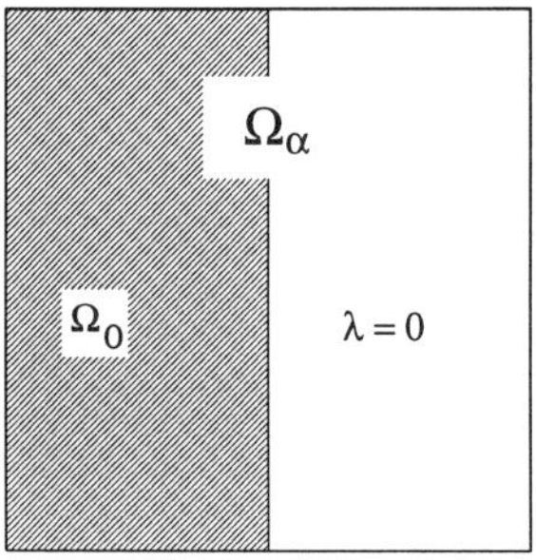

Figure 10.11. Example of partition of Ω_α into $\Omega_0(\alpha)$ and $\Omega_\alpha \setminus \Omega_0(\alpha)$

to be active provided the corresponding components of the KKT vector are zero. In the numerical examples this typically occurred very close to the boundary of $\Omega_0(\alpha)$ (this was, in fact, a sign of optimality).

For the purpose of this section, we will say that (σ, λ) *solves the state problem* (10.21), if σ is the minimizer for (10.21) and λ the part of the KKT vector corresponding to the inequality constraints in $\mathcal{M}(\Omega)$. The optimum design problem reads as

$$\begin{aligned} &\text{minimize} && \mathcal{J}(\sigma(\alpha), \lambda(\alpha)) := \text{meas}\, \Omega_\alpha + r \int_{\Omega_\alpha \setminus \Omega_0(\alpha)} (\lambda(\alpha)) \mathrm{d}\xi \\ &\text{subject to} && \\ & && (\sigma(\alpha), \lambda(\alpha)) \text{ solves (10.21) with } \Omega := \Omega_\alpha \\ & && \alpha \in U_{ad}. \end{aligned} \tag{10.22}$$

Here $r > 0$ is a penalty parameter.

10.3.2 Discretization

The discretization process is analogous to that for elastic-perfectly plastic problem; cf. Sections 10.1.2 and 10.2.2.

What is specific to the masonry structure problem is the definition of the set of admissible stress fields $\mathcal{M}(\Omega_\alpha)$ (10.20). The discrete version of this set is

$$\mathbb{M}(\boldsymbol{\alpha}) = \mathbb{M} := \{\boldsymbol{\sigma} \in \mathbb{R}^{3M} \mid \boldsymbol{\sigma}_{22,k} \leq 0,\ k = 1, 2, \ldots, M\};$$

the reader remembers that $\boldsymbol{\sigma}_{22,k}$ is the 22−component of the discrete stress tensor in the kth element.

The discrete state problem for masonry materials is a convex quadratic program:

$$\begin{aligned} &\text{minimize} \quad \frac{1}{2}(\boldsymbol{F}(\boldsymbol{\alpha})\boldsymbol{\sigma})^T\boldsymbol{\sigma} \\ &\text{subject to} \\ &\qquad\qquad \boldsymbol{\sigma} \in \mathbb{E}(\boldsymbol{\alpha}) \cap \mathbb{M}. \end{aligned} \tag{10.23}$$

In analogy to the continuum case, we say that $(\boldsymbol{\sigma}, \boldsymbol{\lambda})$ solves the state problem (10.23) if $\boldsymbol{\sigma} \in \mathbb{R}^{3M}$ is a minimizer for (10.23) and $\boldsymbol{\lambda} \in \mathbb{R}^M$ is the KKT vector corresponding to inequalities in $\mathbb{M}$.

Denote by $\mathcal{D}_0$ the set of indices of elements lying in $\Omega(u) \setminus \Omega_0(u)$ (i.e., in the region where λ should vanish). The discrete counterpart to the shape optimization problem (10.22) becomes

$$\begin{aligned} &\text{minimize} \quad \boldsymbol{J}(\boldsymbol{\alpha}, \boldsymbol{\sigma}, \boldsymbol{\lambda}) := \text{meas}\, \Omega_h(\boldsymbol{\alpha}) + rh^2 \sum_{i \in \mathcal{D}_0} \boldsymbol{\lambda}^i \\ &\text{subject to} \\ &\qquad\qquad (\boldsymbol{\sigma}, \boldsymbol{\lambda}) \text{ solves (10.23)} \\ &\qquad\qquad \boldsymbol{u} \in \mathbf{U_{ad}}. \end{aligned} \tag{10.24}$$

Recall that h is a discretization parameter and r a penalty parameter.

10.3.3 Numerical method

Once again, (10.24) is MPEC of type (7.1). The assumptions for using the numerical method from Section 7.2 are almost identical with those for the elastic-perfectly plastic problem. Again, the equilibrium problem (10.23) is a quadratic program that can be written as a generalized equation of type (5.24):

$$0 \in \begin{bmatrix} \boldsymbol{F}(\boldsymbol{\alpha})\boldsymbol{\sigma} + \boldsymbol{\mu}\boldsymbol{A}(\boldsymbol{\alpha}) + \boldsymbol{\lambda} \\ \boldsymbol{A}(\boldsymbol{\alpha})\boldsymbol{\sigma} - \boldsymbol{f}(\boldsymbol{\alpha}) \\ -\boldsymbol{\sigma}_{22,1} \\ \vdots \\ -\boldsymbol{\sigma}_{22,M} \end{bmatrix} + N_{\mathbb{R}^{3M} \times \mathbb{R}^{2m} \times \mathbb{R}^M_+}(\boldsymbol{\sigma}, \boldsymbol{\mu}, \boldsymbol{\lambda}). \tag{10.25}$$

We have to show that with some $\tilde{A} \supset \mathbf{U_{ad}}$

(A1) $\boldsymbol{J}$ is continuously differentiable on $\tilde{A} \times \mathbb{R}^{3M}$;

(A2) for all $\alpha \in \tilde{A}$ the GE (10.25) has a unique solution $S(\alpha)$;

(A3) the GE (10.25) is strongly regular at all points $(\alpha, \sigma, \mu, \lambda)$ with $\alpha \in \tilde{A}, \sigma = S_1(\alpha), \mu = S_2(\alpha), \lambda = S_3(\alpha)$;

cf. assumptions (A1)–(A3) from Section 6.2. The verification of these assumptions is just the same as in Section 10.2.3 dealing with elastic-perfectly plastic materials.

The second task is to compute a subgradient from $\partial\Theta(\alpha)$. First, for given control α, we have to solve the adjoint problem, which in our situation amounts to solving the quadratic program (cf. Theorem 7.3; note that the objective J depends on the KKT vector λ)

$$
\begin{array}{ll}
\text{minimize} & \frac{1}{2}\langle p, F(\alpha)p\rangle \\
\text{subject to} & \\
& A(\alpha)p = 0 \\
& p_{3j-1} = \left\langle \begin{array}{ll} rh^2 & \text{if } j \in \mathcal{D}_0 \\ 0 & \text{if } j \notin \mathcal{D}_0 \end{array} \right. \qquad j \in I^+(\alpha,\sigma) \cup M_i(\alpha,\sigma)
\end{array}
$$

with

$$
\begin{aligned}
I(\alpha,\sigma) &= \{i \in \{1,2,\dots,m\} \mid \sigma^i = 0\} \\
I^+(\alpha,\sigma) &= \{i \in I(\alpha,\sigma) \mid \lambda^i > 0\} \\
I^0(\alpha,\sigma) &= I(\alpha,\sigma) \setminus I^+(\alpha,\sigma).
\end{aligned}
$$

$M_i(\alpha,\sigma)$ is a suitably chosen subset of $I^0(\alpha,\sigma)$. The solution of the adjoint problem is a triple (p, q, r); again, q and r, respectively, form the KKT vector associated with the first and second set of the equality constraints. The subgradient associated with M_i is computed as

$$
\nabla_\alpha J(\alpha,\sigma,\lambda) - \left[\mathcal{J}_\alpha(F(\alpha)\sigma) + \sum_{i=1}^{2m} \mu^i \mathcal{J} A^i(\alpha)\right]^T p - [\mathcal{J}_\alpha(A(\alpha)\sigma) - \mathcal{J} f(\alpha)]^T q;
$$

cf. Proposition 7.14.

10.3.4 Examples

Example 10.2 Consider an initial domain $\Omega(\alpha_0)$ as depicted in Figure 10.12. Let $\Gamma_u = \{(\xi_1,\xi_2) \mid 0 \le \xi_1 \le \overline{u}, \xi_2 = 0\}$ be the part of the boundary $\partial\Omega$ with given zero displacements, and let the constant surface traction $T = (5,0)$ act on the left vertical part of $\partial\Omega$, i.e., on the segment $\{(\xi_1,\xi_2) \mid \xi_1 = 0, 0 \le \xi_2 \le 1\}$. The surface traction on the rest of $\partial\Omega$ is prescribed zero and the internal force $F = (0,-10)$ corresponds to the body weight. As penalty parameter $r = 1$ is chosen.. The domain is discretized by triangles defined by 9×9 grid of nodes. The set of admissible controls U_{ad} is characterized by numbers

$$
c_1 = 0.6,\ c_2 = 3.0,\ c_3 = 5.0,\ c_4 = 4.0.
$$

Thus we have 384 state variables (128 elements $\times$ 3 unknowns per element) and 9 design variables (ξ_1–coordinates of principal moving nodes). The index set $\mathcal{D}_0$ associated with the set Ω_0 contains indices of elements lying in the left half of Ω. The discretized state (nonlinear programming) problems (10.23) were solved by the SQP code NLPQL by Schittkowski, 1986, while the nonsmooth optimization problems (10.24) by the code BT, in particular, by

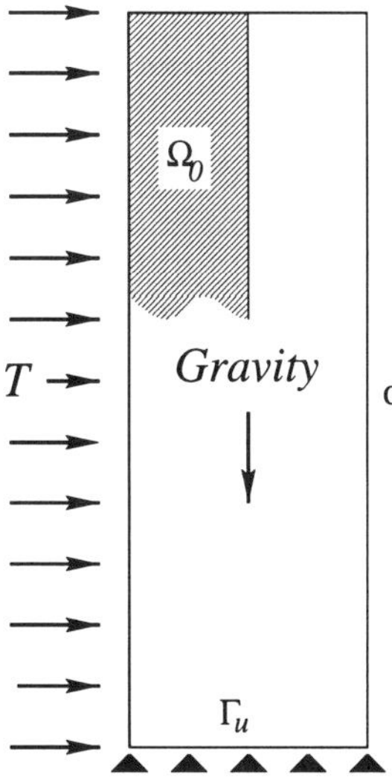

Figure 10.12. Initial data for Example 10.2

its nonconvex version BTNCLC which can handle linear constraints (Schramm and Zowe, 1991). Figure 10.13 shows the optimum shape obtained after 123 BT iterations together with the values of the "vertical" components $\boldsymbol{\sigma}^{3i-1}$ of the state vector (figure (a)) and the corresponding components of the KKT vector $\boldsymbol{\lambda}^i$ (figure (b)). The initial objective function value was $\boldsymbol{J}_{ini} = 174.5166$ and the final one $\boldsymbol{J}_{opt} = 0.9978$. △

Example 10.3 All data are the same as in the previous example, apart from the surface traction T, which now applies only to the upper half of the left vertical part of $\partial\Omega$. The results are shown in Figure 10.14. This time we needed 144 BT iterations. The initial objective function value was $\boldsymbol{J}_{ini} = 6.1524$ and the final one $\boldsymbol{J}_{opt} = 0.8739$. In this example, the resulting shape corresponds very well to the real-world structure shown in Figure 10.10. △

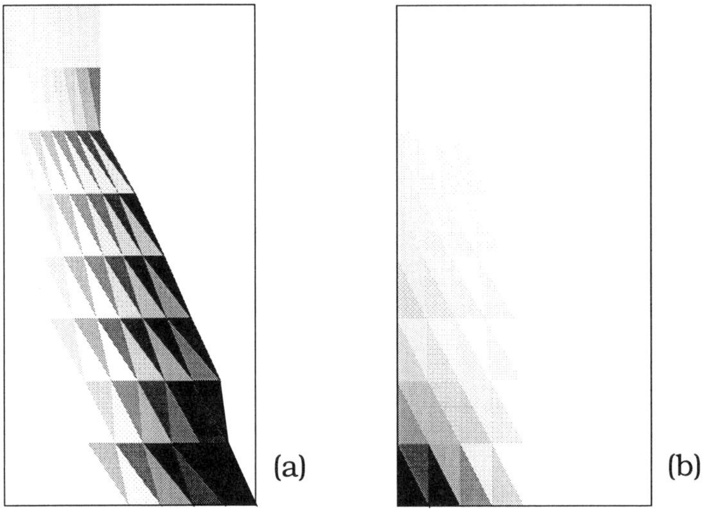

Figure 10.13. Optimal shape, stresses (a) and KKT vector (b) for Example 10.2

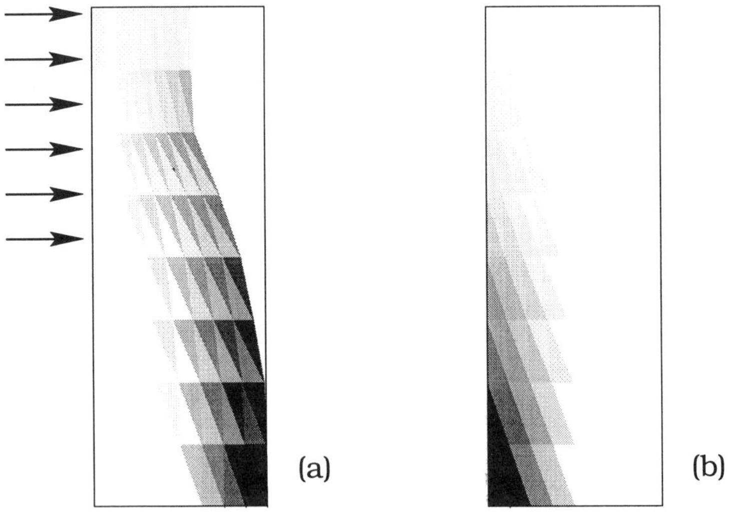

Figure 10.14. Optimal shape, stresses (a) and KKT vector (b) for Example 10.3

11 CONTACT PROBLEM WITH COULOMB FRICTION

In the previous two chapters we have introduced problems with obstacles and the problem of linear elasticity. The goal of this chapter is to combine those two and to introduce probably the most important "nonsmooth" problem of mechanics: the problem of an elastic body in contact with a rigid obstacle. A simple way how to define this problem is to restrict the normal displacement of certain boundary points in the elasticity problem by means of the unilateral contact constraints known from Chapter 9. This leads to a formulation known as contact problem without friction; in most cases, this formulation does not fully reflect the physical reality. To achieve this, one has to take into account the friction between the body and the obstacle—this friction restricts the tangential component of the displacement on the "contact boundary". There are several models of contact problems with friction (cf., e.g., Klarbring, 1986; Lemaitre and Chaboche, 1994). The most realistic one is probably the model of Coulomb friction. In the following sections we will introduce this model, derive its reciprocal (dual) formulation which leads to a quasi-variational inequality and apply the nonsmooth Newton's method of Chapter 3 to the numerical solution of a suitable discrete approximation. We further show that the discretized problem can be formulated as a linear complementarity problem which, again, can be solved by the nonsmooth Newton's method. Finally, we formulate a design problem; the control (design) variable is the coefficient of friction (a property of the material) and the goal to maximize the tangential adhesion between the body and the obstacle. This design problem is an MPEC with an LCP as equilibrium constraint.

Usually, the solution of the contact problem with Coulomb friction is defined as a fixed point of certain mapping; the evaluation of this mapping requires to solve the variational inequality corresponding to the contact problem with given friction. The existence of this fixed point was proved by Nečas et al., 1980 while the finite element approximation and solution techniques were analyzed by Hlaváček et al., 1988. The reciprocal formulation of the problem was studied by Haslinger and Panagiotopoulos, 1984 (cf. also Hlaváček et al., 1988). More recently, the existence of solution to the problem with Coulomb friction has been proved by Eck and Jarušek, 1997a; Eck and Jarušek, 1997b. The authors use penalty technique to obtain existence results for a fairly large class of problems.

11.1 PROBLEM FORMULATION

Let $\Omega \subset \mathbb{R}^2$ be a bounded domain with a Lipschitz boundary $\partial\Omega$. Let $\Gamma_g, \Gamma_u, \Gamma_c \subset \partial\Omega$ be mutually disjoint nonempty sets open in $\partial\Omega$ such that $\mathrm{cl}\Gamma_g \cup \mathrm{cl}\Gamma_u \cup \mathrm{cl}\Gamma_c = \partial\Omega$. Part Γ_g is associated with Neumann boundary conditions (the external force applies here), part Γ_u with homogeneous Dirichlet boundary conditions (the body is fixed here) and the *contact* part Γ_c with unilateral boundary conditions (here the body is in contact with a rigid obstacle); cf. Figure 11.1. We assume, for simplicity, that the contact part Γ_c is a straight-line segment.

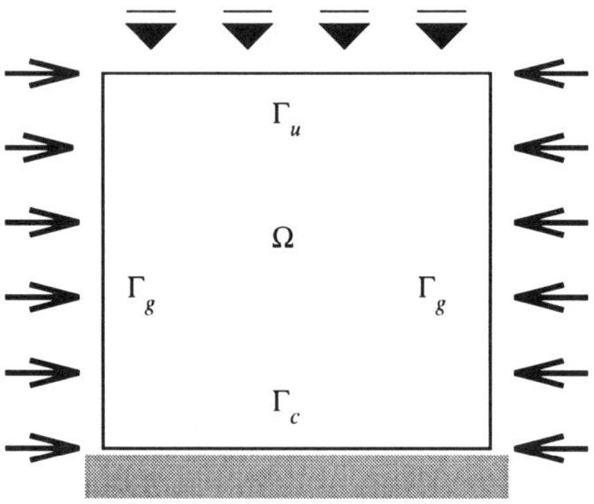

Figure 11.1. Domain Ω and parts of its boundary

Let n denote the outer normal to $\partial\Omega$ (at points where it exists). For a vector function $w : \Gamma_c \to \mathbb{R}^2$ we denote by $w_n = \sum_{i=1}^2 w_i n_i$ its normal and by $w_t = \sum_{i=1}^2 w_i t_i$, $t = (t_1, t_2) = (-n_2, n_1)$, its tangential component.

To define a solution of the contact problem with Coulomb friction, we first have to consider a problem with *given* friction. A function $u[\Omega \to \mathbb{R}^2]$ representing the displacement of an elastic body Ω *solves the contact problem with given friction* if it solves the variational inequality (cf. Hlaváček et al., 1988):

$$\left.\begin{aligned} &\text{Find } u \in K := \{v \in \mathcal{U}(\Omega) \mid \sum_{i=1}^{2} v_i n_i \le 0 \text{ on } \Gamma_c\} \text{ such that} \\ &\mathbf{a}(u, v - u) + \int_{\Gamma_c} \gamma(|v_t| - |u_t|)\mathrm{d}\Gamma_c \ge \int_{\Gamma_g} \langle F, v - u\rangle \mathrm{d}\Gamma_g \quad \text{for all } v \in K, \end{aligned}\right\} \tag{11.1}$$

where $\mathcal{U}(\Omega) := \{v \in (H^1(\Omega))^2 \mid v = 0 \text{ on } \Gamma_u\}$ and K is the closed convex set of admissible displacements. The bilinear form $\mathbf{a}$ is defined as

$$\mathbf{a}(u, v) = \int_\Omega \langle \mathcal{H}e(u), e(v)\rangle \mathrm{d}\xi,$$

where $e(w)$ is the small-strain tensor and $\mathcal{H}$ a bounded, symmetric and elliptic mapping expressing the Hooke's law (cf. (10.4),(10.5)). Function $F \in (H^{-\frac{1}{2}}(\Gamma_g))^2$ is an external force on Γ_g and γ a given *friction function*, $\gamma \geq 0$ on Γ_c. (We ask the reader to recall the notions of displacement vector, strain and stress tensor and Hooke's law introduced in Chapter 10, Section 10.1.)

It can be shown (cf. Hlaváček et al., 1988) that the solution of (11.1) satisfies the contact conditions with given friction:

$$u_n \leq 0, \quad T_n \leq 0, \quad u_n T_n = 0 \qquad \text{on } \Gamma_c \tag{11.2}$$

$$|T_t| \leq \gamma, \quad (\gamma - |T_t|)u_t = 0, \quad u_t T_t \leq 0 \qquad \text{on } \Gamma_c, \tag{11.3}$$

where T denotes the boundary stress vector on Γ_c, T_n its normal and T_t the tangential component. Relations (11.2) are the standard unilateral contact conditions well-known from Chapter 9. Conditions (11.3) say: if the tangential component of the stress vector $|T_t|$ is smaller than γ, then the body does not move along Γ_c; if $|T_t|$ reaches the value of γ, then the body starts to slide along the obstacle.

In the model of *Coulomb friction*, the function γ is not a priori given; in (11.3) it is replaced by $\Phi|T_n|$, where Φ is a *coefficient of friction* characterizing the physical properties of the contact surface. That means, the higher is the normal force on the contact boundary, the higher may be the tangential one before the body starts to slide along the obstacle. The solution of the *contact problem with Coulomb friction* is defined as a fixed point of the mapping

$$\mathcal{F} : \gamma \mapsto \Phi|T_n(u(\gamma))|, \tag{11.4}$$

where $u(\gamma)$ is a solution to (11.1). The existence of a fixed point of $\mathcal{F}$ for small values of the friction coefficient Φ was proved in Nečas et al., 1980 (cf. also Jarušek, 1983), while a numerical algorithm for finding this fixed point was proposed in Haslinger and Panagiotopoulos, 1984. In this algorithm, basically a method of successive approximations, in order to compute a new value of γ, one has to solve (11.1) with the old $\gamma-$value to get the displacement $u(\gamma)$ and then to compute the normal force T_n associated with this displacement. The algorithm, however, might converge intolerably slowly (actually, no convergence proof has been given). Moreover, we obtain the displacements whereas we are more interested in finding the contact stresses. Additionally, this approach might cause numerical difficulties while computing the stresses from the approximated solution of (11.1)—it is well-known that such computation might be quite inaccurate if one uses irregular finite element mesh. Therefore we turn our attention to the *reciprocal* variational formulation of the problem.

11.2 NUMERICAL SOLUTION

11.2.1 Reciprocal formulation of the problem

In this section we follow Haslinger and Panagiotopoulos, 1984. We start with reformulation of the problem with *given* friction. Let

$$\begin{aligned} \Lambda_1 &= \left\{ \mu_1 \in H^{-\frac{1}{2}}(\Gamma_c) \mid \langle\!\langle \mu_1, v_n \rangle\!\rangle \geq 0 \text{ for all } v \in K \right\} \\ \Lambda_2 &= \{ \mu_2 \in L_2(\Gamma_c) \mid |\mu_2| \leq \gamma \text{ a.e. on } \Gamma_c \} \end{aligned} \tag{11.5}$$

where $\langle\!\langle \cdot, \cdot \rangle\!\rangle$ is the duality pairing between $H^{-\frac{1}{2}}(\Gamma_c)$ and $H^{\frac{1}{2}}(\Gamma_c)$. We denote by $\mathcal{G} : \mathcal{U}'(\Omega) \to \mathcal{U}(\Omega)$ the Green's operator corresponding to $\mathcal{U}(\Omega)$ and $\mathbf{a}(\cdot, \cdot)$, which assigns

$\varphi \in (H^{-\frac{1}{2}}(\Gamma_{c,g}))^2$ the unique solution of the problem

$$\text{Find } u \in \mathcal{U}(\Omega) \text{ such that } \quad \mathbf{a}(u,v) = \langle\!\langle \varphi, v \rangle\!\rangle \quad \text{for all } v \in \mathcal{U}(\Omega),$$

and by $\mathbf{b} : (H^{-\frac{1}{2}}(\Gamma_c))^2 \times (H^{-\frac{1}{2}}(\Gamma_c))^2 \to \mathbb{R}$ the bilinear form

$$\mathbf{b}(\mu, \nu) = \langle\!\langle \mu_1, \mathcal{G}(\nu_1, \nu_2)n \rangle\!\rangle + \langle\!\langle \mu_2, \mathcal{G}(\nu_1, \nu_2)t \rangle\!\rangle.$$

Let $\mathbf{g} : (H^{-\frac{1}{2}}(\Gamma_c))^2 \to \mathbb{R}$ be given by

$$\mathbf{g}(\mu) = -\langle\!\langle \mu_1, \mathcal{G}(F)n \rangle\!\rangle - \langle\!\langle \mu_2, \mathcal{G}(F)t \rangle\!\rangle.$$

The problem of finding $\lambda = (\lambda_1, \lambda_2) \in \Lambda_1 \times \Lambda_2$ such that

$$\mathbf{b}(\lambda, \mu - \lambda) - \mathbf{g}(\mu - \lambda) \geq 0 \quad \text{for all } \mu = (\mu_1, \mu_2) \in \Lambda_1 \times \Lambda_2 \tag{11.6}$$

is called *reciprocal (dual) variational formulation* of the contact problem with *given* friction. The following existence result was proved in Haslinger and Panagiotopoulos, 1984:

Theorem 11.1 *There exists a unique solution* $\lambda = (\lambda_1, \lambda_2)$ *of* (11.6). *Moreover,*

$$\lambda_1 = T_n(u), \quad \lambda_2 = T_t(u)$$

where $u \in K$ *is the solution of the primal problem* (11.1).

Let us return to the problem of Coulomb friction. For each $\mu \in (H^{-\frac{1}{2}}(\Gamma_c))^2$ we introduce a closed convex set

$$\Xi_\Phi(\mu) := \left\{ \nu = (\nu_1, \nu_2) \in (H^{-\frac{1}{2}}(\Gamma_c))^2 \mid \nu_1 \leq 0, |\nu_2| \leq \Phi|\mu_1| \right\}.$$

The contact problem with Coulomb friction can now be formulated as a quasi-variational inequality:

$$\left.\begin{array}{l} \text{Find } \lambda \in \Xi_\Phi(\lambda) \text{ such that} \\ \mathbf{b}(\lambda, \mu - \lambda) - \mathbf{g}(\mu - \lambda) \geq 0 \quad \text{for all } \mu = (\mu_1, \mu_2) \in \Xi_\Phi(\lambda). \end{array}\right\} \tag{11.7}$$

To be able to solve this problem with the tools introduced in Chapter 3, we need to approximate it. We cannot discretize $\mathbf{b}(\cdot,\cdot)$ and $\mathbf{g}(\cdot)$ directly, as the explicit form of the Green's operator is known only in special cases. Instead, we discretize the primal formulation, i.e., the bilinear form $\mathbf{a}(\cdot,\cdot)$ and the right-hand side $F(\cdot)$, by bilinear quadrilateral elements. In this way, for a given fixed discretization parameter h, we obtain a stiffness matrix $\boldsymbol{A}$ and a right-hand side vector $\boldsymbol{f}$ (cf. Chapters 9 and 10). As an approximation $\boldsymbol{G}$ of $\mathcal{G}$ we consider the inverse of $\boldsymbol{A}$. We need not to compute the full inverse; only the elements of $\boldsymbol{G}$ associated with nodes on Γ_c are needed. Therefore we can eliminate all unknowns related to the "internal" nodes of the triangulation.

Let us suppose that the unknown variables associated with nodes on Γ_c have indices $1, 2, \ldots, m$. Let m_T be the total number of unknowns and $m_I = m_T - m$ the number of "internal" unknowns. Denote by P_C and P_I, respectively, the $m_T \times m$ and $m_T \times m_I$ matrices

$$P_C = \begin{pmatrix} \mathrm{E}_m \\ 0 \end{pmatrix}, \qquad P_I = \begin{pmatrix} 0 \\ \mathrm{E}_{m_I} \end{pmatrix}.$$

These matrices define a block partitioning of $\boldsymbol{A}$:

$$\boldsymbol{A} = \begin{pmatrix} P_C^T \boldsymbol{A} P_C & P_C^T \boldsymbol{A} P_I \\ P_I^T \boldsymbol{A} P_C & P_I^T \boldsymbol{A} P_I \end{pmatrix}.$$

The reduced stiffness matrix $\boldsymbol{A}_C$ and the corresponding right-hand side vector $\boldsymbol{f}_C$ are obtained by block elimination of "internal" unknowns:

$$\begin{aligned} \boldsymbol{A}_C &= P_C^T \boldsymbol{A} P_C - P_C^T \boldsymbol{A} P_I (P_I^T \boldsymbol{A} P_I)^{-1} P_I^T \boldsymbol{A} P_C, \\ \boldsymbol{f}_C &= P_C^T \boldsymbol{f} - P_C^T \boldsymbol{A} P_I (P_I^T \boldsymbol{A} P_I)^{-1} P_I^T \boldsymbol{f}. \end{aligned}$$

Finally, we define the approximation of $\mathbf{b}(\cdot,\cdot)$ and $\mathbf{g}(\cdot)$, respectively, by means of $\langle \boldsymbol{B}\cdot,\cdot\rangle$ and $\langle \boldsymbol{g},\cdot\rangle$ with an $m \times m$ matrix and $m-$vector

$$\boldsymbol{B} := \boldsymbol{A}_C^{-1} \quad \text{and} \quad \boldsymbol{g} := \boldsymbol{A}_C^{-1} \boldsymbol{f}_C.$$

To approximate the set Ξ_Φ, we assume such numbering of contact unknowns that the tangential component of the stress vector at a contact point $i \in \{1, 2, \ldots, m/2\}$ has an odd number $2i-1$ and the normal component has an even number $2i$. The approximated version of problem (11.7) reads as

$$\left. \begin{array}{l} \text{Find } \boldsymbol{\lambda} \in \Xi_\Phi(\boldsymbol{\lambda}) := \{\boldsymbol{\mu} \in \mathbb{R}^m \mid \boldsymbol{\mu}^{2i} \le 0, |\boldsymbol{\mu}^{2i-1}| \le \Phi|\boldsymbol{\lambda}^{2i}|,\ i = 1, \ldots, m/2\} \\ \text{such that} \\ \qquad \langle \boldsymbol{B}\boldsymbol{\lambda}, \boldsymbol{\mu} - \boldsymbol{\lambda}\rangle - \langle \boldsymbol{g}, \boldsymbol{\mu} - \boldsymbol{\lambda}\rangle \ge 0 \quad \text{for all } \boldsymbol{\mu} \in \Xi_\Phi(\boldsymbol{\lambda}). \end{array} \right\} \tag{11.8}$$

This is a quasi-variational inequality of a special structure (one could call it "mixed ICP"). A suitable choice of boundary conditions guarantees that the matrix $\boldsymbol{A}$ is positive definite, cf. forthcoming examples. Then also $P_I^T \boldsymbol{A} P_I$ and $\boldsymbol{A}_C$ (the Schur complement) are positive definite (Horn and Johnson, 1985, Thm. 7.7.6). Hence $\boldsymbol{B}$ is positive definite and there exists a solution to (11.8) (cf. Kočvara and Outrata, 1995a). The problem can be written as a NSE

$$\boldsymbol{\lambda} - \mathrm{Proj}_{\Xi_\Phi(\boldsymbol{\lambda})}(\boldsymbol{\lambda} - \boldsymbol{B}\boldsymbol{\lambda} + \boldsymbol{g}) = 0 \tag{11.9}$$

to which we can apply the Newton's method from Section 3.4, due to the special structure of the projection (see the code below). The semismoothness of function in (11.9) follows from the analysis at the end of Section 6.3. To guarantee the strong BD-regularity at a solution $\overline{\boldsymbol{\lambda}}$, one has to show the nonsingularity of matrices $E - \boldsymbol{Z}$ for all $\boldsymbol{Z} \in \partial_B \mathrm{Proj}_{\Xi_\Phi(\overline{\boldsymbol{\lambda}})}(\overline{\boldsymbol{\lambda}} - \boldsymbol{B}\overline{\boldsymbol{\lambda}} + \boldsymbol{g})$. An implementation of the method coded in MATLAB is listed here.

```
function [y] = newton(B,g,nc,y,Phi)

% B   ... matrix from (11.8)
% g   ... RHS vector from (11.8)
% nc  ... number of contact unknowns
% y   ... on input the initial approximation
%         on output the solution
% Phi ... coefficient of friction

lo = zeros(nc,1); up = zeros(nc,1); h  = zeros(nc,1);

% the main loop
```

```
for icount=1:1000

   % setting upper and lower bounds

   for i=2:2:nc
      up(i) =  0; lo(i) = -10000000;
   end

   for i=1:2:nc-1
      up(i) = -Phi*y(i+1); lo(i) =  Phi*y(i+1);
   end

   hlo = y-up; hup = y-lo; heq = B*y-g;

   % finding the active constraints

   for i=1:nc
      if(heq(i) >= hup(i))
         h(i) = hup(i); ind(i) = 1;
      elseif(heq(i) <= hlo(i))
         h(i) = hlo(i); ind(i) = 2;
      else
         h(i) = heq(i); ind(i) = 3;
      end
   end

   % stopping test

   test = sqrt(h'*h)
   if(test<0.000000001), break, end

   V=0;

   % forming the Newton matrix V

   for i=1:nc
      if(ind(i)==3)
         V(i,1:nc) = B(i,:);
      elseif(ind(i)==1)
         V(i,i) = 1;
         if(rem(i,2)~=0)
            V(i,i+1) = -Phi;
         end
      elseif(ind(i)==2)
         V(i,i) = 1;
         if(rem(i,2)~=0)
            V(i,i+1) =  Phi;
         end
      end
   end

   % Newton update

   y = y - V\h;

end
```

11.2.2 LCP formulation of the problem

Here we partly follow Al-Fahed et al., 1991. Consider the block partitioning of "contact" matrices and vectors from the previous section according to the normal and tangential components, i.e.,

$$\boldsymbol{B} = \begin{pmatrix} \boldsymbol{B}_{nn} & \boldsymbol{B}_{nt} \\ \boldsymbol{B}_{tn} & \boldsymbol{B}_{tt} \end{pmatrix}, \quad \boldsymbol{v} = \begin{pmatrix} \boldsymbol{v}_n \\ \boldsymbol{v}_t \end{pmatrix}, \quad \text{etc.}$$

Let $\boldsymbol{\lambda}$ be a solution to (11.8). This problem can be written as a GE (cf. Chapter 4)

$$0 \in \boldsymbol{B}\boldsymbol{\lambda} - \boldsymbol{g} + N_{\Xi_\Phi(\boldsymbol{\lambda})}(\boldsymbol{\lambda}).$$

Using Corollary 2.25, we obtain the KKT conditions for (11.8). These conditions guarantee the existence of a vector $\boldsymbol{u}$ (interpreted as contact displacement) such that

$$\boldsymbol{B}(\boldsymbol{\lambda} - \boldsymbol{f}) = \boldsymbol{u} \tag{11.10}$$
$$\boldsymbol{\lambda}_n \le 0, \quad \boldsymbol{u}_n \le 0, \quad \boldsymbol{\lambda}_n^T \boldsymbol{u}_n = 0 \tag{11.11}$$
$$-\Phi\boldsymbol{\lambda}_n - |\boldsymbol{\lambda}_t| \ge 0, \quad \boldsymbol{\lambda}_t^T \boldsymbol{u}_t \le 0, \quad \boldsymbol{u}_t^T(-\Phi\boldsymbol{\lambda}_n - |\boldsymbol{\lambda}_t|) = 0, \tag{11.12}$$

where $\boldsymbol{f} = \boldsymbol{B}^{-1}\boldsymbol{g}$. The trick of rewriting (11.8) as LCP lies in the introduction of a new variable $\boldsymbol{\rho} \in \mathbb{R}^{2m}$ defined by

$$\boldsymbol{\rho} := -\boldsymbol{W}_n^T \boldsymbol{\lambda}_n + \boldsymbol{W}_t^T \boldsymbol{\lambda}_t \tag{11.13}$$

with

$$\boldsymbol{W}_n = \begin{pmatrix} \Phi & \Phi & & & & \\ & & \Phi & \Phi & & \\ & & & & \ddots & \\ & & & & & \Phi \quad \Phi \end{pmatrix}, \quad \boldsymbol{W}_t = \begin{pmatrix} 1 & -1 & & & & \\ & & 1 & -1 & & \\ & & & & \ddots & \\ & & & & & 1 \quad -1 \end{pmatrix}.$$

Clearly, there exists $\boldsymbol{\mu} \in \mathbb{R}^{2m}$, such that

$$\boldsymbol{u}_t = \boldsymbol{W}_t \boldsymbol{\mu}, \quad \boldsymbol{\mu} \ge 0. \tag{11.14}$$

From (11.12),(11.13) and (11.14) we have the following complementarity conditions

$$\boldsymbol{\rho} \ge 0, \quad \boldsymbol{\mu} \ge 0, \quad \boldsymbol{\rho}^T \boldsymbol{\mu} = 0. \tag{11.15}$$

Now, using (11.10) and (11.14), we can evaluate the tangential stress $\boldsymbol{\lambda}_t$ by means of $\boldsymbol{\lambda}_n$ and $\boldsymbol{\mu}$:

$$\boldsymbol{\lambda}_t = \boldsymbol{B}_{tt}^{-1}(\boldsymbol{W}_t\boldsymbol{\mu} - \boldsymbol{B}_{tn}\boldsymbol{\lambda}_n) + \boldsymbol{B}_{tt}^{-1}\boldsymbol{B}_{tn}\boldsymbol{f}_n + \boldsymbol{f}_t.$$

Inserting this into (11.10) and (11.13), we get, respectively,

$$\boldsymbol{u}_n + (\boldsymbol{B}_{nt}\boldsymbol{B}_{tt}^{-1}\boldsymbol{B}_{tn} - \boldsymbol{B}_{nn})\boldsymbol{\lambda}_n - \boldsymbol{B}_{nt}\boldsymbol{B}_{tt}^{-1}\boldsymbol{W}_t\boldsymbol{\mu} = -(\boldsymbol{B}_{nn} - \boldsymbol{B}_{nt}\boldsymbol{B}_{tt}^{-1}\boldsymbol{B}_{tn})\boldsymbol{f}_n$$

and

$$\boldsymbol{\rho} + (\boldsymbol{W}_t^T\boldsymbol{B}_{tt}^{-1}\boldsymbol{B}_{tn} + \boldsymbol{W}_n^T)\boldsymbol{\lambda}_n - \boldsymbol{W}_t^T\boldsymbol{B}_{tt}^{-1}\boldsymbol{W}_t\boldsymbol{\mu} = \boldsymbol{W}_t^T(\boldsymbol{B}_{tt}^{-1}\boldsymbol{B}_{tn}\boldsymbol{f}_n + \boldsymbol{f}_t).$$

Summing up the above lines, we can formulate a linear complementarity problem in variables $\boldsymbol{w} = \begin{pmatrix} \boldsymbol{u}_n \\ \boldsymbol{\rho} \end{pmatrix}$ and $\boldsymbol{z} = \begin{pmatrix} \boldsymbol{\lambda}_n \\ \boldsymbol{\mu} \end{pmatrix}$:

$$\begin{aligned} &\text{Find } \boldsymbol{w}, \boldsymbol{z} \in \mathbb{R}^{3m} \text{ such that} \\ &\boldsymbol{w} - \boldsymbol{M}\boldsymbol{z} = \boldsymbol{b}, \quad \boldsymbol{w} \geq 0, \quad \boldsymbol{z} \geq 0, \quad \boldsymbol{w}^T\boldsymbol{z} = 0 \end{aligned} \tag{11.16}$$

with

$$\boldsymbol{M} = \begin{pmatrix} \boldsymbol{B}_{nn} - \boldsymbol{B}_{nt}\boldsymbol{B}_{tt}^{-1}\boldsymbol{B}_{tn} & \boldsymbol{B}_{nt}\boldsymbol{B}_{tt}^{-1}\boldsymbol{W}_t \\ -\boldsymbol{W}_n^T - \boldsymbol{W}_t^T\boldsymbol{B}_{tt}^{-1}\boldsymbol{B}_{tn} & \boldsymbol{W}_t^T\boldsymbol{B}_{tt}^{-1}\boldsymbol{W}_t \end{pmatrix}, \quad \boldsymbol{b} = \begin{pmatrix} (\boldsymbol{B}_{nt}\boldsymbol{B}_{tt}^{-1}\boldsymbol{B}_{tn} - \boldsymbol{B}_{nn})\boldsymbol{f}_n \\ \boldsymbol{W}_t^T(\boldsymbol{B}_{tt}^{-1}\boldsymbol{B}_{tn}\boldsymbol{f}_n + \boldsymbol{f}_t) \end{pmatrix}.$$

The matrix $\boldsymbol{M}$ is nonsymmetric and singular and, so far, we are not able to say much about its properties. However, we know from the above derivation that there exists at least one solution to (11.16), composed of the solution to (11.8) and of its KKT vector.

11.2.3 Examples

Example 11.1 Let $\Omega = (0,1) \times (0,1)$. The partition of the boundary Γ is shown in Figure 11.1. The bilinear form $\mathbf{a}(\cdot,\cdot)$ is defined by the plane-stress Hooke's law with Young's modulus $E = 1.0$ and Poisson's ratio $\nu = 0.3$. On the contact boundary Γ_c we consider Coulomb friction with three different values of the friction coefficient $\Phi = 0.3, 1.0, 10.0$, respectively. Note that $\Phi = 0$ corresponds to the frictionless problem where $T_t = 0$, while the high values of Φ approximate the homogeneous Dirichlet boundary condition $u_t = 0$ on Γ_c. The problem was discretized by bilinear quadrilateral finite elements. The reported results refer to the mesh of 31×31 nodes and to the QVI formulation (11.8). The number of unknowns in the Newton's method on page 207 was $m = 62$. The results are depicted in Figure 11.2. Let $\boldsymbol{\lambda}$ be the solution of (11.8) and recall that $\boldsymbol{T}_t^i = \boldsymbol{\lambda}^{2i-1}$ and $\boldsymbol{T}_n^i = \boldsymbol{\lambda}^{2i}, i = 1, \ldots, m/2$, are interpreted as tangential and normal stress components on Γ_c. For three different values of Φ, the figures (a),(c),(e) on the left-hand side show $\boldsymbol{T}_n$ (solid line), $\boldsymbol{T}_t$ (dashed line) and the tangential component of the displacement vector u_t (dotted line). The displacement vector u was obtained by solving the problem $Bu = g - \boldsymbol{\lambda}$. Figures (b),(d),(f) on the right-hand side show the distribution of $\Phi|\boldsymbol{T}_n|$ (solid line) and of $|\boldsymbol{T}_t|$ (dashed line) to give an idea where the constraints are active. △

Example 11.2 In this example, $\Omega = (0,4) \times (0,1)$ and the distribution of forces and boundary conditions are shown in Figure 11.3. Other data are the same as in the previous example. The results for a mesh of 41×11 nodes (82 unknowns in the Newton's method) are depicted in Figure 11.4. Again, figures (a),(c),(e) on the left-hand side show the distribution of $\boldsymbol{T}_n$ (solid line), $\boldsymbol{T}_t$ (dashed line), u_t (dotted line) and also of the normal displacements u_n (dash-dot line) which were equal to zero in Example 11.1. Figures (b),(d),(f) on the right-hand side show the distribution of $\Phi|\boldsymbol{T}_n|$ (solid line) and of $|\boldsymbol{T}_t|$ (dashed line). △

In both examples, the Newton's method (applied to the QVI formulation) proved to be very efficient but quite sensitive to the choice of the initial approximation. The method either converged in about five steps or did not converge at all. Typically, for low values of Φ (≤ 0.5) the method did not converge from the starting point zero but it did converge from the point obtained as a solution of a problem with higher value of Φ. Solving the LCP formulation (11.16), the Newton's method was slightly more robust; on the other hand, the problem has three times higher dimension.

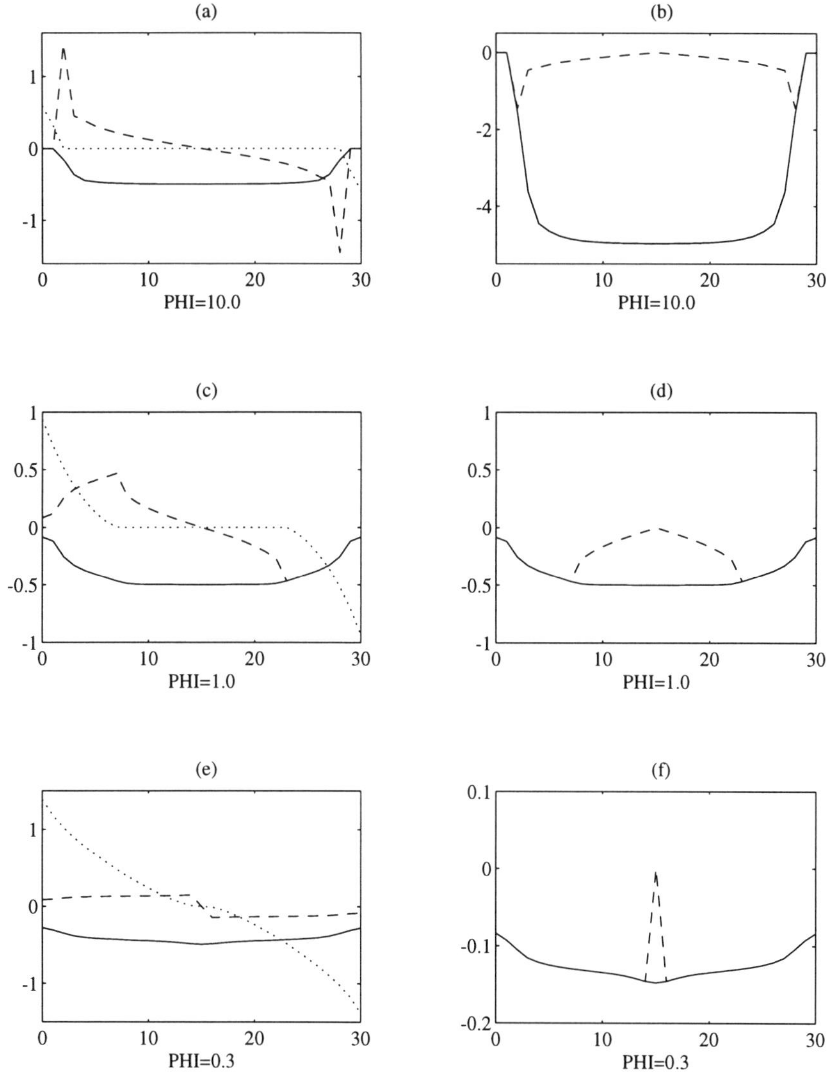

Figure 11.2. Results for Example 11.1

11.3 CONTROL OF FRICTION COEFFICIENTS

11.3.1 Problem formulation

Once we can formulate and solve the contact problem with Coulomb friction as an LCP, we may control it using the tools from the first part of the book. The question arises, what could be a practically useful control (design) variable. Unfortunately, in our approach it cannot be the shape of the elastic body Ω, since the dependence of the matrix M on a

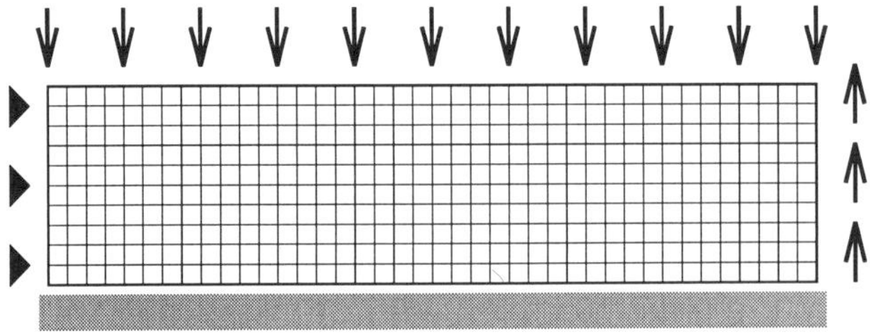

Figure 11.3. Domain Ω for Example 11.2 and the finite element mesh

shape parameter is not tractable; recall that $\boldsymbol{B}$ is the inverse of the Schur complement to the stiffness matrix $\boldsymbol{A}$ and that in $\boldsymbol{M}$ we have a Schur complement of $\boldsymbol{B}$. There is, however, another parameter that might be useful to control—the friction coefficient Φ. We may ask how to distribute Φ along the boundary Γ_c, so that we maximize the tangential adhesion while minimizing a norm of Φ. The second objective reflects the following fact: the higher the friction coefficient, the higher the costs of the material. This problem has applications, e.g., in constructing contact parts of robot and manipulator grippers. So, introducing the matrix

$$\boldsymbol{W}_n(\Phi) = \begin{pmatrix} \Phi_1 & \Phi_1 & & & & & \\ & & \Phi_2 & \Phi_2 & & & \\ & & & & \ddots & & \\ & & & & & \Phi_m & \Phi_m \end{pmatrix},$$

our parametrized state problem reads:

$$\begin{aligned} &\text{For a given } \Phi \in \mathbb{R}^m \text{ find } \boldsymbol{w}, \boldsymbol{z} \in \mathbb{R}^{3m} \text{ such that} \\ &\boldsymbol{w} - \boldsymbol{M}(\Phi)\boldsymbol{z} = \boldsymbol{b}, \quad \boldsymbol{w} \geq 0, \quad \boldsymbol{z} \geq 0, \quad \boldsymbol{w}^T\boldsymbol{z} = 0 \end{aligned} \tag{11.17}$$

with

$$\boldsymbol{M}(\Phi) = \begin{pmatrix} \boldsymbol{B}_{nn} - \boldsymbol{B}_{nt}\boldsymbol{B}_{tt}^{-1}\boldsymbol{B}_{tn} & \boldsymbol{B}_{nt}\boldsymbol{B}_{tt}^{-1}\boldsymbol{W}_t \\ \boldsymbol{W}_n^T(\Phi) - \boldsymbol{W}_t^T\boldsymbol{B}_{tt}^{-1}\boldsymbol{B}_{tn} & \boldsymbol{W}_t^T\boldsymbol{B}_{tt}^{-1}\boldsymbol{W}_t \end{pmatrix}$$

and $\boldsymbol{b}$ from (11.16).

The control problem can be formulated as follows:

$$\begin{aligned} &\text{minimize} \quad \boldsymbol{J}(\Phi, \boldsymbol{u}_t) := r \sum_{i=1}^{m} (\boldsymbol{u}_t)_i^2 + \sum_{i=1}^{m} \Phi_i^2 \\ &\text{subject to} \\ &\qquad\qquad \boldsymbol{u}_t \quad \text{“solves” LCP (11.17)} \\ &\qquad\qquad 0 \leq \Phi_i \leq \Phi_{max}, \qquad i = 1, \ldots, m, \end{aligned} \tag{11.18}$$

with a weight parameter $r > 0$ (recall that $\boldsymbol{u}_t$ is computed by (11.14) from the vector $\boldsymbol{\mu}$, part of the solution to (11.17)).

11.3.2 Numerical method

Problem (11.18) is an MPEC of type (7.1) and can be solved by the nonsmooth technique introduced in Chapter 7. However, unlike in all other examples in the book, we are not able

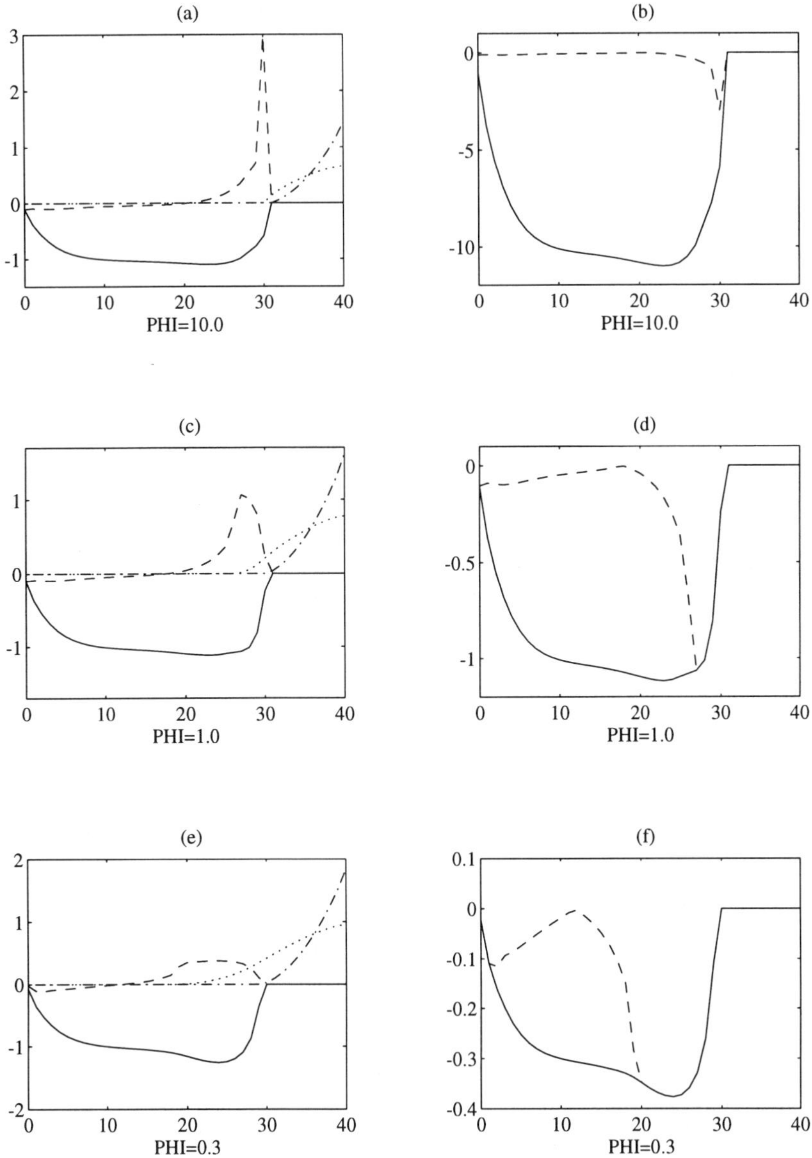

Figure 11.4. Results for Example 11.2

to verify the assumptions (A1)–(A3) from Section 7.2. The assumption (A1) (continuous differentiability of $\boldsymbol{J}$) is still easy to verify. But we are able to guarantee neither the uniqueness of the solution to (11.17) (Assumption (A2)), nor the strong regularity of the

GE associated with (11.17) (Assumption (A3)). Nevertheless, in the solved numerical examples we have never observed any serious difficulties connected with this question.

To compute a subgradient from $\partial\Theta(\boldsymbol{\alpha})$, we first have to solve—for a given control Φ—the adjoint problem, which in our situation amounts to solving the linear system in variable $\boldsymbol{p}$ (cf. Theorem 7.7)

$$\boldsymbol{M}^T_{L\cup(I^0\setminus M_i)}(\Phi)\boldsymbol{p} = (\nabla_\Phi \boldsymbol{J}(\Phi, \boldsymbol{z}))_{L\cup(I^0\setminus M_i)}$$

with

$$\begin{aligned} L(\Phi, \boldsymbol{z}) &= \left\{i \in \{1, 2, \ldots, m\} | \boldsymbol{z}^i > 0\right\}, \\ I^0(\Phi, \boldsymbol{z}) &= \left\{i \in \{1, 2, \ldots, m\} | (\boldsymbol{M}(\Phi)\boldsymbol{z} + \boldsymbol{b})^i = 0,\ \boldsymbol{z}^i = 0\right\}. \end{aligned}$$

$M_i(\Phi, \boldsymbol{z})$ is a suitably chosen subset of $I^0(\Phi, \boldsymbol{z})$. The subgradient associated with M_i can be computed from

$$\nabla_\Phi \boldsymbol{J}(\Phi, \boldsymbol{z}) - [\mathcal{J}_\Phi(\boldsymbol{M}(\Phi)\boldsymbol{z} - \boldsymbol{b})_{L\cup(I^0\setminus M_i)}]^T \boldsymbol{p}$$

cf. Proposition 7.17.

Before we report on the numerical results of a test example, we should mention the software used in the computations. A natural environment for matrix manipulations involved in the formulation and solution of the state problem (computations with Schur complements and matrix inverses, composition of matrix from rows of other matrices, etc.) is MATLAB or a MATLAB-like program. We used a code developed in INRIA and named SCILAB[1]. The advantage of this particular code is that it is equipped with a nonsmooth optimization routine callable directly in its language. (So one does not have to link the code with a routine written, e.g., in FORTRAN. This linking is virtually available but not yet robust enough, in particular while working with large FORTRAN subroutines that call other SCILAB functions.) This SCILAB nonsmooth optimization routine (based on the bundle algorithm and written by C. Lemaréchal) did a good job.

11.3.3 Example

Example 11.3 This example is a continuation of Example 11.1 from Section 11.2.3. Again, let $\Omega = (0, 1) \times (0, 1)$. The partition of the boundary Γ is shown in Figure 11.1. The reported results refer to the discretization of the contact boundary by 31 nodes, i.e., the number of design variables was 31 and the number of state variables was $3 \times 31 = 93$. The results are depicted in Figure 11.5. The first two figures show results of the state problem for two given constant values of the coefficient of friction $\Phi = 0.6$ and $\Phi = 0.9$. Line (1) shows the (constant) distribution of Φ, line (2) the distribution of the tangential component $\boldsymbol{\lambda}_t$ of the contact stress, line (3) its normal component $\boldsymbol{\lambda}_n$ and line (4) the tangential component $\boldsymbol{u}_t$ of the contact displacement vector. Obviously, for $\Phi = 0.9$ (middle figure), the region with zero tangential displacements is larger than that for $\Phi = 0.6$ (upper figure). In the lower figure we see the results of the MPEC. The value of the weight coefficient from (11.18) was set to $r = 5$. Line (1) shows the optimal distribution of the friction coefficient, lines (2)–(4) have the same meaning as above. The region with zero tangential displacements

[1]SCILAB-2.3.1, SCILAB Group (INRIA), can be obtained by anonymous ftp from `ftp.inria.fr`, directory `INRIA/Projects/Meta2/Scilab`

(line (4)) is larger than the one in the two other figures (this is the first, more important part of the upper-level objective). On the other hand, the integral of line (1) (the second, less important part of the objective, the cost of the material) is less than that in the middle figure with $\Phi = 0.9$. $\triangle$

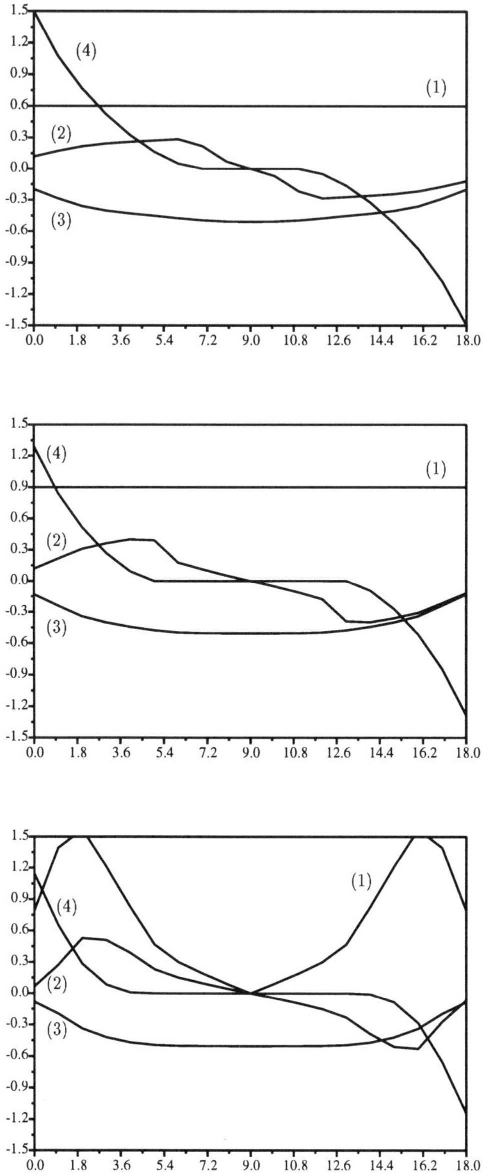

Figure 11.5. Results of Example 11.3

12 ECONOMIC APPLICATIONS

Central to many aspects of economic analysis (and even political theory) is the concept of equilibrium. Similarly as in mechanics, formal characterizations of this concept typically come as complementarity problems or variational inequalities.

Example 12.1 Given a finite number n of tradable goods, let $p \in \mathbb{R}^n$ be the price vector and $E(\mathbb{R}^n \to \mathbb{R}^n)$ be the *excess demand* given by

$$E(p) := D(p) - S(p),$$

where $D(p)$ is the *demand function* and $S(p)$ is the sum of initial endowment and production supply. E is generated by price-taking behaviour of households and firms. The price vector $p \in \mathbb{R}^n$ is declared a *perfectly competitive equilibrium*, if and only if

$$p \geq 0, \quad E(p) \leq 0 \quad \text{and} \quad \langle p, E(p) \rangle = 0. \tag{12.1}$$

The NCP (12.1) captures the simple idea that the ith good is either free ($p^i = 0$) or the respective excess demand component $E^i(p)$ will be nil. This approach to economic theory was initiated by Adam Smith (1776), formalized by Walras, 1954 and rigorously analyzed by Arrow and Debreu, 1954 and McKenzie, 1959. △

Noncooperative game theory is of more recent origin than competitive analysis. Founded mainly by von Neumann (von Neumann and Morgenstern, 1944), and Nash, 1951, it deals with agents who interact less anonymously than via markets. Specifically, suppose that individual $i \in N := \{1, 2, \ldots, n\}$ seeks, without any collaboration, to maximize his utility (payoff) $u_i(y_i, y_{-i})$ with respect to his own set of feasible strategies Y_i. Here $y_{-i} = (y_j)_{j \in N \setminus \{i\}}$ is a short and convenient notation for the strategy profile taken by the rivals of player i. Then y^* is declared a *Nash equilibrium* if neither player $i \in N$ can gain (in terms of increasing $u_i(\cdot, y^*_{-i})$) by unilaterally changing his strategy y^*_i. Formally, such equilibrium satisfies the conditions

$$y_i^* \in \operatorname*{argmax}_{y_i \in Y_i} u_i(y_i, y^*_{-i}) \quad \text{for all } i \in N. \tag{12.2}$$

When each Y_i is a nonempty, closed and convex subset of some Euclidean space $\mathbb{R}^{m_i}$ ($m_i \geq 1$) and each u_i is concave and differentiable with respect to y_i, then (12.2) can be replaced by the first-order optimality conditions

$$\langle \mathcal{M}_i(y^*), y_i - y_i^* \rangle \leq 0 \quad \text{for all } y_i \in Y_i,\ i \in N. \tag{12.3}$$

Here $\mathcal{M}_i(y) = \nabla_{y_i} u_i(y_i, y_{-i})$ signifies the *marginal utility* of i. Letting

$$\mathcal{M}(y) := \begin{bmatrix} \mathcal{M}_1(y_1, y_{-1}) \\ \mathcal{M}_2(y_2, y_{-2}) \\ \vdots \\ \mathcal{M}_n(y_n, y_{-n}) \end{bmatrix}, \tag{12.4}$$

we infer that (under the above assumptions) y^* is a Nash equilibrium if and only if it is a solution of the variational inequality:

$$\left.\begin{array}{l} \text{Find } y \in Y \text{ such that} \\ \qquad \langle -\mathcal{M}(y), v - y \rangle \geq 0 \quad \text{for all } v \in Y, \end{array}\right\} \tag{12.5}$$

where $Y = \times_{i \in N} Y_i$.

Admittedly, the Nash equilibrium concept is quite demanding. Such an outcome has two characteristics, both - at a first glance - presupposing much knowledge, foresight and planning. Indeed, to reach equilibrium, each agent must

- entertain correct beliefs about the actions of his rivals, and
- respond optimally.

Clearly, in practice, one cannot always assume that the agents will play equilibrium strategies right away. They must rather be offered time to learn and possibilities to adapt. We shall not touch upon this important issue how (which and whether) equilibrium may be approached or reached. Instead, we shall study an MPEC with the equilibrium constraint generated by so-called Cournot equilibrium which is the Nash equilibrium in a special economic model. It will be shown that under reasonable assumptions the method from Section 7.2 can be applied to its numerical solution. In Section 12.2 we look at equilibria characterized by (12.5), where, however, Y is a proper subset of $\times_{i \in N} Y_i$. Such cases, in which agents are subject to additional coupling constraints, yield what is called generalized (or constrained) Nash equilibria (GNE).

12.1 THE COURNOT OLIGOPOLY

In this section we reconsider the classic Cournot (1838) model in which individual (firm) $i \in N := \{1, 2, \ldots, n\}$ furnishes a quantity $y_i \in Y_i \subset \mathbb{R}_+$ of some homogeneous good[1] to a common market. Doing so he incurs differentiable cost $f_i(y_i)$ and obtains revenues $y_i p(T)$, where

$$T := \sum_{i \in N} y_i$$

denotes total supply and $p(\vartheta)$ is the price at which (price-taking) consumers are willing to demand. The function $p[\mathrm{int}\mathbb{R}_+ \to \mathrm{int}\mathbb{R}_+]$ is assumed differentiable and usually called *inverse demand curve.* It turns out that the Nash equilibrium concept is well-suited to the computation of "equilibrium productions" for the individuals (firms) acting at this oligopolistic market. This was already observed by Cournot in the case of duopoly and therefore one speaks sometimes of Cournot equilibrium. In terms of the preceding notation

$$u_i(y) = y_i p(T) - f_i(y_i), \quad i \in N.$$

This simple model remains a workhorse within economics, and directly suggests, how Cournot equilibria can be analyzed and computed.

Assume that Y_i, $i \in N$, are nonempty closed intervals in $\mathbb{R}_+$. The respective VI (12.5), written as generalized equation, attains the form

$$0 \in -\mathcal{M}(y) + N_Y(y), \tag{12.6}$$

where

$$\mathcal{M}_i(y) := \nabla_{y_i} u_i(y) = p(T) + y_i \nabla p(T) - \nabla f_i(y_i)$$

is the *marginal profit* of individual i. To ensure that (12.6) is equivalent to the corresponding relations (12.2) it is, as said, convenient to have each u_i concave with respect to y_i. For that purpose suppose, quite naturally, that all functions f_i are convex (i.e., marginal production costs are nondecreasing). For the revenue terms we shall invoke the following result from Murphy et al., 1982.

Lemma 12.1 *Assume that*

(i) p is twice continuously differentiable and strictly convex on $\mathrm{int}\mathbb{R}_+$;

(ii) $\vartheta p(\vartheta)$ is a concave function of ϑ.

Then for each $K > 0$ the function $g(\vartheta) := \vartheta p(\vartheta + K)$ is strictly concave on $\mathrm{int}\mathbb{R}_+$.

Proof. Assume by contradiction that g is not strictly concave, which implies $\nabla^2 g(\vartheta) \geq 0$ for some $\vartheta > 0$. Then one has

$$\begin{aligned} 0 &\leq \nabla^2 g(\vartheta) \\ &= 2\nabla p(\vartheta + K) + \vartheta \nabla^2 p(\vartheta + K) \\ &= \left[2\nabla p(\vartheta + K) + (\vartheta + K)\nabla^2 p(\vartheta + K)\right] - K\nabla^2 p(\vartheta + K) \\ &\leq -K\nabla^2 p(\vartheta + K) \end{aligned}$$

[1] y_i is the ith component of the production vector y and therefore it should actually be denoted by y^i. However, we prefer to write y_i to be consistent with the notation, introduced at the beginning of this chapter.

in virtue of the concavity of $\vartheta p(\vartheta)$. Since $K > 0$, this implies that $\nabla^2 p(\vartheta + K) \leq 0$, which contradicts the strict convexity of p. ∎

Granted the hypotheses of Lemma 12.1 and the convexity of each f_i, $i \in N$, the GE (12.6) provides a full characterization of Cournot equilibrium.

We specialize now to the important (and presumably frequent) situation, where one distinguished individual, say 1, has a temporal advantage over the others. Specifically, he is able to commit his action y_1 *before* the others are allowed to follow suit. This player, called the Leader, cannot renege on his choice y_1; the latter must be credible and irrevocable. Let the intervals $Y_i, i = 2, 3, \ldots, n$, be bounded and β^i be their respective upper production bounds. Concerning the Leader, we will suppose that

(H) $\mathcal{M}_1(0, \beta^2, \beta^3, \ldots, \beta^n) = p(\sum_{i=2}^n \beta^i) - \nabla f_1(0) > 0.$

Since p is typically decreasing, this hypothesis implies that for the Leader the zero production is definitely not an optimal one. (H) thus enables to replace interval Y_1 by another interval $\tilde{Y}_1$ with a positive (arbitrary small) lower bound. In this way, the sum T of the productions is always positive and $p(T)$ is well-defined.

To be consistent with the notation in the first part of the book, we put

$$\begin{aligned} x &:= y_1 \\ z &:= (y_2, y_2, \ldots, y_n), \end{aligned}$$

and assume that individuals $2, \ldots, n$ apply also in this situation their Cournot strategies. This means that they solve for the given strategy of the Leader the perturbed GE

$$0 \in F(x, y_2, y_3, \ldots, y_n) + N_{Y_2 \times Y_3 \times \cdots \times Y_n}(y_2, y_3, \ldots, y_n), \tag{12.7}$$

where $F = (F^2, F^3, \ldots, F^n)$ with

$$F^i(x, y_2, y_3, \ldots, y_n) = \nabla f_i(y_i) - p(T) - y_i \nabla p(T), \quad i = 2, 3, \ldots, n,$$

and $T = x + \sum_{i=2}^n y_i$. Suppose that (12.7) has for all $x \in Y_1$ a unique solution. Then the Leader, optimizing his profit, has to solve a *Stackelberg problem*

$$\begin{aligned} &\text{minimize} \quad f_1(x) - xp\left(x + \textstyle\sum_{i=2}^n y_i\right) \\ &\text{subject to} \\ &\qquad\qquad (y_2, y_3, \ldots, y_n) \in S(x) \\ &\qquad\qquad x \in \tilde{Y}_1, \end{aligned} \tag{12.8}$$

where S is the solution map generated by the GE (12.7). The resulting production x is called the *Stackelberg strategy* of individual 1.

Problem (12.8) is an MPEC of type (7.1), where the GE can be converted to the form (5.24). The numerical method of Chapter 7 can be applied to its solution provided the assumptions (A1),(A2) and (A3) can be verified. To this purpose we will assume that

(i) the functions f_i, $i = 1, 2, \ldots, n$, are convex and twice continuously differentiable, and

(ii) the assumptions of Lemma 12.1 are fulfilled.

Then, in virtue of (H), we have no difficulties with (A1). To verify (A2),(A3), we prove the following crucial result.

Lemma 12.2 *Let hypothesis (H) hold true and the above assumptions (i),(ii) be fulfilled. Assume that* $x \in \tilde{Y}_1$ *and* $y_i \in Y_i$, $i = 2, 3, \ldots, n$. *Then* $\mathcal{J}_z F(x, z)$ *is positive definite.*

Proof. Instead of $\mathcal{J}_z F(x, z)$, we can certainly check the positive definiteness of the symmetrized partial Jacobian

$$\begin{aligned}
&\tfrac{1}{2}\left(\mathcal{J}_z F(x,z) + (\mathcal{J}_z F(x,z))^{\mathrm{T}}\right) \\
&= \begin{bmatrix}
\nabla^2 f_2(y_2) - 2\nabla p(T) - y_2 \nabla^2 p(T) & -\nabla p(T) - \frac{y_2+y_3}{2}\nabla^2 p(T) & \cdots & \cdots & -\nabla p(T) - \frac{y_2+y_n}{2}\nabla^2 p(T) \\
-\nabla p(T) - \frac{y_2+y_3}{2}\nabla^2 p(T) & \vdots & \vdots & \vdots & \vdots \\
\vdots & \vdots & \vdots & \vdots & \vdots \\
-\nabla p(T) - \frac{y_2+y_n}{2}\nabla^2 p(T) & \cdots & \cdots & \cdots & \nabla^2 f_n(y_n) - 2\nabla p(T) - y_n \nabla^2 p(T)
\end{bmatrix}
\end{aligned}$$

$$\left.\begin{aligned}
&= \operatorname{diag}\left\{\nabla^2 f_2(y_2), \ldots, \nabla^2 f_n(y_n)\right\} \\
&+ \begin{bmatrix}
-2\nabla p(T) - \frac{y_2+y_2}{2}\nabla^2 p(T) & \cdots & -\nabla p(T) - \frac{y_2+y_n}{2}\nabla^2 p(T) \\
\vdots & & \vdots \\
-\nabla p(T) - \frac{y_n+y_2}{2}\nabla^2 p(T) & \cdots & -2\nabla p(T) - \frac{y_n+y_n}{2}\nabla^2 p(T)
\end{bmatrix}.
\end{aligned}\right\} \quad (12.9)$$

By the assumptions, the diagonal matrix in (12.9) is positive semidefinite and so it suffices to prove the positive definiteness of the second matrix in (12.9). To this purpose we multiply it by $\frac{2}{T\nabla^2 p(T)}$, which does not influence its definiteness ($\nabla^2 p(T) > 0$). Further, for the sake of simplicity, put

$$\sigma := \frac{-2\nabla p(T)}{T\nabla^2 p(T)}, \quad \tilde{x} := \frac{x}{T} \quad \text{and} \quad \tilde{y}_i := \frac{y_i}{T}, \quad i = 2, 3, \ldots, n.$$

We observe that $\tilde{x} + \sum_{i=2}^{n} \tilde{y}_i = 1$ and $\sigma \geq 1$. Indeed, as in the proof of Lemma 12.1, one has (for $\vartheta > 0$)

$$\nabla^2(\vartheta p(\vartheta)) = 2\nabla p(\vartheta) + \vartheta \nabla^2 p(\vartheta) \leq 0,$$

which implies

$$\frac{-2\nabla p(T)}{T\nabla^2 p(T)} \geq 1.$$

Using this notation, we have to prove that the matrix

$$\Delta := \begin{bmatrix}
2\sigma - (\tilde{y}_2 + \tilde{y}_2) & \sigma - (\tilde{y}_2 + \tilde{y}_3) & \cdots & \sigma - (\tilde{y}_2 + \tilde{y}_n) \\
\sigma - (\tilde{y}_3 + \tilde{y}_2) & 2\sigma - (\tilde{y}_3 + \tilde{y}_3) & \cdots & \sigma - (\tilde{y}_3 + \tilde{y}_n) \\
\vdots & \vdots & & \vdots \\
\sigma - (\tilde{y}_n + \tilde{y}_2) & \sigma - (\tilde{y}_n + \tilde{y}_3) & \cdots & 2\sigma - (\tilde{y}_n + \tilde{y}_n)
\end{bmatrix}$$

is positive definite. If $y_i = \tilde{y}_i = 0$ for $i = 1, 2, \ldots, n$, then

$$\Delta = \sigma \begin{bmatrix} 2 & 1 & \cdots & 1 \\ 1 & 2 & \cdots & 1 \\ \vdots & \vdots & & \vdots \\ 1 & 1 & \cdots & 2 \end{bmatrix}$$

and the positive definiteness is straightforward. Indeed, for any nonzero vector $h \in \mathbb{R}^{n-1}$ one has

$$\langle h, \Delta h \rangle = \sigma \left(\sum_{i=1}^{n-1} (h^i)^2 + \left(\sum_{i=1}^{n-1} h^i \right)^2 \right) > 0.$$

Hence we can confine ourselves with the case $z = (y_2, y_3, \ldots, y_n) \neq 0$. We use that a symmetric matrix is positive definite if and only if all its eigenvalues are positive (Horn and Johnson, 1985). Let λ be an eigenvalue and u be a corresponding eigenvector, i.e.,

$$(\Delta u)^i = \sigma u^i - \tilde{y}_{i+1} \sum_{j=1}^{n-1} u^j - \sum_{j=1}^{n-1} \tilde{y}_{j+1} u^j + \sigma \sum_{j=1}^{n-1} u^j = \lambda u^i, \quad i = 1, 2, \ldots, n-1. \quad (12.10)$$

In the following we will distinguish the cases

$$\sum_{j=1}^{n-1} u^j = 0 \quad (12.11)$$

and

$$\sum_{j=1}^{n-1} u^j \neq 0. \quad (12.12)$$

If (12.11) holds, then

$$\sigma u^i - \sum_{j=1}^{n-1} \tilde{y}_{j+1} u^j = \lambda u^i, \quad i = 1, 2, \ldots, n-1. \quad (12.13)$$

By adding these equations over i, one obtains

$$\sigma \sum_{i=1}^{n-1} u^i - (n-1) \sum_{j=1}^{n-1} \tilde{y}_{j+1} u^j = \lambda \sum_{i=1}^{n-1} u^i,$$

which implies $\sum_{j=1}^{n-1} \tilde{y}_{j+1} u^j = 0$. Equations (12.13) is thus reduced to

$$\sigma u^i = \lambda u^i, \quad i = 1, 2, \ldots, n-1.$$

Since $u \neq 0$, we have $\lambda = \sigma \geq 1 > 0$ and we are done.

In case of (12.12) we assume without any loss of generality that $\sum_{j=1}^{n-1} u^j = 1$. Then equations (12.10) attain the form

$$\sigma u^i + \sigma - \tilde{y}_{i+1} - \sum_{j=1}^{n-1} \tilde{y}_{j+1} u^j = \lambda u^i, \quad i = 1, 2, \ldots, n-1. \quad (12.14)$$

We add over i and arrive at

$$\sigma + (n-1)\sigma - \sum_{i=1}^{n-1} \tilde{y}_{i+1} - (n-1)\sum_{j=1}^{n-1} \tilde{y}_{j+1}u^j = \lambda,$$

that is,

$$n\sigma - 1 + \tilde{x} - (n-1)\sum_{j=1}^{n-1} \tilde{y}_{j+1}u^j = \lambda.$$

If $\sum_{j=1}^{n-1} \tilde{y}_{j+1}u^j \leq 0$, one has

$$\lambda \geq n\sigma - 1 + \tilde{x} \geq n - 1 + \tilde{x} > 0,$$

and so it remains to consider the case $\sum_{j=1}^{n-1} \tilde{y}_{j+1}u^j > 0$. We multiply the ith equation (12.14) by $\tilde{y}_{i+1}$ (note that at least one of these numbers is nonzero) and sum them up over i. This yields

$$\sigma \sum_{i=1}^{n-1} \tilde{y}_{i+1}u^i + \sigma(1-\tilde{x}) - \sum_{i=1}^{n-1} (\tilde{y}_{i+1})^2 - \sum_{j=1}^{n-1} \tilde{y}_{j+1}u^j(1-\tilde{x}) = \lambda \sum_{i=1}^{n-1} \tilde{y}_{i+1}u^i,$$

that is,

$$\sigma(1-\tilde{x}) - \sum_{i=1}^{n-1} (\tilde{y}_{i+1})^2 = (\lambda - \sigma + 1 - \tilde{x}) \sum_{i=1}^{n-1} \tilde{y}_{i+1}u^i. \tag{12.15}$$

Since $\tilde{y}_{i+1} \in [0,1)$ for $i = 1, 2, \ldots, n-1$, we have

$$\sum_{i=1}^{n-1} (\tilde{y}_{i+1})^2 \leq \sum_{i=1}^{n-1} \tilde{y}_{i+1} = 1 - \tilde{x}.$$

This implies that

$$\lambda - \sigma + 1 - \tilde{x} \geq 0,$$

since the left-hand side of (12.15) is non-negative. It follows that also in this case all eigenvalues λ are positive. The proof is complete. ■

The preceding result implies that under the assumptions (i),(ii) and under (H) the GE in (12.7) possesses a unique solution for each x from an open set $\tilde{\mathcal{A}}$ containing $\tilde{Y}_1$ due to Theorem 4.1 and Theorem 4.4(i). Further, in virtue of Theorem 5.8, this GE is strongly regular at all pairs (x, z) with $x \in \tilde{\mathcal{A}}$ and y being the corresponding Cournot equilibrium. It follows that in this case the assumptions (A2),(A3) are fulfilled. We now take a concrete form of the functions f_i and p from Murphy et al., 1982 to be able to perform test computations. Let

$$f_i(\xi) = c_i\xi + \frac{\delta_i}{\delta_i + 1} K_i^{-1/\delta_i} (\xi)^{(1+\delta_i)/\delta_i}, \tag{12.16}$$

where $c_i, \delta_i, K_i,\ i = 1, 2, \ldots, n$, are given positive parameters; further let

$$p(T) = 5000^{1/\gamma} T^{-1/\gamma}, \tag{12.17}$$

with a positive parameter γ called *demand elasticity*. The production cost functions (12.16) are clearly twice continuously differentiable on $\mathbb{R}$ and convex on $\mathbb{R}_+$. The inverse demand curve (12.17) is twice continuously differentiable and strictly convex on $\text{int}\mathbb{R}_+$. We also observe that the so-called *industry revenue curve*

$$\vartheta p(\vartheta) = 5000^{1/\gamma}\vartheta^{(\gamma-1)/\gamma}$$

is concave on $\text{int}\mathbb{R}_+$ for $\gamma \geq 1$. The numerical method of Chapter 7 can thus be applied to the solution of the Stackelberg problem (12.8), defined by the functions (12.16),(12.17) with $\gamma \geq 1$.

We turn our attention to the computation of a subgradient of the respective function Θ at $\overline{x} \in \widetilde{Y}_1$. In the simple case $n = 2$ the assumptions of Corollary 7.12 are satisfied, so the choice of a suitable $i \in \mathbb{K}(\overline{x}, \overline{z})$ is an easy job ($\overline{z}$ is the unique solution of the GE (12.7) with $x = \overline{x}$). In the more interesting case $n > 2$, Corollary 7.13 applies provided just one constraint is weakly active.

Proposition 12.3 *Consider problem (12.8) with the functions f_i, $i = 1, 2, \ldots, n$, and p given by (12.16),(12.17) and assume that $\gamma \geq 1$ and hypothesis (H) holds true. Let $(\overline{x}, \overline{z}) \in \widetilde{Y}_1 \times \bigtimes_{i=2}^{n} Y_i$, where $\overline{z}$ is the (unique) solution of the GE (12.7) for $x = \overline{x}$, and denote $\overline{T} := \overline{x} + \sum_{i=2}^{n} \overline{y}_i$. Further suppose that for $\ell \in \{1, 2, \ldots, n-1\}$*

(i) either the lower or the upper bound on $y_{\ell+1}$ is weakly active, and

(ii) $$1 - \frac{\overline{y}_{\ell+1}}{\overline{T}}\left(\frac{1}{\gamma} + 1\right) \neq 0.$$

Then any $i \in \mathbb{K}(\overline{x}, \overline{z})$ generates a subgradient of the corresponding composite objective Θ at $\overline{x}$.

Proof. We suppose without loss of generality that $\overline{y}_2, \ldots, \overline{y}_\ell$ are interior points of the corresponding intervals and $\overline{y}_{\ell+2}, \ldots, \overline{y}_n$ are boundary points. By Corollary 7.13 it suffices to show that the $\ell \times \ell$ matrix

$$\left[\mathcal{J}_x F_{L\cup I^0}(\overline{x}, \overline{z}), \mathcal{J}_y F_{L\cup I^0, L}(\overline{x}, \overline{z})\right] \tag{12.18}$$

is nonsingular, which is the same as showing the nonsingularity of its transpose

$$\begin{bmatrix} (\mathcal{J}_y F_{L\cup I^0, L}(\overline{x}, \overline{z}))^{\mathrm{T}} \\ (\mathcal{J}_x F_{L\cup I^0}(\overline{x}, \overline{z}))^{\mathrm{T}} \end{bmatrix}$$

$$= \left[\begin{matrix} \nabla^2 f_2(\overline{y}_2) - 2\nabla p(\overline{T}) - \overline{y}_2\nabla^2 p(\overline{T}) & \cdots \\ -\nabla p(\overline{T}) - \overline{y}_2\nabla^2 p(\overline{T}) & \vdots \\ \vdots & \vdots \\ -\nabla p(\overline{T}) - \overline{y}_2\nabla^2 p(\overline{T}) & \cdots \end{matrix}\right.$$

$$\left.\begin{matrix} \cdots & -\nabla p(\overline{T}) - \overline{y}_\ell\nabla^2 p(\overline{T}) & -\nabla p(\overline{T}) - \overline{y}_{\ell+1}\nabla^2 p(\overline{T}) \\ \vdots & \vdots & -\nabla p(\overline{T}) - \overline{y}_{\ell+1}\nabla^2 p(\overline{T}) \\ \vdots & \nabla^2 f_\ell(\overline{y}_\ell) - 2\nabla p(\overline{T}) - \overline{y}_\ell\nabla^2 p(\overline{T}) & \vdots \\ \cdots & -\nabla p(\overline{T}) - \overline{y}_\ell\nabla^2 p(\overline{T}) & -\nabla p(\overline{T}) - \overline{y}_{\ell+1}\nabla^2 p(\overline{T}) \end{matrix}\right].$$

Hence suppose $h \in \mathbb{R}^\ell$ is such that

$$\begin{bmatrix} A \\ -\nabla p(\overline{T}) - \overline{y}_2 \nabla^2 p(\overline{T}) \cdots - \nabla p(\overline{T}) - \overline{y}_\ell \nabla^2 p(\overline{T}) \end{bmatrix} \begin{bmatrix} h^1 \\ h^2 \\ \vdots \\ h^{\ell-1} \end{bmatrix}$$

$$+ \begin{bmatrix} -\nabla_p(\overline{T}) - \overline{y}_{\ell+1} \nabla^2 p(\overline{T}) \\ \vdots \\ -\nabla_p(\overline{T}) - \overline{y}_{\ell+1} \nabla^2 p(\overline{T}) \end{bmatrix} h^\ell = 0, \qquad (12.19)$$

where

$$A = \begin{bmatrix} \nabla^2 f_2(\overline{y}_2) - 2\nabla p(\overline{T}) - \overline{y}_2 \nabla^2 p(\overline{T}) & \cdots & -\nabla p(\overline{T}) - \overline{y}_\ell \nabla^2 p(\overline{T}) \\ -\nabla p(\overline{T}) - \overline{y}_2 \nabla^2 p(\overline{T}) & \cdots & -\nabla p(\overline{T}) - \overline{y}_\ell \nabla^2 p(\overline{T}) \\ \vdots & & \vdots \\ -\nabla p(\overline{T}) - \overline{y}_2 \nabla^2 p(\overline{T}) & \cdots & \nabla^2 f_\ell(\overline{y}_\ell) - 2\nabla p(\overline{T}) - \overline{y}_\ell \nabla^2 p(\overline{T}) \end{bmatrix}.$$

Lemma 12.2 implies that matrix A is non-singular and assumption (ii) ensures that $-\nabla p(\overline{T}) - \overline{y}_{\ell+1} \nabla^2 p(\overline{T}) \neq 0$. Therefore, in any non-zero solution of (12.19) one has $h^\ell \neq 0$. Thus, (12.19) has a non-zero solution if and only if the vector $(-\nabla p(\overline{T}) - \overline{y}_2 \nabla^2 p(\overline{T}), \ldots, -\nabla p(\overline{T}) - \overline{y}_\ell \nabla^2 p(\overline{T}))^{\mathrm{T}}$ is a linear combination of the rows of A with coefficients α_i, such that $\sum_{i=1}^{\ell-1} \alpha_i = 1$. However, looking at the structure of A, we immediately see that this is impossible, because $-\nabla p(\overline{T}) > 0$ and $\nabla^2 f_i(\overline{y}_i) \geq 0,\ i = 2, 3, \ldots, \ell$. Thus, the matrix (12.18) is nonsingular and the assertion has been proved. ■

Of course, the above criterion does not cover the case of more than one weakly active constraint when the verification of the assumptions of Theorem 7.10 becomes substantially more difficult.

We performed numerical tests for $n = 5$, $\gamma \in [1, 2]$ and the parameters of the production cost functions shown in Table 12.1. Tables 12.2 and 12.3 present the (locally optimal) productions and profits of Firm 1, applying alternatively the Cournot and the Stackelberg strategy for different values of the demand elasticity. The GEs were solved by the method discussed in Fukushima, 1992 using a sequential quadratic programming code NLPQL (Schittkowski, 1986) for the resulting optimization problems. For the minimization of the corresponding function Θ over $\tilde{Y}_1$, we used the bundle-trust algorithm BT sketched in Chapter 3. The chosen accuracies were $\varepsilon = 1.0 \times 10^{-12}$ in NLPQL and $\varepsilon = 5.0 \times 10^{-4}$ in BT. NSIM indicates the number of evaluations of Θ for the starting vector x set to be a half of the upper production bound.

Alternatively, to promote the application of the Stackelberg strategy, we set $c_1 = 2$ (instead of 10). For these data, Tables 12.4 and 12.5 show the productions and profits of all firms for $\gamma = 1$. The upper production bound in Table 12.5 is only imposed on Firms 2–5. The results indicate that, even if the Cournot equilibrium production of Firm 1 clearly dominates on the market, its profit improvement connected with the change to the Stackelberg strategy does not exceed 2%. This improvement becomes negligible, whenever the remaining firms are subject to some production limitations (Table 12.5).

Table 12.1. Parameter specification for the production costs

	Firm 1	*Firm 2*	*Firm 3*	*Firm 4*	*Firm 5*
c_i	10	8	6	4	2
K_i	5	5	5	5	5
δ_i	1.2	1.1	1.0	0.9	0.8

Table 12.2. Upper production bound = 150 for all firms

γ		1.0	1.1	1.3	1.5	1.7
Firm 1	Production	47.811	36.9325	21.2179	10.9588	4.2421
Cournot	Profit	337.3078	199.9345	67.21	18.9205	3.1268
Firm 1	Production	55.5483	42.5354	24.142	12.388	4.7536
Stackelberg	Profit	343.3453	203.1551	68.1356	19.1540	3.1612
NSIM		6	9	9	12	9

Table 12.3. Different upper production bounds (50,40,30,25,20)

γ		1.0	1.1	1.3	1.5	1.7
Firm 1	Production	48.2346	38.5177	22.8567	11.8907	5.4829
Cournot	Profit	345.8091	223.8366	80.5432	22.6793	5.3116
Firm 1	Production	50.0	39.7924	24.2706	13.0107	6.0018
Stackelberg	Profit	346.8933	224.0375	80.7860	22.8377	5.3492
NSIM		5	6	12	5	5

12.2 GENERALIZED NASH EQUILIBRIUM

Strategic interaction among noncooperative agents takes manifold forms. Classically, the influence of any single player on his rivals is left merely at the level of their utility functions. Quite often, though, individual actions directly modify the feasible sets of others (pollution or non-coordinated use of common resources being two important examples). The overall strategy space Y is thus a (still closed, convex) *strict* subset of $\times_{i \in N} Y_i$. Correspondingly, $y^* \in Y$ is then named a *generalized Nash equilibrium* (GNE), if

$$y_i^* \in \operatorname*{argmax}_{(y_i, y_{-i}^*) \in Y} \; u_i(y_i, y_{-i}^*) \quad \text{for all } i \in N. \tag{12.20}$$

Table 12.4. Productions and profits—upper production bound = 150

		Firm 1	*Firm 2*	*Firm 3*	*Firm 4*	*Firm 5*
Cournot	Production	86.0903	46.1366	47.2898	45.37	41.0537
Equilibrium	Profit	943.2682	316.1257	381.7337	422.3705	436.1552
Firm 1	Production	99.5329	44.3804	45.8893	44.2806	40.2357
Stackelberg	Profit	958.6347	284.683	350.5039	393.2799	410.5319
NSIM		6				

Table 12.5. Productions and profits—upper production bound = 44 for Firms 2–5

		Firm 1	*Firm 2*	*Firm 3*	*Firm 4*	*Firm 5*
Cournot	Production	86.8613	44.0	44.0	44.0	41.6005
Equilibrium	Profit	982.2808	326.2177	387.0534	435.1105	454.1278
Firm 1	Production	89.8377	44.0	44.0	44.0	41.3293
Stackelberg	Profit	983.1496	317.5352	378.3709	426.428	445.1253
NSIM		10				

Granted appropriate differentiability of utilities, (12.20) entails the system

$$\langle \mathcal{M}_i(y^*), y_i - y_i^* \rangle \le 0 \quad \text{for all } y_i \in \{v \in \mathbb{R}^{m_i} \mid (v, y_{-i}^*) \in Y\},\ i \in N \tag{12.21}$$

of individual variational inequalities (as before $\mathcal{M}_i(y) = \nabla_{y_i} u_i(y_i, y_{-i})$ denotes the marginal utility). We strongly emphasize that any solution of the GE

$$0 \in -\mathcal{M}(y) + N_Y(y) \tag{12.22}$$

with $\mathcal{M}$ given by (12.4) is a GNE. There may, however, be solutions to (12.21) which do not fit (12.22). Solutions y^* to (12.22) are termed *normal equilibria* (Rosen, 1965). Since (12.22) is mathematically a more tractable object than system (12.21), normal equilibria can be computed by a number of effective methods; cf. Harker and Pang, 1990. However, in economic terms there are a priori no reasons why normal equilibria should be more legitimate than other GNE. Therefore, we shall focus on the latter, for which the computational challenge is notably greater.

To make progress we need explicit representations of all *feasible sections*

$$K_i(y_{-i}) := \{v \in \mathbb{R}^{m_i} \mid (v, y_{-i}) \in Y\}, \quad i \in N.$$

Namely, we will assume that for $i \in N$

$$K_i(y_{-i}) = Y_i \cap \{v \in \mathbb{R}^{m_i} \mid g^j(v, y_{-i}) \le 0,\ j = 1, 2, \ldots, s,\} \tag{12.23}$$

and

$$Y_i = \left\{v \in \mathbb{R}^{m_i} \mid \tilde{g}_i^j(v) \le 0,\ j = 1, 2, \ldots, \ell_i\right\}, \tag{12.24}$$

where with $m := \sum_{i \in N} m_i$ the functions $g^j[\mathbb{R}^m \to \mathbb{R}]$, $j = 1, 2, \ldots, s$, and $\tilde{g}_i^j[\mathbb{R}^{m_i} \to \mathbb{R}]$, $j = 1, 2, \ldots, \ell_i$, $i \in N$, are twice continuously differentiable and convex. Denote

$$g(y) = (g^1(y), g^2(y), \ldots, g^s(y))^T$$

and

$$\tilde{g}_i(y_i) = (\tilde{g}_i^1(y_i), \tilde{g}_i^2(y_i), \ldots, \tilde{g}_i^{\ell_i}(y_i))^T, \; i \in N.$$

Further we shall suppose that with a GNE, say y^*, the constraint system

$$\begin{bmatrix} g(y_1, y^*_{-1}) \\ \vdots \\ g(y_n, y^*_{-n}) \\ \tilde{g}_1(y_1) \\ \vdots \\ \tilde{g}_n(y_n) \end{bmatrix} \leq 0 \tag{12.25}$$

satisfies (ESCQ) at y^*.

Then we infer from (12.21) that (y^*, y^*) solves the problem:

$$\begin{aligned} &\text{Find } (x, y) \in \mathbb{R}^m \times \mathbb{R}^m \text{ such that} \\ &x = y \\ &0 \in \begin{bmatrix} \mathcal{L}(x, y, \lambda) \\ -G(x, y) \end{bmatrix} + N_{\mathbb{R}^m \times \mathbb{R}_+^{ns+t}}(y, \lambda), \end{aligned} \tag{12.26}$$

where $t := \sum_{i \in N} \ell_i$, λ is the KKT vector associated with the constraints (12.25),

$$\mathcal{L}(x, y, \lambda) := \begin{bmatrix} -\nabla_{y_1} u_1(y) \\ -\nabla_{y_2} u_2(y) \\ \vdots \\ -\nabla_{y_n} u_n(y) \end{bmatrix} + (\nabla_y G(x, y))^{\mathrm{T}} \lambda$$

is the D-Lagrangian and G is defined by

$$G(x, y) := \begin{bmatrix} g(y_1, x_{-1}) \\ \vdots \\ g(y_n, x_{-n}) \\ \tilde{g}_1(y_1) \\ \vdots \\ \tilde{g}_n(y_n) \end{bmatrix}.$$

Our idea of the numerical solution of (12.26) is based on its reformulation to the form

$$\begin{aligned} &\text{minimize} \quad \|x - y\| \\ &\text{subject to} \\ &0 \in \begin{bmatrix} \mathcal{L}(x, y, \lambda) \\ -G(x, y) \end{bmatrix} + N_{\mathbb{R}^m \times \mathbb{R}_+^{ns+t}}(y, \lambda) \end{aligned} \tag{12.27}$$

which is an MPEC of form (7.1). It is clear that a pair $(\hat{x}, \hat{y})$ is a solution of (12.26) if and only if it is such a solution to (12.27) for which the objective attains zero value (i.e. $\hat{x} = \hat{y}$). The objective in (12.27) is, however, not differentiable at the solutions of (12.26) so that the assumption (A1) from Section 7.2 is not fulfilled. Nevertheless, since the Euclidean norm is differentiable everywhere apart from 0, we have no difficulties in applying the numerical method of Chapter 7: either $x \neq y$ at a current iterate (x, y) and the objective is differentiable, or $x = y$ and we have a solution.

To apply the method of Section 7.2 to the numerical solution of (12.27), we still have to ensure the assumptions (A2),(A3). Suppose for a while that the map

$$F(y) := \begin{bmatrix} -\nabla_{y_1} u_1(y) \\ -\nabla_{y_2} u_2(y) \\ \vdots \\ -\nabla_{y_n} u_n(y) \end{bmatrix} \tag{12.28}$$

is strongly monotone, over $\mathbb{R}^m$. Then, by Theorem 4.4, the GE in (12.27) possesses a solution whenever $\times_{i \in N} K_i(x_{-i})$ is nonempty. Moreover, the y-component of this solution is unique. We can thus write $y = S_1(x)$ and need just to ensure (ELICQ) at all pairs $(x, S_1(x))$. This constraint qualification, however, can be easily violated due to a collision between the inequalities defining Y_i and the inequalities $g(y_i, x_{-i}) \leq 0,\ i \in N$, for certain values of x. This is illustrated by the following example taken from Harker, 1991.

Example 12.2 Consider the generalized Nash equilibrium given by $n = 2,\ m = 2, s = 1$,

$$\begin{aligned} u_1(y_1, y_2) &= -(y_1)^2 - \tfrac{8}{3} y_1 y_2 + 34 y_1 \\ u_2(y_1, y_2) &= -(y_2)^2 - \tfrac{5}{4} y_1 y_2 + \tfrac{97}{4} y_2 \\ Y_1 &= Y_2 = [0, 10] \end{aligned}$$

and

$$g(y) = y_1 + y_2 - 15.$$

Then the map

$$\begin{bmatrix} -\nabla_{y_1} u_1(y_1, y_2) \\ --\nabla_{y_2} u_2(y_1, y_2) \end{bmatrix} = \begin{bmatrix} 2y_1 + \frac{8}{3} y_2 - 34 \\ 2y_2 + \frac{5}{4} y_1 - \frac{97}{4} \end{bmatrix}$$

is evidently strongly monotone over $\mathbb{R}^2$. Unfortunately,

$$G(x, y) = \begin{bmatrix} y_1 + x_2 - 15 \\ y_2 + x_1 - 15 \\ -y_1 \\ y_1 - 10 \\ -y_2 \\ y_2 - 10 \end{bmatrix}$$

so that (ELICQ) may not hold (e.g. at $x_1 = y_1 = 10,\ x_2 = y_2 = 5$). △

In such a case, the strong regularity cannot be ensured. If, however, at least the *reduced constraint system*

$$\begin{bmatrix} g(y_1, x_{-1}) \\ \vdots \\ g(y_n, x_{-n}) \end{bmatrix} \leq 0 \tag{12.29}$$

satisfies (ELICQ) at all (x, y), $x \in \Omega := \times_{i \in N} Y_i$, $y = S_1(x)$, then we can decouple the constraints generated by Y_i, $i \in N$, and (12.29) and arrive at the "modified" problem:

Find $(x, y) \in \Omega \times \mathbb{R}^m$ such that

$$\begin{aligned} & x = y \\ & 0 \in \begin{bmatrix} \tilde{\mathcal{L}}(x, y, \tilde{\lambda}) \\ -\tilde{G}(x, y) \end{bmatrix} + N_{\mathbb{R}^m \times \mathbb{R}^{ns}_+}(y, \tilde{\lambda}), \end{aligned} \tag{12.30}$$

where $\tilde{\lambda}$ is the KKT vector associated with the constraints (12.29),

$$\tilde{\mathcal{L}}(x, y, \tilde{\lambda}) := \begin{bmatrix} -\nabla_{y_1} u_1(y) \\ -\nabla_{y_2} u_2(y) \\ \vdots \\ -\nabla_{y_n} u_n(y) \end{bmatrix} + (\nabla_y \tilde{G}(x, y))^{\mathrm{T}} \tilde{\lambda} \tag{12.31}$$

is the D-Lagrangian and $\tilde{G}$ is defined by

$$\tilde{G}(x, y) := \begin{bmatrix} g(y_1, x_{-1}) \\ \vdots \\ g(y_n, x_{-n}) \end{bmatrix}.$$

Proposition 12.4 *Under the above assumptions on (12.29) each solution of (12.30) is a solution of (12.27).*

Proof. Let $(\hat{y}, \hat{y})$ be a solution of (12.30) and $\tilde{\lambda} \in \mathbb{R}^{ns}_+$ its associated KKT vector. Then $\hat{y} \in \Omega$ and we easily verify that with

$$\lambda = \begin{pmatrix} \tilde{\lambda} \\ 0 \end{pmatrix} \in \mathbb{R}^{ns+t}_+$$

the triple $(\hat{y}, \hat{y}, \lambda)$ is a solution of the GE in (12.27). ∎

Unfortunately, we can also lose some GNE in this way, but this loss does not have to cause any harm as shown, for instance, in Example 12.3. Taking this loss as acceptable, the numerical method from Section 7.2 is applicable to the computation of the remaining GNE provided

(i) F (defined by (12.28)) is strongly monotone;

(ii) the inequalities (12.29) satisfy (ELICQ) at all pairs (x, y) with $x \in \Omega$, $y = S_1(x)$.

Sometimes also the following useful condition holds true.

(G) $\nabla_y \tilde{G}(x, y)$ does not depend on x and at each pair (x, y) with $x \in \Omega$, $y = S_1(x)$ the partial gradients $\nabla_{x_{-i}} g^j(y_i, x_{-i})$ are linearly independent for the active inequalities.

Then Corollary 7.11 applies and for each $i \in \mathbb{K}(x, y)$ formula (7.32) provides us with a subgradient of the corresponding composite objective Θ.

Example 12.3 (Example 12.2 continued) In this problem, conditions (i),(ii) and (G) are fulfilled and we could easily solve it by the proposed method. As in the examples of Section 12.1, we used the codes BT and NLPQL (with the same accuracies). The set of GNE can also be computed analytically (Harker, 1991) and one gets

$$\{(5, 9)\} \cup [(9, 6), (10, 5)].$$

Table 12.6 shows that our method reaches all extreme points of this set from suitable starting points.

Table 12.6.

Starting vector		GNE		NSIM
0	0	5	9	13
5	5	5	9	6
10	10	9	6	43
10	0	10	5	7
0	10	5	9	10

△

Assumption (i) can be weakened. It suffices to require strong monotonicity of

$$\tilde{F}(x, \cdot) = \begin{bmatrix} -\nabla_{y_1} u_1(\cdot, x_{-1}) \\ -\nabla_{y_2} u_2(\cdot, x_{-2}) \\ \vdots \\ -\nabla_{y_n} u_n(\cdot, x_{-n}) \end{bmatrix}$$

over $\mathbb{R}^m$ for all $x \in \Omega$. In this case, we have to replace the D-Lagrangian (12.31) by

$$\widehat{\mathcal{L}}(x, y, \tilde{\lambda}) := \begin{bmatrix} -\nabla_{y_1} u_1(y_1, x_{-1}) \\ -\nabla_{y_2} u_2(y_2, x_{-2}) \\ \vdots \\ -\nabla_{y_n} u_n(y_n, x_{-n}) \end{bmatrix} + (\nabla_y \tilde{G}(x, y))^{\mathrm{T}} \tilde{\lambda}.$$

Alternatively, a generalized variant of this method proposed in Dempe, 1995; Dempe, 1998, could be applied directly to the MPEC (12.27).

Let us return to the Cournot equilibrium from the previous section. Our aim is to find a production vector $(y_1^*, y_2^* \dots, y_n^*)$ such that each y_i^*, $i = 1, 2, \dots, n$, solves the optimization problem

$$\begin{aligned} &\text{minimize} \quad f_i(y_i) - y_i p\left(y_i + \sum_{\substack{j \in N \\ j \neq i}} y_i^*\right) \\ &\text{subject to} \\ &\qquad\qquad y_i \in Y_i \; . \\ &\qquad\qquad y_i + \sum_{\substack{j \in N \\ j \neq i}} y_j^* \leq P, \end{aligned} \tag{12.32}$$

with the joint upper production bound P. Due to adding the coupling constraint $y_i + \sum_{\substack{j \in N \\ j \neq i}} y_j^* \leq P$ we obtain a GNE, where (in the previous notation) $s = 1$ and

$$g(y) = \sum_{i \in N} y_i - P. \tag{12.33}$$

Assume that hypothesis (H) holds true and $Y_i = [\alpha^i, \beta^i]$ with $\alpha^i \geq 0$ and $\beta^i \geq P$ for $i \in N$. Then problem (12.30) becomes:

$$\begin{aligned} &\text{Find} \quad (x, y) \in \mathbb{R}^n \times \mathbb{R}^n \quad \text{such that} \\ &\qquad x = y \\ &0 \in \begin{bmatrix} \nabla f_1(y_1) - p(T) - y_1 \nabla p(T) - \omega^1 + \kappa^1 \\ \vdots \\ \nabla f_n(y_n) - p(T) - y_n \nabla p(T) - \omega^n + \kappa^n \\ y_1 - \alpha^1 \\ \vdots \\ y_n - \alpha^n \\ P - \sum_{i=2}^{n} x_i - y_1 \\ \vdots \\ P - \sum_{i=1}^{n-1} x_i - y_n \end{bmatrix} + N_{\mathbb{R}^n \times \mathbb{R}^n_+ \times \mathbb{R}^n_+}(y, \omega, \kappa). \end{aligned} \tag{12.34}$$

In (12.34) ω denotes a KKT vector associated with the constraints $y^i \geq \alpha^i$ and κ is a KKT vector associated with the coupling constraints. Since all constraints are affine, the GE in (12.34) with $x = y$ amounts to the necessary and sufficient optimality conditions for programs (12.32). This implies that each solution $x = y = y^*$ of (12.34) is a "generalized Cournot equilibrium". For fixed $x \in \mathbb{R}^n$, the GE in (12.34) describes a Cournot equilibrium in which the production intervals of single firms are given by $K_i(x_{-i}) = \left[\alpha^i, P - \sum_{j \in N \setminus \{i\}} x_j\right]$, $i \in N$. Theorem 4.1 tells us that this GE has a solution whenever the sets $K_i(x_{-i})$ are nonempty. Moreover, (ELICQ) holds at all pairs (x, y), $y = S_1(x)$, at which no interval $K_i(x_{-i})$ shrinks to a singleton. Thus, by Theorem 4.4(i), Proposition 4.5(i) and Lemma 12.2, the GE in (12.34) possesses a unique solution $S(x) = (y, \omega, \kappa)$ at each x for which

$$P - \sum_{j \in N \setminus \{i\}} x_j > \alpha^i, \quad i = 1, 2, \ldots, n. \tag{12.35}$$

We further observe that, by Theorem 5.8, the GE in (12.34) is strongly regular at $(x, S(x))$ provided (12.35) holds. In this way we can guarantee single-valuedness of S and strong regularity on somewhat smaller sets than those mentioned in (A2),(A3). Nevertheless, the numerical method of Section 7.2 has good chances to converge, provided we start close to an equilibrium, at which the lower bounds $y_i \geq \alpha^i$ are not active. Moreover, Corollary 7.11 applies at all points at which the lower bounds $y_i \geq \alpha^i$ are not active.

In the test runs we took the data from Section 12.1, i.e. $n = 5$, and the functions f_i and p given by (12.16) and (12.17) with parameters from Table 12.1. The demand elasticity γ was set to 1.1. Again we used the codes BT and NLPQL with the accuracy in BT slightly decreased to 5.0×10^{-4}. In Table 12.7 we collected the productions and profits of single firms at equilibria, corresponding to different joint production limits P and the starting vector $(10, 10, 10, 10, 10)^{\mathrm{T}}$. These results exhibit a good conformity between the x- and the y-values. For comparison, the standard Cournot equilibrium is also displayed. As expected, the use of different starting vectors leads to different solutions.

Table 12.7. Productions and profits—starting vector $(10, 10, 10, 10, 10)$

		Firm 1	*Firm 2*	*Firm 3*	*Firm 4*	*Firm 5*	NSIM
GNE with	Production	35.9963	41.1087	43.0847	41.9888	37.8214	
$P = 200$	Profit	209.9161	291.9367	359.7178	403.8771	419.142	32
GNE with	Production	27.5761	30.444	31.5525	30.9377	29.4895	
$P = 150$	Profit	330.1226	411.8522	475.7894	514.9803	535.2218	24
GNE with	Production	19.0198	19.9758	20.3453	20.1403	20.5184	
$P = 100$	Profit	444.5986	503.2123	549.3577	580.219	624.59	28
GNE with	Production	14.6847	14.6847	14.6847	14.6847	16.2613	
$P = 75$	Profit	501.7835	530.3274	558.6158	586.5221	675.9499	21
no coupling	Production	36.9325	41.8181	43.7066	42.6592	39.179	
constraint	Profit	199.9345	279.7157	346.5898	391.2786	410.3566	—

The above approach is especially recommendable, when a "small" value of $\|x - y\|$ implies that x (and thus also y) is close to an $\hat{x}$ solving (12.30). Let Ξ denote the set of these solutions and put

$$\Psi(x, y, \tilde{\lambda}) := \begin{bmatrix} x - y \\ \tilde{\mathcal{L}}(x, y, \tilde{\lambda}) \\ -\tilde{G}(x, y) + N_{\mathbb{R}^{ns}_{+}}(\tilde{\lambda}) \end{bmatrix}$$

so that problem (12.30) can be written in the form:

$$\text{Find} \quad (x, y) \in \Omega \times \mathbb{R}^m \quad \text{such that}$$
$$0 \in \Psi(x, y, \tilde{\lambda})$$

Proposition 12.5 *Assume that* $\Xi \neq \emptyset$ *and all sets* Y_i *are polyhedral. Let the utilities* u_i, $i = 1, 2, \ldots, n$, *be quadratic functions and* $\tilde{G}$ *an affine map (in both variables* x *and* y*). Then there exist reals* $\varepsilon > 0$ *and* $\rho \geq 0$ *such that*

$$dist_{\Xi}(x) \leq \rho \|x - y\|, \tag{12.36}$$

whenever $\|x - y\| \leq \varepsilon$, $x \in \Omega$ *and* (x, y) *satisfies (with some* $\tilde{\lambda}$*) the GE in (12.30).*

Proof. In virtue of our assumptions, the multifunction $\Psi[\mathbb{R}^m \times \mathbb{R}^m \times \mathbb{R}^{ns}_+ \to \mathbb{R}^m \times \mathbb{R}^m \times \mathbb{R}^{ns}]$ is polyhedral. Consequently, the multifunction $\Psi^{-1}(\cdot) \cap (\Omega \times \mathbb{R}^m \times \mathbb{R}^{ns}_+)$ is also polyhedral and, by Theorem 2.4, locally upper Lipschitz with modulus $\rho \geq 0$ at each $(z_1, z_2, z_3) \in \mathrm{Dom}\Psi^{-1}$. Hence there exists a neighbourhood $\mathcal{O}$ of $(0, 0, 0) \in \mathbb{R}^m \times \mathbb{R}^m \times \mathbb{R}^{ns}$ such that

$$\Psi^{-1}(z_1, z_2, z_3) \cap (\Omega \times \mathbb{R}^m \times \mathbb{R}^{ns}_+) \subset \Psi^{-1}(0, 0, 0) \cap (\Omega \times \mathbb{R}^m \times \mathbb{R}^{ns}_+) + \rho \|z\| \mathbb{B}$$
$$\text{for all } z \in \mathcal{O}.$$

We infer that

$$\Psi^{-1}(x - y, 0, 0) \cap (\Omega \times \mathbb{R}^m \times \mathbb{R}^{ns}_+) \subset \Psi^{-1}(0, 0, 0) \cap (\Omega \times \mathbb{R}^m \times \mathbb{R}^{ns}_+) + \rho \|x - y\| \mathbb{B}, \tag{12.37}$$

whenever $(x - y, 0, 0) \in \mathcal{O}$. Since Ξ is the canonical projection of $\Psi^{-1}(0, 0, 0) \cap (\Omega \times \mathbb{R}^m \times \mathbb{R}^{ns}_+)$ onto $\mathbb{R}^m$, inequality (12.36) follows directly from (12.37). ■

The assumptions of Proposition 12.5 are satisfied in Example 12.2, but they do not hold for our generalized Cournot equilibrium problem.

Under suitable assumptions, the above approach can also be applied to consistency problems of the general form

$$\text{Find} \quad (x, y, \lambda) \in U_{\text{ad}} \times \mathbb{R}^m \times \mathbb{R}^s_+ \quad \text{such that}$$
$$\varphi(x, y, \lambda) = 0 \tag{12.38}$$
$$(y, \lambda) \in S(x),$$

where U_{ad} is a nonempty closed set, φ is a continuously differentiable function defined on $\mathbb{R}^n \times \mathbb{R}^m \times \mathbb{R}^s$ and S is given by the GE (5.24). This model incorporates a class of quasi-variational inequalities and also the so-called equilibrium programming problems introduced in Flåm and Antipin, 1997. Further, in this way we may look, e.g., for those solutions of perturbed variational inequalities at which some prescribed constraints are active (inactive).

Bibliographical notes

The Cournot equilibrium was thoroughly studied in Murphy et al., 1982; Harker, 1984; Cach, 1996 in connection with algorithms for its computation. Lemmas 12.1,12.2 come from

Murphy et al., 1982 and Cach, 1996, respectively. The Stackelberg problem (12.8) was used in Harker and Choi, 1987 to test a special penalty method for the numerical solution of MPECs. The results of Section 12.1 originate from Outrata and Zowe, 1995b, but the satisfaction of (A3) was only conjectured there. Here we were able to verify (A3) on the basis of Lemma 12.2.

The GNE was investigated in Debreu, 1952; Ichiishi, 1983; Harker, 1991; Flåm and Kummer, 1992 both from the theoretical and the numerical point of view. The idea to compute these equilibria via a suitable MPEC was proposed and briefly sketched in Outrata and Zowe, 1995b. Here we analyzed it in more detail. In Outrata and Zowe, 1995a and Kočvara and Outrata, 1995a an effective numerical approach, based on the nonsmooth Newton's method from Chapter 3, was proposed for the solution of a class of quasi-variational inequalities. This approach, however, cannot be applied to the computation of GNE, since the essential regularity assumption is violated at all equilibria at which the coupling constraints are active.

Appendices

Appendix A
Cookbook

In this appendix we give a summary of formulas needed for the numerical solution of various MPECs.

A.1 PROBLEM

We solve an MPEC

$$\begin{aligned}
&\text{minimize} \quad f(x, z) \\
&\text{subject to} \\
&\qquad z = S(x) \\
&\qquad x \in U_{\text{ad}},
\end{aligned} \tag{A.1}$$

where $x \in \mathbb{R}^n$, $z \in \mathbb{R}^k$, f maps $\mathbb{R}^n \times \mathbb{R}^k$ into $\mathbb{R}$, and U_{ad} is a nonempty and closed subset of $\mathbb{R}^n$. The map S is the solution map of an equilibrium problem given by a GE of type (1.1). We assume that this equilibrium problem has a unique solution for each control $x \in U_{\text{ad}}$, i.e., that S is single-valued.

A.2 ASSUMPTIONS

Let U_{ad} be bounded and $\tilde{A}$ be an open set containing U_{ad}. In order to be able to apply the implicit programming approach, the following assumptions must be verified:

(A1) f is continuously differentiable on $\tilde{A} \times \mathbb{R}^k$;

(A2) S is single-valued on $\tilde{A}$;

(A3) the considered GE is strongly regular at all points (x, z) with $x \in \tilde{A}$, $z = S(x)$.

The verification of (A1) is a standard mathematical task.

The single-valuedness of S on $\tilde{A}$ (the uniqueness of the solution to the equilibrium problem) can be verified by using the theorems in Section 4.2. Typically, one needs Theorem 4.7 for LCPs and a combination of Theorem 4.8 with Proposition 4.5(ii) for nonlinear programs and variational inequalities.

As for the strong regularity assumption (A3), the reader will use theorems in Section 5.3. Theorem 5.8 is particularly useful for nonlinear programs and variational inequalities.

A.3 FORMULAS

To solve the MPEC (A.1), we write it as an optimization problem

$$\begin{array}{ll} \text{minimize} & \Theta(x) := f(x, S(x)) \\ \text{subject to} & \\ & x \in U_{\text{ad}} \end{array} \tag{A.2}$$

with a nonsmooth objective $\Theta(x)$. The nonsmooth code BT described in Chapter 3 is the proper tool to deal with it. At each iterate, say x_k, the code needs

- the function value $\Theta(x_k)$

and

- one element (subgradient) of the generalized Jacobian $\partial\Theta(x_k)$.

The computation of the function value is straightforward: for the given control x^k, one solves the equilibrium (state) problems and plugs the solution into f. To obtain the subgradient information, one has to solve another problem—the adjoint problem. There are general formulas for generally formulated state problems, but these can be simplified if the state problem has simpler structure. From this reason, we give here the adjoint problems and subgradients formulas for several typical equilibrium problems of graded complexity.

A.3.1 QP, constraints do not depend on control

We assume that the equilibrium problem is a quadratic program and that only the objective function depend on the control x (not the constraints). We further assume that the upper-level objective f only depends on the solution of the QP, not on the associated multipliers. Examples of such MPECs can be found in Sections 9.2 and 9.4.

Equilibrium problem. Let $C[\mathbb{R}^n \to \mathbb{R}^{m\times m}]$ and $b[\mathbb{R}^n \to \mathbb{R}^m]$ be maps that assign a control x a square symmetric matrix $C(x)$ and a vector $b(x)$, respectively. Further, let $A \in \mathbb{R}^{\ell\times m}, c \in \mathbb{R}^\ell, B \in \mathbb{R}^{s\times m}$ and $d \in \mathbb{R}^s$. The equilibrium problem reads

$$\begin{array}{ll} \text{minimize} & \frac{1}{2}\langle y, C(x)y\rangle - \langle b(x), y\rangle \\ \text{subject to} & \\ & Ay = c \\ & By \le d. \end{array}$$

Denote by μ and λ the components of the KKT vector corresponding to the equality and inequality constraints, respectively.

Adjoint problem. (Corollary 7.5) For a given control x and the associated solution y of the equilibrium problem, the adjoint problem is the following quadratic program in variable p:

$$\begin{array}{ll} \text{minimize} & \frac{1}{2}\langle p, C(x)p\rangle - \langle \nabla_y f(x,y), p\rangle \\ \text{subject to} & \\ & Ap = 0 \\ & B^j p = d^i, \quad j \in I^+(x,y) \cup M_i(x,y) \end{array}$$

with

$$\begin{aligned} I(x,y) &= \{i \in \{1,2,\ldots,m\} \mid \langle B^i, y\rangle = 0\} \\ I^+(x,y) &= \{i \in I(x,y) \mid \lambda^i > 0\} \\ I^0(x,y) &= I(x,y) \setminus I^+(x,y). \end{aligned}$$

$M_i(x,y)$ is a subset of $I^0(x,y)$, i.e., an element of the family $\mathcal{P}(I^0(x,y))$.

Subgradient formula. (Proposition 7.14) Let x be a given control, y the solution of the equilibrium problem and p the solution of the adjoint problem. Then, for a "suitable" M_i,

$$\nabla_x f(x,y) - [\mathcal{J}_x(C(x)y - b(x))]^T p$$

is an element of the generalized gradient $\partial\Theta(x)$.

A.3.2 QP, constraints depend on control

We consider the same problem as in the previous section, but now the constrains may depend on the control x, too. We again assume that the upper-level objective f only depends on the solution of the QP, not on the multipliers. An example of such MPEC can be found in Section 10.2.

Equilibrium problem. Let $C[\mathbb{R}^n \to \mathbb{R}^{m\times m}]$ and $b[\mathbb{R}^n \to \mathbb{R}^m]$ be maps that assign a control x a square symmetric matrix $C(x)$ and a vector $b(x)$, respectively. Further, let $A[\mathbb{R}^n \to \mathbb{R}^{\ell\times m}], c[\mathbb{R}^n \to \mathbb{R}^\ell], B[\mathbb{R}^n \to \mathbb{R}^{s\times m}]$ and $d[\mathbb{R}^n \to \mathbb{R}^s]$ be maps defining the constraints. The equilibrium problem in y reads

$$\begin{aligned} &\text{minimize} \quad \tfrac{1}{2}\langle y, C(x)y\rangle - \langle b(x), y\rangle \\ &\text{subject to} \\ &\qquad\qquad A(x)y = c(x) \\ &\qquad\qquad B(x)y \le d(x). \end{aligned}$$

Denote by μ and λ the components of the KKT vector corresponding to the equality and inequality constraints, respectively.

Adjoint problem. (Corollary 7.5) For a given control x and the associated solution y of the equilibrium problem, the adjoint problem is the following quadratic program in variable p:

$$\begin{aligned} &\text{minimize} \quad \tfrac{1}{2}\langle p, C(x)p\rangle - \langle \nabla_y f(x,y), p\rangle \\ &\text{subject to} \\ &\qquad\qquad A(x)p = 0 \\ &\qquad\qquad \langle B^j(x), p\rangle = 0, \quad j \in I^+(x,y) \cup M_i(x,y) \end{aligned}$$

with

$$\begin{aligned} I(x,y) &= \{i \in \{1,2,\ldots,m\} \mid \langle B^i(x), y\rangle = d^i(x)\} \\ I^+(x,y) &= \{i \in I(x,y) \mid \lambda^i > 0\} \\ I^0(x,y) &= I(x,y) \setminus I^+(x,y). \end{aligned}$$

$M_i(x,y)$ is a subset of $I^0(x,y)$, i.e., an element of the family $\mathcal{P}(I^0(x,y))$. The solution of the adjoint problem is a triple (p,q,r), where q and r are the components of the KKT vector associated with the first and second set of the equality constraints, respectively.

Subgradient formula. (Proposition 7.14) Let x be a given control, (y,μ,λ) the solution of the equilibrium problem and (p,q,r) the solution of the adjoint problem. Then, for a "suitable" M_i,

$$\nabla_x f(x,y) - [\mathcal{J}_x(C(x)y - b(x)) + \sum_{j=1}^{\ell} \mu^j \mathcal{J} A^j(x) + \sum_{j=1}^{s} \lambda^j \mathcal{J} B^j(x)]^T p$$
$$-[\mathcal{J}_x(A(x)y - c(x))]^T q$$
$$- \sum_{j \in I^+(x,y) \cup M_i(x,y)} \mathcal{J}_x(\langle B^j(x), y\rangle - d^j(x)) r^j$$

is an element of the generalized gradient $\partial\Theta(x)$.

A.3.3 QP, constraints depend on control, upper-level objective depends on multipliers

We consider the same problem as in the previous paragraph, with constrains depending on the control x. Assume now that the upper-level objective f also depends on the multipliers μ, λ, not only on the solution of the QP. An example of such an MPEC can be found in Section 10.3.

Equilibrium problem. Let $C[\mathbb{R}^n \to \mathbb{R}^{m\times m}]$ and $b[\mathbb{R}^n \to \mathbb{R}^m]$ be maps that assign a control x a square symmetric matrix $C(x)$ and a vector $b(x)$, respectively. Further, let $A[\mathbb{R}^n \to \mathbb{R}^{\ell\times m}], c[\mathbb{R}^n \to \mathbb{R}^\ell], B[\mathbb{R}^n \to \mathbb{R}^{s\times m}]$ and $d[\mathbb{R}^n \to \mathbb{R}^s]$ be maps defining the constraints. The equilibrium problem in y reads

$$\begin{array}{ll} \text{minimize} & \frac{1}{2}\langle y, C(x)y\rangle - \langle b(x), y\rangle \\ \text{subject to} & \\ & A(x)y = c(x) \\ & B(x)y \le d(x). \end{array}$$

Denote by μ and λ the components of the KKT vector corresponding to the equality and inequality constraints, respectively.

Adjoint problem. (Theorem 7.3) For a given control x and the associated solution (y,μ,λ) of the equilibrium problem, the adjoint problem is the following quadratic program in variable p:

$$\begin{array}{ll} \text{minimize} & \frac{1}{2}\langle p, C(x)p\rangle - \langle \nabla_y f(x,y,\mu,\lambda), p\rangle \\ \text{subject to} & \\ & A(x)p = \nabla_\mu f(x,y,\mu,\lambda) \\ & \langle B^j(x), p\rangle = \dfrac{\partial f(x,y,\mu,\lambda)}{\partial \lambda^j}, \qquad j \in I^+(x,y) \cup M_i(x,y) \end{array}$$

with

$$I(x,y) \quad = \quad \{i \in \{1,2,\ldots,m\} \mid \langle B^i(x), y\rangle = d^i(x)\}$$

$$
\begin{aligned}
I^+(x,y) &= \{i \in I(x,y) \mid \lambda^i > 0\} \\
I^0(x,y) &= I(x,y) \setminus I^+(x,y).
\end{aligned}
$$

$M_i(x,y)$ is a subset of $I^0(x,y)$, i.e., an element of the family $\mathcal{P}(I^0(x,y))$. The solution of the adjoint problem is a triple (p,q,r), where q and r are the components of the KKT vector associated with the first and second set of the equality constraints, respectively.

Subgradient formula. (Proposition 7.14) Let x be a given control, (y,μ,λ) the solution of the equilibrium problem and (p,q,r) the solution of the adjoint problem. Then, for a "suitable" M_i,

$$
\begin{aligned}
\nabla_x f(x,y,\mu,\lambda) - [\mathcal{J}_x(C(x)y - b(x)) + \sum_{j=1}^{\ell} \mu^j \mathcal{J}A^j(x) + \sum_{j=1}^{s} \lambda^j \mathcal{J}B^j(x)]^T p \\
-[\mathcal{J}_x(A(x)y - c(x))]^T q \\
- \sum_{j \in I^+(x,y) \cup M_i(x,y)} \mathcal{J}_x(\langle B^j(x), y\rangle - d^j(x)) r^j
\end{aligned}
$$

is an element of the generalized gradient $\partial\Theta(x)$.

A.3.4 VI, constraints do not depend on control

Let the state problem be a variational inequality. Assume that only the operator defining the VI depends on the control x. Further suppose that the upper-level objective f only depends on the control x and on the solution y of the VI.

Equilibrium problem. Let $F[\mathbb{R}^{n\times m} \to \mathbb{R}^m]$ be a continuous mapping and Ω a set given by a system of equations and inequalities:

$$
\Omega = \{v \in \mathbb{R}^m | h^i(v) = 0,\ i = 1,2,\ldots,\ell,\ g^j(v) \le 0,\ j = 1,2,\ldots,s\};
$$

here the functions $h^i[\mathbb{R}^m \to \mathbb{R}]$, $i = 1,2,\ldots,\ell$, are affine and $g^j[\mathbb{R}^m \to \mathbb{R}]$, $j = 1,2,\ldots,s$, are convex and continuously differentiable. The equilibrium problem reads

$$
\begin{aligned}
&\text{Find } y \in \Omega \text{ such that} \\
&\qquad \langle F(x,y), v - y\rangle \ge 0 \quad \text{for all } v \in \Omega.
\end{aligned}
$$

Denote by

$$
\mathcal{L}(x,y,\mu,\lambda) := F(x,y) + \sum_{i=1}^{\ell} \mu^i \nabla h^i(y) + \sum_{i=1}^{s} \lambda^i \nabla g^i(y)
$$

the D-Lagrangian.

Adjoint problem. (Corollary 7.4) For a given control x and the associated solution y of the equilibrium problem, the adjoint problem is a linear variational inequality in variable p:

$$
\begin{aligned}
&\text{Find } p \in L_i(x,y) \text{ such that} \\
&\qquad \langle \mathcal{J}_y\mathcal{L}(x,y,\mu,\lambda)^T p - \nabla_y f(x,y), z - p\rangle \ge 0 \quad \text{for all } z \in L_i(x,y),
\end{aligned}
$$

with

$$
L_i(x,y) := \{z \mid \mathcal{J}H(y)z = 0, \langle \nabla g^i(y), z\rangle = 0,\ j \in I^+(x,y) \cup M_i(x,y)\}
$$

and

$$\begin{aligned} I(x,y) &= \{i \in \{1,2,\ldots,m\} \mid g^i(y) = 0\} \\ I^+(x,y) &= \{i \in I(x,y) \mid \lambda^i > 0\} \\ I^0(x,y) &= I(x,y) \setminus I^+(x,y). \end{aligned}$$

$M_i(x,y)$ is a subset of $I^0(x,y)$, i.e., an element of the family $\mathcal{P}(I^0(x,y))$.

Subgradient formula. (Proposition 7.14) Let x be a given control, y the solution of the equilibrium problem and p the solution of the adjoint problem. Then, for a "suitable" M_i,

$$\nabla_x f(x,y) - (\mathcal{J}_x F(x,y))^T p$$

is an element of the generalized gradient $\partial\Theta(x)$.

A.3.5 VI, constraints depend on control, upper-level objective depends on multipliers

Let us consider the same problem as in the previous paragraph (VI as state problem) but let now the constraints that define the set Ω depend on the control x and the upper-level objective depend on the "multipliers" associated with these constraints. An example of such an MPEC can be found in Section 10.3

Equilibrium problem. Let $F[\mathbb{R}^{n\times m} \to \mathbb{R}^m]$ be a continuous mapping and Ω a set given by a system of equations and inequalities:

$$\Omega(x) = \{v \in \mathbb{R}^m \,|\, h^i(x,v) = 0,\ i = 1,2,\ldots,\ell, \text{ and } g^j(x,v) \le 0,\ j = 1,2,\ldots,s\};$$

here the functions $h^i[\mathbb{R}^{n\times m} \to \mathbb{R}]$, $i = 1,2,\ldots,\ell$, are affine and $g^j[\mathbb{R}^{n\times m} \to \mathbb{R}]$, $j = 1,2,\ldots,s$, are convex and continuously differentiable for all x. The equilibrium problem reads:

$$\begin{aligned} &\text{Find } y \in \Omega(x) \text{ such that} \\ &\quad \langle F(x,y), v - y\rangle \ge 0 \quad \text{for all } v \in \Omega(x), \end{aligned}$$

Denote again by

$$\mathcal{L}(x,y,\mu,\lambda) := F(x,y) + \sum_{i=1}^{\ell} \mu^i \nabla h^i(x,y) + \sum_{i=1}^{s} \lambda^i \nabla g^i(x,y)$$

the D-Lagrangian.

Adjoint problem. (Theorem 7.3) For a given control x and the associated solution (y,μ,λ) of the equilibrium problem, the adjoint problem is the following linear system in variables p,q,r:

$$\begin{aligned} (\mathcal{J}_y\mathcal{L}(x,y,\mu,\lambda))^T p + (\mathcal{J}_y H(x,y))^T q - (\mathcal{J}_y G_{I^+\cup M_i}(x,y))^T r &= \nabla_y f(x,y,\mu,\lambda) \\ \mathcal{J}_y H(x,y) p &= \nabla_\mu f(x,y,\mu,\lambda) \end{aligned}$$

$$\langle \nabla_y g^j(x,y), p\rangle = \frac{\partial f(x,y,\mu,\lambda)}{\partial \lambda^j} \quad \text{for } j \in I^+(x,y) \cup M_i(x,y),$$

with

$$\begin{aligned} I(x,y) &= \{i \in \{1,2,\ldots,m\} \mid g^i(x,y) = 0\} \\ I^+(x,y) &= \{i \in I(x,y) \mid \lambda^i > 0\} \\ I^0(x,y) &= I(x,y) \setminus I^+(x,y). \end{aligned}$$

$M_i(x,y)$ is a subset of $I^0(x,y)$, i.e., an element of the family $\mathcal{P}(I^0(x,y))$.

Subgradient formula. (Proposition 7.14) Let x be a given control, (y,μ,λ) the solution of the equilibrium problem and (p,q,r) the solution of the adjoint problem. Then, for a "suitable" M_i,

$$\nabla_x f(x,y) - (\mathcal{J}_x \mathcal{L}(x,y,\mu,\lambda))^T p_i - (\mathcal{J}_x H(x,y))^T q_i + (\mathcal{J}_x G_{I^+ \cup M_i}(x,y))^T r_i$$

is an element of the generalized gradient $\partial\Theta(x)$.

A.3.6 LCP

Let the equilibrium problem be a linear complementarity problem. Examples of such MPECs can be found in Sections 9.2, 9.4 and 11.3.

Equilibrium problem. Let $A[\mathbb{R}^n \to \mathbb{R}^{m\times m}]$ and $b[\mathbb{R}^n \to \mathbb{R}^m]$ be maps that assign a control x a square symmetric matrix $A(x)$ and a vector $b(x)$, respectively. Further, let $\Psi \in \mathbb{R}^m$ be given. The equilibrium problem reads

Find $y \in \mathbb{R}^m$ such that

$$A(x)y + b(x) \geq 0, \quad y - \Psi \geq 0, \quad \langle A(x)y + b(x), y - \Psi\rangle = 0.$$

Adjoint problem. (Theorem 7.7) For a given control x and the associated solution y of the equilibrium problem, the adjoint problem is the following system of linear equations in variable p:

$$A^T_{L\cup(I^0\setminus M_i),L\cup(I^0\setminus M_i)}p = (\nabla_y f(x,y))_{L\cup(I^0\setminus M_i)}$$

with

$$\begin{aligned} L(x,y) &= \left\{i \in \{1,2,\ldots,m\} | y^i > \Psi^i\right\}, \\ I^0(x,y) &= \left\{i \in \{1,2,\ldots,m\} | (A(x)y + b(x))^i = 0,\ y^i = \Psi^i\right\}. \end{aligned}$$

$M_i(x,y)$ is a subset of $I^0(x,y)$, i.e., an element of the family $\mathcal{P}(I^0(x,y))$. Recall that for an $s \times s$ matrix M, an $s-$vector v and index set $I \subset \{1,2,\ldots,s\}$, $M_{I,I}$ denotes a submatrix of M with elements $M^{ij}, i,j \in I$, while b_I is a subvector with elements $b^i, i \in I$.

Subgradient formula. (Proposition 7.17) Let x be a given control, y the solution of the equilibrium problem and p the solution of the adjoint problem. Then, for a "suitable" M_i,

$$\nabla_x f(x,y) - [\mathcal{J}_x (A(x)y + b(x))_{L\cup(I^0\setminus M_i)}]^T p$$

is an element of the generalized gradient $\partial\Theta(x)$.

A.3.7 NCP

Let us now assume that the equilibrium problem is a nonlinear complementarity problem.

Equilibrium problem. Assume that $F[\mathbb{R}^{n\times m} \to \mathbb{R}^m]$ is a continuous mapping and $\Psi \in \mathbb{R}^m$ is a given vector. The equilibrium problem reads

Find $y \in \mathbb{R}^m$ such that

$$F(x,y) \geq 0, \quad y - \Psi \geq 0, \quad \langle F(x,y), y - \Psi \rangle = 0.$$

Adjoint problem. (Theorem 7.7) For a given control x and the associated solution y of the equilibrium problem, the adjoint problem is the following system of linear equations in variable p:

$$(\mathcal{J}_y F_{L\cup(I^0\setminus M_i),L\cup(I^0\setminus M_i)}(x,y))^T p = (\nabla_y f(x,y))_{L\cup(I^0\setminus M_i)}$$

with

$$\begin{aligned} L(x,y) &= \left\{i \in \{1,2,\ldots,m\} | y^i > \Psi^i\right\}, \\ I^0(x,y) &= \left\{i \in \{1,2,\ldots,m\} | F^i(x,y) = 0, \ y^i = \Psi^i\right\}. \end{aligned}$$

$M_i(x,y)$ is a subset of $I^0(x,y)$, i.e., an element of the family $\mathcal{P}(I^0(x,y))$.

Subgradient formula. (Proposition 7.17) Let x be a given control, y the solution of the equilibrium problem and p the solution of the adjoint problem. Then, for a "suitable" M_i,

$$\nabla_x f(x,y) - [\mathcal{J}_x F_{L\cup(I^0\setminus M_i)}(x,y)]^T p$$

is an element of the generalized gradient $\partial\Theta(x)$.

A.3.8 ICP

Finally, let the equilibrium problem be an implicit complementarity problem. An example of such an MPEC can be found in Section 9.3

Equilibrium problem. Assume that $F[\mathbb{R}^{n\times m} \to \mathbb{R}^m]$ and $\Phi[\mathbb{R}^{n\times m} \to \mathbb{R}^m]$ are continuously differentiable functions. We consider the equilibrium problem

Find $y \in \mathbb{R}^m$ such that

$$F(x,y) \geq 0, \quad y - \Phi(x,y) \geq 0, \quad \langle F(x,y), y - \Phi(x,y) \rangle = 0.$$

Adjoint problem. (Theorem 7.6) For a given control x and the associated solution y of the equilibrium problem, the adjoint problem is the following system of linear equations in variable p:

$$\begin{bmatrix} -\mathcal{J}_y F_{L\cup(I^0\setminus M_i)}(x,y) \\ E_{I^+\cup M_i} - \mathcal{J}_y \Phi_{I^+\cup M_i}(x,y) \end{bmatrix}^T p = \nabla_y f(x,y)$$

with

$$\begin{aligned} I^+(x,y) &= \left\{i \in \{1,2,\ldots,m\} | F^i(x,y) > 0\right\} \\ L(x,y) &= \left\{i \in \{1,2,\ldots,m\} | y^i > \Phi^i(x,y)\right\} \\ I^0(x,y) &= \left\{i \in \{1,2,\ldots,m\} | F^i(x,y) = 0, \ y^i = \Phi^i(x,y)\right\}. \end{aligned}$$

$M_i(x,y)$ is a subset of $I^0(x,y)$, i.e., an element of the family $\mathcal{P}(I^0(x,y))$.

Subgradient formula. (Proposition 7.16) Let x be a given control, y the solution of the equilibrium problem and p the solution of the adjoint problem. Then, for a "suitable" M_i,

$$\nabla_x f(x,y) + \begin{bmatrix} -\mathcal{J}_x F_{L\cup(I^0\setminus M_i)}(x,y) \\ \mathcal{J}_x \Phi_{I^+\cup M_i}(x,y) \end{bmatrix}^T p$$

is an element of the generalized gradient $\partial\Theta(x)$.

Appendix B
Basic facts on elliptic boundary value problems

We introduce basic notation and facts on elliptic boundary value problems; in particular, the notion of distribution, Sobolev spaces $H^k(\Omega)$, 2-nd order boundary value problem and variational inequality. We give three theorems on existence and uniqueness of the solution to elliptic boundary value problems. We only speak of linear differential operators of the second order, as all the examples in the book fall into this category. We present here only very basic facts; details can be found in many books on this subject, e.g., Dautray and Lions, 1988; Fučík and Kufner, 1980; Kinderlehrer and Stampaccia, 1980; Lions and Magenes, 1972; Nečas, 1967; Oden and Reddy, 1976. In this text we mainly follow selected parts from Dautray and Lions, 1988 and Oden and Reddy, 1976.

B.1 DISTRIBUTIONS

Test Functions. We introduce a special class of real-valued functions defined on $\mathbb{R}$, called the class of *test functions* on $\mathbb{R}$ and denoted by $\mathcal{D}(\mathbb{R})$, which has the following properties:

(i) Each $\phi(\xi) \in \mathcal{D}(\mathbb{R})$ is in $C_0^\infty(\mathbb{R})$ (infinitely differentiable functions with compact support).

(ii) There is a special topology associated with $\mathcal{D}(\mathbb{R})$ in which the convergence of sequences of test functions is defined in the following way. A sequence $\{\phi_n(\xi)\}_{n=1}^\infty \in \mathcal{D}(\mathbb{R})$ *converges* to a test functions $\phi(\xi) \in \mathcal{D}(\mathbb{R})$ if

  (ii.1) The supports of all $\phi_n(\xi)$ are contained in a fixed compact subset of $\mathbb{R}$.

  (ii.2) The derivative of *any* given order r of the $\phi_n(\xi)$ converge uniformly, as $n \to \infty$, to the corresponding derivative of the order r of $\phi(\xi)$.

Distribution. A functional q on the space $\mathcal{D}(\mathbb{R})$ is called a *distribution* or a *generalized function* if and only if it is linear and continuous. The dual space $\mathcal{D}(\mathbb{R})'$ of the space of test functions is the *space of distributions*. We list some basic properties of $\mathcal{D}(\mathbb{R})'$:

(i) If $q \in \mathcal{D}(\mathbb{R})'$, we write

$$q(\phi) := \langle q, \phi \rangle \qquad \phi \in \mathcal{D}(\mathbb{R}),$$

where $\langle \cdot, \cdot \rangle$ is the duality pairing, i.e., a bilinear map of $\mathcal{D}(\mathbb{R})' \times \mathcal{D}(\mathbb{R})$ into $\mathbb{R}$.

(ii) $\mathcal{D}(\mathbb{R})'$ is a linear space.

(iii) The space $\mathcal{D}(\mathbb{R})'$ is endowed with the weak star (weak*) topology, in which a sequence of distributions $\{q_n\}$ converges to a distribution q if, for any $\phi \in \mathcal{D}(\mathbb{R})$,

$$\lim_{n\to\infty} \langle q_n, \phi \rangle = \langle q, \phi \rangle.$$

Locally Integrable Functions. A function f is said to be *locally integrable* if the Lebesgue integral

$$\int_a^b |f(\xi)| \mathrm{d}\xi$$

exists for every compact interval $[a, b]$. Obviously, every locally integrable function f defines a distribution. We say that f *generates* the distribution F and do not distinguish between f and F; henceforth we write

$$F(\phi) = \langle f, \phi \rangle = \int_{-\infty}^{\infty} f(\xi)\phi(\xi)\mathrm{d}\xi.$$

A distribution q that can be generated from a locally integrable function is called a *regular distribution*. If a distribution is not regular, it is said to be *singular*.

Derivatives of Distributions. Let q be an arbitrary distribution. The functional p defined by

$$\langle p, \phi \rangle = -\langle q, \phi' \rangle \qquad \text{for all } \phi \in \mathcal{D}(\mathbb{R})$$

is called *distributional* or *generalized derivative* of q, and we use the notation

$$p = q'$$

that is,

$$\langle q', \phi \rangle = -\langle q, \phi' \rangle \qquad \text{for all } \phi \in \mathcal{D}(\mathbb{R}).$$

The quantity q' is also a distribution.

The functional p defined by

$$\langle p, \phi \rangle = (-1)^k \left\langle q, \frac{\partial^k \phi}{\partial \xi^k} \right\rangle \qquad \text{for all } \phi \in \mathcal{D}(\mathbb{R})$$

is called the kth *distributional derivative* of q, and we use the notation $p = q^{(k)}$.

Distributions in several dimensions. The extension of all the ideas covered thus far to functions defined on $\mathbb{R}^n$ is straightforward.

Let $\alpha = (\alpha_1, \ldots, \alpha_n)$ be a *multi-index* and $|\alpha| = \alpha_1 + \ldots + \alpha_n$. For a function $q : \mathbb{R}^n \to \mathbb{R}$, we denote

$$D^\alpha q = \frac{\partial^{|\alpha|} q}{\partial \xi_1^{\alpha_1} \ldots \partial \xi_n^{\alpha_n}}.$$

The continuous linear functional p given by

$$\langle p, \phi \rangle = (-1)^{|\alpha|} \langle q, D^\alpha \phi \rangle \qquad \text{for all } \phi \in \mathcal{D}(\mathbb{R}^n),$$

where q is a distribution, is called α-th *distributional partial derivative* of q, and we use the notation

$$p = D^\alpha q.$$

B.2 SOBOLEV SPACES

Let $\Omega \in \mathbb{R}^n$ be a bounded domain (open connected set).

Definition B.1 *We say that Ω has a* Lipschitz boundary $\partial\Omega$ *if there exist constants $\alpha > 0$ and $\beta > 0$ and a finite number of local coordinate systems and local maps $a_r, 1 \le r \le R$ which are Lipschitz continuous on their respective domains of definitions $\{\hat{\xi}^r \in \mathbb{R}^{n-1} \mid |\hat{\xi}^r| \le \alpha\}$, such that*

$$\begin{aligned}
&\partial\Omega = \bigcup_{r=1}^{R} \{(\xi_1^r, \hat{\xi}^r) \mid \xi_1^r = a_r(\hat{\xi}^r), |\hat{\xi}^r| < \alpha\}, \\
&\{(\xi_1^r, \hat{\xi}^r) \mid a_r(\hat{\xi}^r) < \xi_1^r < a_r(\hat{\xi}^r) + \beta, |\hat{\xi}^r| < \alpha\} \subset \Omega, \qquad 1 \le r \le R, \\
&\{(\xi_1^r, \hat{\xi}^r) \mid a_r(\hat{\xi}^r) - \beta < \xi_1^r < a_r(\hat{\xi}^r), |\hat{\xi}^r| < \alpha\} \subset compl\overline{\Omega}, \quad 1 \le r \le R,
\end{aligned}$$

where $\hat{\xi}^r = (\xi_2^r, \dots, \xi_n^r)$ and complA is a complement of A (see Figure B.1).

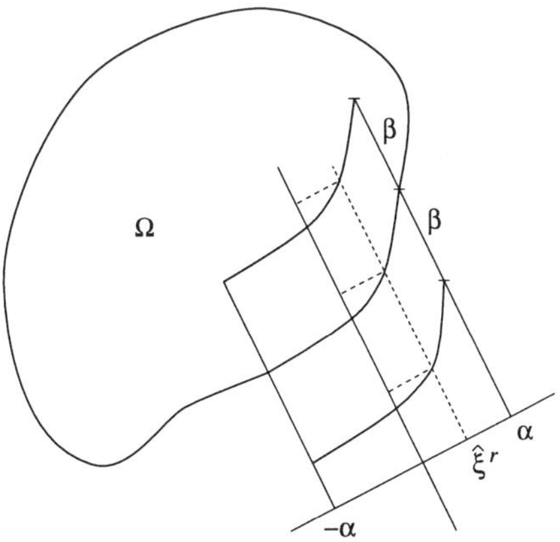

Figure B.1. Domain with a Lipschitz boundary

We assume that our Ω always has a Lipschitz boundary.

Sobolev spaces $H^m(\Omega)$. Let us recall the definition of the L_2 space:

$$L_2(\Omega) = \{v \mid \int_\Omega |v|^2 \mathrm{d}\xi = \|v\|_{L_2(\Omega)}^2 < +\infty\}$$

with the inner product

$$(u, v)_{L_2(\Omega)} = \int_\Omega uv\mathrm{d}\xi \qquad u, v \in L_2(\Omega).$$

It is well-known that $L_2(\Omega)$ is a Hilbert space.

The *Sobolev space* $H^m(\Omega)$ is defined as follows:

$$H^m(\Omega) = \{v \mid D^\alpha v \in L_2(\Omega) \quad \text{for all } |\alpha| \le m\}.$$

It can be shown that $H^m(\Omega)$ is again a Hilbert space with the inner product

$$(u,v)_{H^m(\Omega)} := (u,v)_{m,\Omega} = \sum_{|\alpha|\le m} \int_\Omega D^\alpha u D^\alpha v \mathrm{d}\xi \tag{B.1}$$

and the norm

$$\|v\|^2_{m,\Omega} = \sum_{|\alpha|\le m} \|D^\alpha v\|^2_{L_2(\Omega)}. \tag{B.2}$$

We define a space $H_0^m(\Omega)$ as a completion of $\mathcal{D}(\Omega)$ in the norm (B.2). Obviously, $H_0^m(\Omega)$ is a subspace of $H^m(\Omega)$.

To be able to speak of the values of function on the boundary of Ω (set of zero Lebesgue measure), we need the following result:

Theorem B.1 *There exists a linear continuous operator $\gamma_0 : H^1(\Omega) \mapsto L_2(\partial\Omega)$ such that*

$$\gamma_0 v = v|_{\partial\Omega} \qquad \textit{for all } v \in C^1(\mathrm{cl}\Omega).$$

The function $\gamma_0 v$ is called *trace* of v (often it is denoted by tr$v$). The meaning of the theorem is the following: if we have two "close" functions $u, v \in H^1(\Omega)$ then, due to continuity, also their traces are "close". (Note that two functions from $H^1(\Omega)$ which are equal in Ω and differ on $\partial\Omega$ have the same $H^1(\Omega)$ norm.) Similarly, there exist operators $\gamma_j, j > 0$, the traces of the normal derivatives $\dfrac{\partial^j v}{\partial n^j}$.

We can now characterize the space $H_0^1(\Omega)$ as follows:

$$H_0^1(\Omega) = \{v \in H^1(\Omega) \mid \gamma_0 v = 0\}.$$

B.3 ELLIPTIC PROBLEMS

Abstract variational problem. Let V be a real Hilbert space, V' its dual, a a continuous bilinear form on $V \times V$ and F a continuous linear form on V (i.e., $F \in V'$). The bilinear form a is said to be *V-elliptic* (or *V-coercive*) if there exists $\alpha \ge 0$ such that

$$a(v,v) \ge \alpha \|v\|^2 \qquad \text{for all } v \in V.$$

We consider the following problem:

$$\begin{aligned} &\text{Find } u \in V \text{ such that} \\ &a(u,v) = F(v) \qquad \text{for all } v \in V, \end{aligned} \tag{B.3}$$

called *abstract variational problem.* The following Lax–Milgram Theorem gives a condition for the existence of a unique solution to (B.3).

Theorem B.2 (Lax–Milgram Theorem) *Let $a(\cdot,\cdot)$ be a continuous bilinear form and F be a continuous linear form on V. If in addition the form $a(\cdot,\cdot)$ is V-elliptic, then the problem (B.3) admits a unique solution.*

We introduce the operator $A : V \to V$ defined by

$$(Au, v) = a(u,v) \qquad \text{for all } v \in V;$$

the existence of Au follows from the Riesz Theorem.

Now let V and H be two Hilbert spaces with $V \subset H$, V dense and continuously embedded in H (for continuous embedding we use the notation $V \hookrightarrow H$). We can identify H and H' and have

$$V \hookrightarrow H = H' \hookrightarrow V'.$$

The bilinear form a is said to be *V-coercive with respect to H* if

$$a(u,v) + \lambda(u,v) \quad \text{is } V\text{-coercive for a suitable } \lambda \in \mathbb{R}.$$

We have the following generalization of the Lax-Milgram Theorem.

Theorem B.3 *Let the embedding $V \hookrightarrow H$ be compact. If the bilinear form $a(\cdot,\cdot)$ is V-coercive with respect to H, then we have the following alternative:*

(i) $kerA = \{0\}$ and A is an isomorphism of $Dom(A)$ onto H; or

(ii) $kerA \neq \{0\}$, then $kerA$ is of finite dimension and the problem

$$Au = f$$

(with $f \in H$ given) has a solution only if f belongs to the image of A.

We now consider a case when

$$\mathcal{D}(\Omega) \hookrightarrow V \hookrightarrow H \hookrightarrow \mathcal{D}'(\Omega),$$

where Ω is an open set in $\mathbb{R}^n$. In addition, we assume that $\mathcal{D}(\Omega)$ is dense in H. This occurs, in particular, when $V = H^1(\Omega)$ and $H = L_2(\Omega)$, see the examples below. Let a be again a continuous bilinear form on $V \times V$. For a fixed $u \in V$, we consider the mapping

$$\varphi \to a(u,\varphi)$$

from $\mathcal{D}(\Omega)$ to $\mathbb{R}$; this is a continuous linear form on $\mathcal{D}(\Omega)$. Thus, there exist $\mathcal{A}u \in \mathcal{D}'(\Omega)$ such that

$$\langle \mathcal{A}u, \varphi\rangle = a(u,\varphi), \qquad \text{for all } \varphi \in \mathcal{D}(\Omega), \tag{B.4}$$

where $\langle\cdot,\cdot\rangle$ denotes the duality pairing between $\mathcal{D}(\Omega)$ and $\mathcal{D}'(\Omega)$. We can now come to the differential operators.

2-nd order differential operators. Now, let Ω be a domain in $\mathbb{R}^n$, $H = L_2(\Omega)$ and V be a closed subspace of $H^1(\Omega)$ containing $H_0^1(\Omega)$. We consider a 2-nd order differential operator A of the form

$$Au = -\sum_{i,j=1}^{n} \frac{\partial}{\partial \xi_i}(a_{ij}\frac{\partial u}{\partial \xi_j}) + a_0 u. \tag{B.5}$$

We call A *formal differential operator*, as it only represents a certain formal writing; it is not clear if the derivatives in (B.5) exist. We associate with A a bilinear form $a(u,v)$ defined by

$$a(u,v) = \int (\sum_{i,j=1}^{n} a_{ij}\frac{\partial u}{\partial \xi_i}\frac{\partial v}{\partial \xi_j} + a_0 uv)d\xi \tag{B.6}$$

with $u, v \in H^1(\Omega)$. We further associate with $a(u, v)$ the operator $\mathcal{A}$ from (B.4). In this sense, we can associate the abstract variational problem (B.3) with the operator equation

$$\mathcal{A}u = F$$

and further with a formal differential equation

$$Au = f,$$

where $F(v) = \int_\Omega f(\xi)v(\xi)\mathrm{d}\xi$. A function $u \in V$ is said to be a *weak solution* of the formal differential equation $Au = f$ if it solves the abstract variational problem (B.3) with a given by (B.6).

Let us assume that $a_{ij}, a_0 \in L_\infty(\Omega)$ (functions bounded a.e. in Ω) and that the coefficients a_{ij} are such that A is *strongly elliptic*, i.e., there is some $\alpha > 0$ such that

$$\sum_{i,j=1}^{n} a_{ij}\zeta_i\zeta_j \geq \alpha \sum_{i=1}^{n} |\zeta_i|^2 \qquad \text{a.e. in } \Omega, \text{ for all } \zeta \neq 0.$$

(If the matrix of coefficients a_{ij} is symmetric, this condition means that this matrix is positive definite and its smallest eigenvalue is bounded below by α.) We further assume that

$$a_0(\xi) \geq \beta \qquad \text{a.e. in } \Omega$$

where β is a prescribed constant. Then the bilinear form $a(u, v)$ is V-elliptic if $\beta > 0$. If $\beta = 0$, the bilinear form $a(u, v)$ is only V-coercive with respect to $L_2(\Omega)$.

Example B.1 (Dirichlet problem) Let $\Omega \subset \mathbb{R}^n$ be a bounded domain with a Lipschitz boundary. Let $V = H_0^1(\Omega)$, $f \in L_2(\Omega)$ be given and $F(v) := \int_\Omega f(\xi)v(\xi)\mathrm{d}\xi$. We consider a formal differential operator

$$Au := -\Delta u = -\sum_{i=1}^{n} \frac{\partial^2}{\partial \xi_i^2}$$

and the associated bilinear form

$$a(u, v) := \int_\Omega \sum_{i=1}^{n} \frac{\partial u}{\partial \xi_i}\frac{\partial v}{\partial \xi_i}\mathrm{d}\xi.$$

Operator A is strongly elliptic, but the coefficient a_0 from (B.5) is zero, hence a is only V-coercive with respect to $L_2(\Omega)$. However, due to Poincaré's inequality, a is even V-elliptic. The variational problem

$$a(u, v) = F(v) \qquad \text{for all } v \in V$$

admits a unique solution $u \in V$.

The corresponding operator equation is

$$-\Delta u = F \qquad \text{in } \mathcal{D}'(\Omega).$$

As $u \in H_0^1$, we further have

$$\gamma_0 u = u|_{\partial\Omega} = 0.$$

The formal differential equation reads

$$-\Delta u = f \qquad \text{in } \Omega$$

with the boundary condition

$$u = 0 \qquad \text{on } \partial\Omega.$$

△

Example B.2 (Neumann problem) Let again $\Omega \in \mathbb{R}^n$ be a bounded domain with a Lipschitz boundary. Let $V = H^1(\Omega)$, $f \in L_2(\Omega)$ and $g \in L_2(\partial\Omega)$ be given and

$$F(v) := \int_\Omega f(\xi)v(\xi)\mathrm{d}\xi + \int_{\partial\Omega} g(\xi)v(\xi)\mathrm{d}\partial\Omega.$$

We consider a formal differential operator

$$Au := -\Delta u + a_0 u = -\sum_{i=1}^{n} \frac{\partial^2}{\partial \xi_i^2} + a_0 u, \quad a_0 > 0$$

and the associated bilinear form

$$a(u,v) := \int_\Omega (\sum_{i=1}^{n} \frac{\partial u}{\partial \xi_i}\frac{\partial v}{\partial \xi_i} + a_0 uv)\mathrm{d}\xi. \tag{B.7}$$

The linear form F is obviously continuous on $H^1(\Omega)$ (due to the Trace Theorem) and the form a is V-elliptic. Hence the variational problem

$$a(u,v) = F(v) \qquad \text{for all } v \in V$$

has a unique solution $u \in H^1(\Omega)$ (Theorem B.2). Note that if a_0 were zero, then the bilinear form a would only be V-coercive with respect to $L_2(\Omega)$ and Theorem B.3 would apply.

Now, if we choose $v = \varphi \in \mathcal{D}(\Omega)$, (B.7) can be written as

$$\langle -\Delta u + a_0 u, \varphi\rangle = \langle f, \varphi\rangle; \tag{B.8}$$

from this we have the associated operator equation

$$-\Delta u + a_0 u = f \qquad \text{in } \mathcal{D}'(\Omega).$$

From the generalized Green's formula

$$\langle \gamma_1 u, v\rangle = \int_\Omega \Delta u \cdot v\mathrm{d}\xi + \int_\Omega \sum_{i=1}^{n} \frac{\partial u}{\partial \xi_i}\frac{\partial v}{\partial \xi_i}\mathrm{d}\xi$$

($\langle\cdot,\cdot\rangle$ denotes here the duality between $H^{\frac{1}{2}}(\partial\Omega)$ and its dual $H^{-\frac{1}{2}}(\partial\Omega)$), we further have the boundary condition implied by (B.8) and (B.7):

$$\gamma_1 u = \left.\frac{\partial u}{\partial n}\right|_{\partial\Omega} = g \qquad \text{in } H^{-\frac{1}{2}}(\partial\Omega).$$

The associated formal boundary value problem reads

$$\begin{aligned} -\Delta u + a_0 u &= f & &\text{in } \Omega \\ \frac{\partial u}{\partial n} &= g & &\text{on } \partial\Omega. \end{aligned}$$

△

Linear variational inequalities. Let us now consider a *closed convex subset K of V*. Let again a be a continuous bilinear form on $V \times V$ and F a continuous linear form on V. The projection w of the solution of the abstract variational problem (B.3) onto K is characterized by

$$\begin{array}{ll} \text{(i)} & w \in K \\ \text{(ii)} & a(w, v-w) \geq F(v-w) \quad \text{for all } v \in K. \end{array}$$

The problem of finding w that satisfies the above conditions is called *linear variational inequality*. The next theorem is a generalization of the Lax-Milgram Theorem for this class of problems.

Theorem B.4 (Lions-Stampacchia) *Let $a(\cdot,\cdot)$ be a continuous bilinear form on $V \times V$ and F be a continuous linear form on V. Let K be a closed convex subset of V. If $a(\cdot,\cdot)$ is V-elliptic, then the variational inequality*

$$\begin{array}{l} \textit{Find } u \in K \textit{ such that} \\ \quad a(u, v-u) \geq F(v-u) \quad \textit{for all } v \in K. \end{array}$$

has a unique solution.

Example B.3 Let $\Omega \subset \mathbb{R}^n$ be a bounded domain with a Lipschitz boundary. Let $V = H_0^1(\Omega)$, $f \in L_2(\Omega)$ be given and $F(v) := \int_\Omega f(\xi)v(\xi)\mathrm{d}\xi$. We consider a formal differential operator

$$Au := -\Delta u = -\sum_{i=1}^{n} \frac{\partial^2}{\partial \xi_i^2}$$

and the associated bilinear form

$$a(u,v) := \int_\Omega \sum_{i=1}^{n} \frac{\partial u}{\partial \xi_i}\frac{\partial v}{\partial \xi_i}\mathrm{d}\xi.$$

Further, let K be a subset of $H_0^1(\Omega)$:

$$K := \{v \in H_0^1(\Omega) \mid v \geq 0 \text{ a.e. in } \Omega\};$$

obviously, K is closed and convex. We already know from Example B.1 that a is V-elliptic. Hence the variational inequality

$$a(u, v-u) \geq L(v-u) \qquad \text{for all } v \in K$$

admits a unique solution $u \in K$.

The classic interpretation of this problem is no longer a formal differential equation but a (formal) linear complementarity problem

$$\left.\begin{array}{r} -\Delta u \geq f, \quad u \geq 0 \\ (\Delta u + f)u = 0 \end{array}\right\} \quad \text{in } \Omega$$
$$u = 0 \qquad \text{on } \partial\Omega.$$

△

Appendix C
Complementarity problems

C.1 PROOF OF THEOREM 4.7

Theorem 4.7 plays a crucial role both in linear complementarity and in the stability theory of variational inequalities. There are different proofs available, cf. Cottle et al., 1992 and the references therein. The proof presented below comes from Cottle et al., 1992 and relies essentially on two important statements: The Frank–Wolfe Theorem from quadratic programming and a characterization of P-matrices due to Fiedler and Pták. For the reader's convenience we start with their formulation.

Theorem C.1 (Frank and Wolfe, 1956) *Consider a quadratic function $f[\mathbb{R}^n \to \mathbb{R}]$ and assume that f is bounded below on a nonempty (convex) polyhedral set Ω. Then the minimization problem*

$$\begin{array}{ll} \textit{minimize} & f(x) \\ \textit{subject to} & \\ & x \in \Omega \end{array}$$

possesses a solution.

Theorem C.2 (Fiedler and Pták, 1962) *Let A be an $[m \times m]$ matrix. Then the following statements are equivalent.*

(i) A is a P-matrix.

(ii) If $z^i(Az)^i \leq 0$ for all $i = 1, 2, \ldots, m$, then $z = 0$ (i.e. A reverses the sign of no nonzero vector).

In the proof we will still need the following lemma which relates the LCP from Theorem 4.7 to the quadratic program

$$\begin{array}{ll} \text{minimize} & \langle y, Ay + b\rangle \\ \text{subject to} & \\ & Ay + b \geq 0 \\ & y \geq 0. \end{array} \tag{C.1}$$

Lemma C.3 *Let the set of feasible points in program (C.1) be nonempty. Then (C.1) has a solution, say $\hat{y}$. Moreover, there exists a KKT vector $\lambda \in \mathbb{R}^m_+$ so that the pair $(\hat{y}, \lambda)$ satisfies the relations*

$$(A + A^T)\hat{y} + b - A^T\lambda \geq 0 \tag{C.2}$$

$$\langle (A + A^T)\hat{y} + b - A^T\lambda, \hat{y}\rangle = 0 \tag{C.3}$$

$$\langle \lambda, A\hat{y} + b\rangle = 0 \tag{C.4}$$

$$(\hat{y} - \lambda)^i(A^T(\hat{y} - \lambda))^i \leq 0 \quad \textit{for all } i = 1, 2, \ldots, m. \tag{C.5}$$

Proof. In virtue of Theorem C.1, problem (C.1) possesses a solution $\hat{y}$. We may now easily verify that the relations (C.2)–(C.4) are the corresponding KKT conditions in which the objective is considered in the form $\frac{1}{2}\langle y, (A + A^T)y\rangle + \langle y, b\rangle$ and λ is the associated KKT vector.

It remains to prove the inequalities (C.5). From (C.3) we deduce that

$$\hat{y}^i (A^T(\hat{y} - \lambda))^i \le 0 \quad \text{for all } i = 1, 2, \dots, m \tag{C.6}$$

since $(A\hat{y})^i + b^i \ge 0$ for all i. Similarly, we may multiply the i-th component of the vector on the left-hand side of (C.2) by λ^i and invoke the complementarity condition

$$\lambda^i (A\hat{y} + b)^i = 0, \quad i = 1, 2, \dots, m.$$

This yields

$$-\lambda^i (A^T(\hat{y} - \lambda))^i \le 0 \quad i = 1, 2, \dots, m. \tag{C.7}$$

The desired inequalities follow by adding of (C.6) and (C.7). ■

After this preparatory work we are now ready to prove Theorem 4.7. Suppose first that A is a P-matrix. Then we claim that

$$\Xi := \{y \in \mathbb{R}^m \mid Ay > 0, y > 0\} \ne \emptyset.$$

Indeed, if this is not the case, then by the Ville's theorem of the alternative (Cottle et al., 1992) there exists a vector $u \ne 0$ such that $u \ge 0$ and $A^T u \le 0$. In such a case, however, A^T reverses the sign of a nonzero vector and by Theorem C.2 A^T is not a P-matrix. By definition, A cannot be a P-matrix either, which contradicts our assumption.

This implies that the feasible set in (C.1) is nonempty for each $b \in \mathbb{R}^m$. To see this, let $\bar{y} \in \Xi$. We observe that

$$\vartheta A\bar{y} = A(\vartheta\bar{y}) \ge -b$$

for a suitably large $\vartheta > 0$ and of course $\vartheta\bar{y} > 0$ as well. Hence $\vartheta\bar{y}$ is a feasible point for the quadratic program (C.1).

By Lemma C.3 the quadratic program (C.3) has a solution $\hat{y}$ and there exists a KKT vector λ such that $(\hat{y}, \lambda)$ satisfies the conditions (C.2)–(C.5). Since A is a P-matrix, the inequalities (C.5) imply that $\hat{y} = \lambda$. Equality (C.4) becomes

$$\langle \hat{y}, A\hat{y} + b\rangle = 0$$

and $\hat{y}$ is thus a solution of our LCP. To show that $\hat{y}$ is a unique solution, assume the existence of a alternate solution $\tilde{y}$. Since $\tilde{y}^i (A\hat{y} + b)^i \ge 0$ and $\hat{y}^i (A\tilde{y} + b)^i \ge 0$ for all i, we have

$$\begin{aligned} (\hat{y} - \tilde{y})^i (A\hat{y} + b)^i &\le 0 \\ -(\hat{y} - \tilde{y})^i (A\tilde{y} + b)^i &\le 0 \end{aligned}$$

for all $i = 1, 2, \dots, m$. By adding these two inequalities, we obtain

$$(\hat{y} - \tilde{y})^i (A(\hat{y} - \tilde{y}))^i \le 0,$$

contradicting the fact that A reverses the sign of no nonzero vector.

Conversely, assume that A is not a P-matrix. Then, by Theorem C.2, there exists a vector $z \ne 0$ such that

$$z^i (Az)^i \le 0 \quad \text{for all } i. \tag{C.8}$$

We express z in the form $z^+ - z^-$ and note that $z^+ \neq z^-$. Similarly we set $u^+ := (Az)^+$, $u^- := (Az)^-$ and

$$\bar{b} := u^+ - A(z^+).$$

Since $z = z^+ - z^-$ and $Az = u^+ - u^-$, one also has

$$\bar{b} := u^- - A(z^-).$$

Due to (C.8),

$$z_i^+ u_i^+ = z_i^- u_i^- = 0 \quad \text{for all } i,$$

and so we observe that both vectors z^+ and z^- solve our LCP for $b = \bar{b}$. Since $z^+ \neq z^-$, we conclude that whenever A is not a P-matrix, there exists a vector b (e.g. $\bar{b}$) for which the LCP (4.32) has at least two distinct solutions. In this way Theorem 4.7 has been established.

C.2 SUPPLEMENT TO PROOF OF THEOREM 4.9

In Theorem 4.9 it is assumed that $F[\mathbb{R}^m \to \mathbb{R}^m]$ is strongly monotone, continuously differentiable and Lipschitz on $\mathbb{R}^m$. In the proof we need the following two properties:

(i) F is a homeomorphism of $\mathbb{R}^m$ onto itself;

(ii) F^{-1} is continuously differentiable and strongly monotone on $\mathbb{R}^m$.

These two useful properties are proved in the sequel. We note first that

$$\|F(x) - F(y)\| \geq \alpha\|x - y\| \quad \text{for all } x, y \in \mathbb{R}^n, \tag{C.9}$$

where α is the coefficient of the strong monotonicity. Indeed, by the Cauchy–Schwarz inequality

$$\|F(x) - F(y)\|\|x - y\| \geq \alpha\|x - y\|^2 \quad \text{for all } x, y \in \mathbb{R}^n,$$

which immediately yields (C.9).

Lemma C.4 *The inverse operator F^{-1} is well-defined on the range of F and one has*

$$\|F^{-1}(u) - F^{-1}(v)\| \leq \frac{1}{\alpha}\|u - v\| \quad \textit{for all } u, v \in \mathcal{R}(F). \tag{C.10}$$

Proof. By (C.9), F is injective on $\mathbb{R}^m$ so that F^{-1} is well-defined on $\mathcal{R}(F)$. To all points $u, v \in \mathcal{R}(F)$ there exist pre-images $x, y \in \mathbb{R}^m$ and

$$\|F^{-1}(u) - F^{-1}(v)\| = \|x - y\| \leq \frac{1}{\alpha}\|F(x) - F(y)\| = \frac{1}{\alpha}\|u - v\| \quad \text{for all } u, v \in \mathcal{R}(F).$$

The assertion has been proved. ∎

Let F be a linear map defined by an $[m \times m]$ matrix A. Then A is invertible, (C.9) amounts to $\|Ad\| \geq \alpha\|d\|$ for all $d \in \mathbb{R}^n$ and (C.10) becomes

$$\|A^{-1}\| \leq \frac{1}{\alpha}. \tag{C.11}$$

As the next step we recall the Hadamard's Theorem from Ortega and Rheinboldt, 1970.

Theorem C.5 *Let $G[\mathbb{R}^n \to \mathbb{R}^n]$ be continuously differentiable and*

$$\|(\mathcal{J}G(x))^{-1}\| \leq \gamma < +\infty \quad \textit{for all } x \in \mathbb{R}^n.$$

Then G is a homeomorphism of $\mathbb{R}^n$ onto itself.

On the basis of inequality (C.11) and Theorem C.5, we are now able to verify property (i). By Proposition 4.5(iii) for all $d \in \mathbb{R}^m$

$$\langle d, \mathcal{J}F(x)d\rangle \geq \alpha\|d\|^2 \quad \text{for all } x \in \mathbb{R}^m,$$

which implies

$$\|\mathcal{J}F(x)d\| \geq \alpha\|d\| \quad \text{for all } x \in \mathbb{R}^m.$$

In virtue of (C.11), we have thus

$$\|(\mathcal{J}F(x))^{-1}\| \leq \frac{1}{\alpha},$$

and Theorem C.5 applies.

To establish (ii), observe first that the continuous differentiability of F^{-1} follows from the chain rule, and

$$\mathcal{J}F^{-1}(x) = (\mathcal{J}F(x))^{-1} \quad \text{for all } x \in \mathbb{R}^m.$$

It remains to prove the strong monotonicity of F^{-1}. Let $x, y \in \mathbb{R}^m, x \neq y$, and put $u := F(x)$, $v := F(y)$. By the strong monotonicity of F

$$\langle u - v, F^{-1}(u) - F^{-1}(v)\rangle \geq \alpha\|F^{-1}(u) - F^{-1}(v)\|^2.$$

Further, due to the Lipschitz continuity of F,

$$\|u - v\| \leq L\|F^{-1}(u) - F^{-1}(v)\|,$$

where L is the Lipschitz modulus of F. Hence,

$$\langle F^{-1}(u) - F^{-1}(v), u - v\rangle \geq \frac{\alpha}{L^2}\|u - v\|^2.$$

In this way also property (ii) has been proved.

References

Al-Fahed, A. M., Stavroulakis, G. E., and Panagiotopoulos, P. D. (1991). Hard and soft fingered robot grippers. The linear complementarity approach. *ZAMM–Zeitschrift für angewandte Mathematik und Mechanik*, 71:257–265.

Arrow, K. and Debreu, G. (1954). Existence of equilibrium for competitive economy. *Econometrica*, 22:265–290.

Aubin, J. P. and Frankowska, H. (1990). *Set-Valued Analysis*. Birkhäuser, Boston.

Axelsson, O. (1996). *Iterative Solution Methods*. Cambridge University Press, Cambridge.

Baiocchi, C. and Capelo, A. (1984). *Variational and Quasi-Variational Inequalities*. J. Wiley & Sons, New York.

Barbu, V. (1984). *Optimal Control of Variational Inequalities*. Research Notes in Mathematics 100. Pitman, London.

Benedict, B., Sokołowski, J., and Zolesio, J. P. (1984). Shape optimization for contact problems. In *System Modelling and Optimization*, Lecture Notes in Control and Information Sciences 59, pages 790–799. Springer-Verlag, Berlin.

Bensoussan, A. and Lions, J. J. (1973). Contrôle impulsionnel et inéquations quasi-variationnels d'évolution. *Comptes Rendus de l'Academie Sciences, Paris, Série A*, 276:1333–1338.

Byrd, R., Hribar, M. B., and Nocedal, J. (July 1997). An interior-point algorithm for large scale nonlinear programming. Report OTC 97/05, Optimization Technology Center, Northwest University, Illinois.

Cach, J. (1996). *A nonsmooth approach to the computation of equilibria (in Czech)*. Diplom Thesis, Charles University Prague, Prague.

Chan, D. and Pang, J.-S. (1982). The generalized quasi-variational problem. *Mathematics of Operations Research*, 7:211–222.

Chaney, R. W. (1990). Piecewise C^k-functions in nonsmooth analysis. *Nonlinear Analysis, Theory, Methods, and Applications*, 15:649–660.

Cheney, E. W. and Goldstein, A. A. (1959). Newton's method for convex programming and Tchebycheff approximation. *Numerische Mathematik*, 1:253–268.

Ciarlet, P. G. (1978). *The Finite Element Method for Elliptic Problems*. North-Holland, Amsterdam, New York, Oxford.

Ciarlet, P. G. (1988). *Mathematical Elasticity, Vol I: Three Dimensional Elasticity*. Studies in Mathematics and its Applications 20. North Holland, Amsterdam.

Clarke, F. F. (1983). *Optimization and Nonsmooth Analysis*. J. Wiley & Sons, New York.

Conn, A. R., Gould, N. I. M., and Toint, P. L. (1992). *LANCELOT: a Fortran Package for Large-Scale Nonlinear Optimization (Release A)*. Lecture Notes in Computation Mathematics 17. Springer-Verlag, Berlin.

Cottle, R. W. (1966). Nonlinear programs with positively bounded Jacobians. *SIAM J. on Applied Mathematics*, 14:147–158.

Cottle, R. W., Pang, J.-S., and Stone, R. E. (1992). *The Linear Complementarity Problem*. Academic Press, Boston.

Cryer, C. W. (1971). The method of Christopherson for solving free boundary problems for infinite journal bearings by means of finite differences. *Mathematics of Computation*, 25:435–444.

Dafermos, S. (1988). Sensitivity analysis in variational inequalities. *Mathematics of Operations Research*, 13:421–434.

Dautray, L. and Lions, J. L. (1988). *Mathematical Analysis and Numerical Methods for Science and Technology. Volume 2: Functional and Variational Methods*. Springer-Verlag, Berlin-Heidelberg.

Debreu, G. (1952). A social equilibrium existence theorem. *Proc. of the National Academy of Sciences of the U.S.A.*, 38:886–893.

DeLuca, T., Facchinei, F., and Kanzow, C. (1996). A semismooth equation approach to the solution of nonlinear complementarity problems. *Mathematical Programming*, 75:407–439.

Dempe, S. (1987). A simple algorithm for the linear bilevel programming problem. *Optimization*, 18:373–385.

Dempe, S. (1995). On generalized differentiability of optimal solutions and its application to an algorithm for solving bilevel optimization problems. In Qi, L., Du, D., and Womersley, R., editors, *Recent Advances in Nonsmooth Optimization*, pages 36–56, Singapore. World Scientific.

Dempe, S. (1998). An implicit function approach to bilevel programming problems. In Migdalas, A., Pardalos, P. M., and Värbrand, P., editors, *Multilevel Optimization: Algorithms and Applications*, pages 273–294. Kluwer Acad. Publishers.

Dempe, S. and Schmidt, H. (1996). On an algorithm solving two-level programming problems with non-unique lower-level solutions. *Computational Optimization and Applications*, 6:227–249.

DeSilva, A. H. (1978). *Sensitivity formulas for nonlinear factorable programming and their application to the solution of an implicitly defined optimization model of US crude oil production*. PhD thesis, George Washington University, Washington, D.C.

Dieudonné, J. (1960). *Foundations of Modern Analysis*. Academic Press, New York.

Dontchev, A. L. (1995). Implicit function theorems for generalized equations. *Mathematical Programming*, 70:91–106.

Dontchev, A. L. and Hager, W. W. (1994). Implicit functions, Lipschitz maps, and stability in optimization. *Mathematics of Operations Research*, 19:753–768.

Dontchev, A. L. and Rockafellar, R. T. (1996). Characterizations of strong regularity for variational inequalities over polyhedral convex sets. *SIAM J. on Optimization*, 7:1087–1105.

Duvaut, G. and Lions, J. L. (1972). *Les inèquations en mécanique et en physique*. Dunod, Paris.

Eck, C. and Jarušek, J. (1997a). Existence results for the static contact problem with Coulomb friction. SFB 404 Bericht Nr. 97/2, Universität Stuttgart.

Eck, C. and Jarušek, J. (1997b). Existence results for the semicoercive static contact problem with Coulomb friction. SFB 404 Bericht Nr. 97/45, Universität Stuttgart.

Edmunds, T. E. and Bard, J. F. (1991). Algorithms for nonlinear bilevel mathematical programs. *IEEE Transactions on Systems, Man and Cybernetics*, 21:83–89.

Ermoliev, Y. Y. (1976). *Stochastic Programming Methods*. Nauka, Moscow.

Facchinei, F., Fischer, A., and Kanzow, C. (1995). A semismooth Newton method for variational inequalities: Theoretical results and preliminary numerical experience. Preprint 102, Institute of Applied Mathematics, University of Hamburg.

Facchinei, F., Jiang, H., and Qi, L. (preprint 1997). A smoothing method for mathematical programs with equilibrium constraints. *SIAM J. on Optimization*. To appear.

Facchinei, F. and Kanzow, C. (1997). A nonsmooth inexact Newton method for the solution of large-scale nonlinear complementarity problems. *Mathematical Programming*, 76:493–512.

Fiacco, A. V. and McCormick, G. P. (1968). *Nonlinear Programming: Sequential Unconstrained Minimization Techniques*. J. Wiley & Sons, New York.

Fiedler, M. (1986). *Special Matrices and Their Application in Numerical Mathematics*. Martinus Nijhoff, Dordrecht.

Fiedler, M. and Pták, V. (1962). On matrices with non-positive off-diagonal elements and positive principal minors. *Czechoslovak Mathematical Journal*, 12:382–400.

Flåm, S. D. and Antipin, A. S. (1997). Equilibrium programming using proximal-like algorithms. *Mathematical Programming*, 78:29–41.

Flåm, S. D. and Kummer, B. (April 1992). Great fish wars and Nash equilibria. WP–No. 0892, Department of Economics, University of Bergen, Norway.

Frank, M. and Wolfe, P. (1956). An algorithm for quadratic programming. *Naval Research Logistics Quarterly*, 3:95–110.

Friesz, T. F., Cho, H.-J., Mehta, N. J., Tobin, R. L., and Anandalingam, G. (1992). A simulated annealing approach to the network design problem with variational inequality constraints. *Transportation Science*, 26:18–26.

Friesz, T. L., Tobin, R. L., Cho, H.-J., and Mehta, N. J. (1990). Sensitivity analysis based heuristic algorithms for mathematical programs with variational inequality constraints. *Mathematical Programming*, 48:265–284.

Fučík, S. and Kufner, A. (1980). *Nonlinear Differential Equations*. Elsevier, Amsterdam.

Fukushima, M. (1992). Equivalent differentiable optimization problems and descent methods for asymptotic variational inequality problems. *Mathematical Programming*, 53:99–110.

Giaquinta, M. and Giusti, E. (1985). Research on the equilibrium of masonry structures. *Archive for Rational Mechanics and Analysis*, 88:359–392.

Gill, P. E., Murray, W., and Saunders, M. A. (December 1997). User's guide for SNOPT 5.3: a Fortran package for large-scale nonlinear programming. Report NA 97-5, Department of Mathematics, University of California, San Diego.

Haraux, A. (1977). How to differentiate the projection on a convex set in Hilbert space. Some applications to variational inequalities. *J. of the Mathematical Society of Japan*, 29:615–631.

Harker, P. (1991). Generalized Nash games and quasi-variational inequalities. *European J. of Operational Research*, 54:81–94.

Harker, P. T. (1984). A variational inequality approach for the determination of oligopolistic market equilibrium. *Mathematical Programming*, 30:105–111.

Harker, P. T. and Choi, S. C. (1987). A penalty function approach to mathematical programs with variational inequality constraints. Technical report, Decision Sciences Department, University of Pennsylvania.

Harker, P. T. and Pang, J.-S. (1988). On the existence of optimal solutions to mathematical program with equilibrium constraints. *Operations Research Letters*, 7:61–64.

Harker, P. T. and Pang, J.-S. (1990). Finite-dimensional variational inequalities and complementarity problems: a survey of theory, algorithms and applications. *Mathematical Programming*, 60:161–220.

Hartman, P. and Stampacchia, G. (1966). On some nonlinear elliptic differential functional equations. *Acta Mathematica*, 115:153–188.

Haslinger, J. and Neittaanmäki, P. (1996). *Finite Element Approximation for Optimal Shape Design: Theory and Applications*. J. Wiley & Sons, Chichester.

Haslinger, J. and Panagiotopoulos, P. D. (1984). The reciprocal variational approach to the Signorini problem with friction, Approximation results. *Proceedings of the Royal Society of Edinburgh, Sect. A*, 98:365–383.

Hiriart-Urruty, J.-B. and Lemaréchal, C. (1993). *Convex Analysis and Minimization Algorithms*. Springer-Verlag, Berlin–Heidelberg.

Hlaváček, I. (1986). Shape optimization of elasto-plastic bodies obeying Hencky's law. *Aplikace Matematiky*, 31:486–499.

Hlaváček, I., Haslinger, J., Nečas, J., and Lovíšek, J. (1988). *Solution of Variational Inequalities in Mechanics*. Springer-Verlag, New York.

Hlaváček, I. and Křížek, M. (1992). Weight minimization of elastic bodies weakly supporting tension. I. Domains with one curved side. *Applications of Mathematics*, 37:201–240.

Hoffmann, A. J. (1952). On approximate solutions of systems of linear inequalities. *J. of Research of the National Bureau of Standards*, 49:263–265.

Hogan, W. W. (1973). Point-to-set maps in mathematical programming. *SIAM Review*, 15:591–603.

Horn, R. and Johnson, C. (1985). *Matrix Analysis*. Cambridge University Press, Cambridge.

Ichiishi, T. (1983). *Game Theory for Economic Analysis*. Academic Press, New York.

Ishizuka, Y. and Aiyoshi, E. (1992). Double penalty method for bilevel optimization problems. In Anandalingam, G. and Friesz, T., editors, *Hierarchical Optimization*, pages 73–88. Annals of Operations Research 34.

Jarušek, J. (1983). Contact problems with bounded friction. Coercive case. *Czechoslovak Mathematical Journal*, 33:237–261.

Jiang, H. (preprint 1997). Local properties of solutions of nonsmooth variational inequalities. *Optimization*. To appear.

Jittorntrum, K. (1984). Solution point differentiability without strict complementarity in nonlinear programming. *Mathematical Programming Study*, 21:127–138.

Josephy, N. J. (1979). Newton's method for generalized equations. Technical Summary Report 1965, Mathematics Research Center, University of Wisconsin-Madison.

Kalashnikov, V. V. and Kalashnikova, N. I. (1996). Solving two-level variational inequality. *J. of Global Optimization*, 8:289–294.

Kanzow, C. and Qi, H.-D. (March 1997). A QP-free constrained Newton-type method for variational inequality problems. Preprint 121, Reihe A, Hamburger Beiträge zur Angewandte Mathematik, Universität Hamburg.

Karamardian, S. (1971). Generalized complementarity problem. *J. of Optimization Theory and Applications*, 8:161–167.

Kelley, J. E. (1960). The cutting plane method for solving convex programs. *Journal of the SIAM*, 8:703–712.

Kinderlehrer, D. and Stampaccia, G. (1980). *An Introduction to Variational Inequalities and Their Application*. Academic Press, New York.

Kiwiel, K. C. (1985). *Methods of Descent for Nondifferentiable Optimization*. Springer-Verlag, Berlin-Heidelberg.

Kiwiel, K. C. (1986). A method for solving certain quadratic programming problems arising in nonsmooth optimization. *IMA Journal of Numerical Analysis*, 6:137–152.

Kiwiel, K. C. (1990). Proximity control in bundle methods for convex nondifferentiable optimization. *Mathematical Programming*, 46:105–122.

Klarbring, A. (1986). A mathematical programming approach to three-dimensional contact problems with friction. *Computer Methods in Applied Mechanics and Engineering*, 58:175–200.

Kočvara, M. (1997). Topology optimization with displacement constraints: A bilevel programming approach. *Structural Optimization*, 14:256–263.

Kočvara, M. and Outrata, J. V. (1994a). A numerical approach to the design of masonry structures. In Henry, J. and Yvon, J.-P., editors, *Systems Modelling and Optimization*, Lecture Notes in Control and Information Sciences 197, pages 195–205, London. Springer-Verlag.

Kočvara, M. and Outrata, J. V. (1994b). On optimization of systems governed by implicit complementarity problems. *Numerical Functional Analysis and Optimization*, 15:869–887.

Kočvara, M. and Outrata, J. V. (1994c). On optimization of systems governed by implicit complementarity problems. *Numerical Functional Analysis and Optimization*, 15:869–887.

Kočvara, M. and Outrata, J. V. (1994d). Shape optimization of elasto-plastic bodies governed by variational inequalities. In Zolésio, J. P., editor, *Boundary Control and Boundary Variation*, Lecture Notes in Pure and Applied Mathematics 168, pages 261–271, New York. M. Dekker.

Kočvara, M. and Outrata, J. V. (1995a). On a class of quasi-variational inequalities. *Optimization Methods & Software*, 5:275–295.

Kočvara, M. and Outrata, J. V. (1995b). On the solution of optimum design problems with variational inequalities. In Du, D., Qi, L., and Womersley, R., editors, *Recent Advances in Nosmooth Optimization*, pages 172–192, Singapore. World Scientific.

Kočvara, M. and Outrata, J. V. (1997). A nonsmooth approach to optimization problems with equilibrium constraints. In Ferris, M. C. and Pang, J.-S., editors, *Complementarity and Variational Problems*, pages 148–164, Philadelphia. SIAM.

Kočvara, M. and Zowe, J. (1994). An iterative two-step algorithm for linear complementarity problems. *Numerische Mathematik*, 68:95–106.

Kočvara, M. and Zowe, J. (1996). How mathematics can help in design of mechanical structures. In Griffiths, D. and Watson, G., editors, *Proc. of the 16th Biennial Conference on Numerical Analysis*, pages 76–93. Longman Scientific and Technical.

Kojima, M. (1980). Strongly stable stationary solutions in nonlinear programs. In Robinson, S. M., editor, *Analysis and Computation of Fixed Points*, pages 93–138, New York. Acad. Press.

Kruskal, J. B. (1969). Two convex counterexamples: A discontinuous envelope function and a nondifferentiable nearest-point mapping. *Proc. of the American Mathematical Society*, 23:697–703.

Kummer, B. (1992). Newton's method based on generalized derivatives for nonsmooth functions: Convergence analysis. In Oettli, W. and Pallaschke, D., editors, *Advances in Optimization*, Lecture Note in Economics and Mathematical Systems 382, pages 171–194, Berlin. Springer-Verlag.

Kummer, B. (1997). Lipschitzian and pseudo-Lipschitzian inverse functions and applications to nonlinear optimization. In Fiacco, A., editor, *Mathematical Programming with Data Perturbations*, Lecture Notes in Pure and Applied Mathematics, Vol. 195, pages 201–222.

Kuntz, L. and Scholtes, S. (1994). A nonsmooth variant of the Mangasarian-Fromowitz constraint qualification. *J. of Optimization Theory and Applications*, 82:59–75.

Kyparisis, J. (1990). Solution differentiability for variational inequalities. *Mathematical Programming*, 48:285–302.

Lemaitre, J. and Chaboche, J.-L. (1994). *Mechanics of Solids Materials*. Cambridge University Press, Cambridge.

Lemaréchal, C. (1974). An algorithm for minimizing convex functions. In Rosenfeld, J. L., editor, *Proc. of the IFIP'74 Congress*, pages 552–556, Amsterdam. North-Holland.

Lemaréchal, C. (1975). An extension of Davidon method to nondifferentiable problems. *Mathematical Programming Study*, 3:95–109.

Lemaréchal, C. (1981). A view of line searches. In Oettli, W. and Stoer, J., editors, *Optimization and Optimal Control*, Lecture Notes in Control and Information Sciences, Berlin–Heidelberg. Springer–Verlag.

Lemaréchal, C. (1989). Nondifferentiable optimization. In Nemhauser, G., Kan, A. R., and Todd, M., editors, *Handbooks in Operations Research and Management Science, Volume 1, Optimization*, Amsterdam. North Holland.

Lemaréchal, C. and Imbert, M. B. (1985). Le Module M1FC1. Technical report, INRIA, Le Chesnay.

Lemaréchal, C. and Sagastizábal, C. (1997). Variable metric bundle methods. From conceptual to implementable forms. *Mathematical Programming*, 76:393–410.

Lemaréchal, C., Strodiot, J.-J., and Bihain, A. (1981). On a bundle algorithm for nonsmooth optimization. In Mangasarian, O. L., Meyer, R. R., and Robinson, S. M., editors, *Nonlinear Programming 4*, New York. Academic Press.

Lions, J. L. and Magenes, E. (1972). *Non-Homogeneous Boundary Value Problems and Applications*. Springer-Verlag, Berlin.

Loridan, P. and Morgan, J. (1989a). New results on approximate solutions in two level optimization. *Optimization*, 20:819–836.

Loridan, P. and Morgan, J. (1989b). A theoretical approximation scheme for Stackelberg problems. *J. of Optimization Theory and Applications*, 61:95–110.

Luenberger, D. G. (1979). *Introduction to Dynamic Systems*. J. Wiley & Sons, New York.

Lukšan, L. and Vlček, J. (1996). PNEW—A bundle-type algorithm for nonsmooth optimization. Technical Report 718, Institute of Computer Science, Prague.

Luo, Z.-Q., Pang, J.-S., and Ralph, D. (1997). *Mathematical Programs with Equilibrium Constraints*. Cambridge University Press, Cambridge.

Luo, Z.-Q., Pang, J.-S., Ralph, D., and Wu, S. Q. (1996). Exact penalization and stationarity conditions of mathematical programs with equilibrium constraints. *Mathematical Programming*, 75:19–76.

Mäkelä, M. M. and Neittaanmäki, P. (1992). *Nonsmooth Optimization*. World Scientific, Singapore.

Mangasarian, O. L. (1994). *Nonlinear Programming*. SIAM Classics in Applied Mathematics 10, Philadelphia.

Mangasarian, O. L. and Shiau, T. H. (1987). Lipschitz continuity of solutions of linear inequalities, programs and complementarity problems. *SIAM J. on Control and Optimization*, 25:583–595.

Marcotte, P. (1986). Network design problem with congestion effects: a case of bilevel programming. *Mathematical Programming*, 34:142–162.

Marcotte, P. and Zhu, D. L. (1996). Exact and inexact penalty methods for the generalized bilevel programming problems. *Mathematical Programming*, 74:141–158.

McKenzie, W. (1959). On the existence of general equilibrium for a competitive market. *Econometrica*, 27:54–71.

Mifflin, R. (1977). Semismooth and semiconvex functions in constrained optimization. *SIAM J. on Control and Optimization*, 15:959–972.

Mifflin, R. (1982). A modification and an extension of Lemaréchal's algorithm for nonsmooth optimization. *Mathematical Programming Study*, 17:77–90.

Mifflin, R. (1996). A quasi-second-order proximal bundle algorithm. *Mathematical Programming*, 73:51–72.

Mifflin, R., Sun, D., and Qi, L. (1996). Quasi-Newton bundle-type methods for nondifferentiable convex optimization. Applied Mathematics Report, University of New South Wales.

Mordukhovich, B. S. (1994). Lipschitzian stability of constrained systems and generalized equations. *Nonlinear Analysis, Theory, Methods and Applications*, 22:173–206.

Mosco, V. (1976). Implicit variational problems and quasi-variational inequalities. In *Lecture Notes in Mathematics*, Vol. 543, pages 83–156, Berlin. Springer Verlag.

Murphy, F. H., Sherali, H. D., and Soyster, A. L. (1982). A mathematical programming approach for determining oligopolistic market equilibrium. *Mathematical Programming*, 24:92–106.

Murty, K. G. (1988). *Linear Complementarity, Linear and Nonlinear Programming*. Heldermann Verlag, Berlin.

Nash, J. (1951). Non-cooperative games. *Annals of Mathematics*, 54:286–295.

Nečas, J., Jarušek, J., and Haslinger, J. (1980). On the solution of the variational inequality to the Signorini-problem with small friction. *Bolletino dela UMI*, 17B:796–811.

Nečas, J. (1967). *Les méthodes directes en théorie des équations elliptiques*. Masson, Paris.

Nečas, J. and Hlaváček, I. (1981). *Mathematical Theory of Elastic and Elasto-Plastic Bodies: An Introduction*. Elsevier, Amsterdam.

Oden, J. T. and Reddy, J. N. (1976). *An Introduction to the Mathematical Theory of Finite Elements*. J. Wiley & Sons, New York.

Ortega, J. M. and Rheinboldt, W. C. (1970). *Iterative Solutions of Nonlinear Equations in Several Variables*. Academic Press, New York.

Outrata, J. V. (1990). On the numerical solution of a class of Stackelberg problems. *Zeitschrift für Operations Research*, 4:255–278.

Outrata, J. V. (1993). On necessary optimality conditions for Stackelberg problems. *J. of Optimization Theory and Applications*, 76:305–320.

Outrata, J. V. (1994). On optimization problems with variational inequality constraints. *SIAM J. on Optimization*, 4:340–357.

Outrata, J. V. (1995). Semismoothness in parametrized quasi-variational inequalities. In Doležal, J. and Fidler, J., editors, *System Modelling and Optimization*, pages 203–210, London. Chapman & Hall.

Outrata, J. V. (1997). On a special class of mathematical programs with equilibrium constraints. In Horst, R., Sachs, E., and Tichatschke, T., editors, *Recent Advances in Optimization*, pages 246–260, Berlin. Springer.

Outrata, J. V. and Zowe, J. (1995a). A Newton method for class of quasi-variational inequalities. *Computational Optimization and Applications*, 4:5–21.

Outrata, J. V. and Zowe, J. (1995b). A numerical approach to optimization problems with variational inequality constraints. *Mathematical Programming*, 68:105–130.

Pang, J.-S. (1981). The implicit complementarity problem. In Mangasarian, O. L., Meyer, R. R., and Robinson, S. M., editors, *Nonlinear Programming*, pages 487–518, New York. Academic Press.

Pang, J.-S. (1990a). Newton's method for B-differentiable equations. *Mathematics of Operations Research*, 15:311–341.

Pang, J.-S. (1990b). Solution differentiabilaity and continuation of Newton's method for variational inequality problems over polyhedral sets. *Journal of Optimization Theory and Applications*, 66:121–135.

Pang, J.-S. (1997). Error bounds in mathematical programming. *Mathematical Programming*, 79:299–332.

Pang, J.-S., Han, S. P., and Rangaraj, N. (1991). Minimization of locally Lipschitzian functions. *SIAM J. on Optimization*, 1:57–82.

Pang, J.-S. and Qi, L. (1993). Nonsmooth equations: Motivations and algorithms. *SIAM J. on Optimization*, 3:443–465.

Pang, J.-S. and Ralph, D. (1996). Piecewise smoothness, local invertibility, and parametric analysis of normal maps. *Mathematics of Operations Research*, 21:401–426.

Poljak, B. T. (1978). Subgradient methods: A survey of Soviet research. In Lemaréchal, C. and Mifflin, R., editors, *Nonsmooth Optimization*, pages 5–29, Oxford. Pergamon Press.

Qi, L. (1993). Convergence analysis of some algorithms for solving nonsmooth equations. *Mathematics of Operations Research*, 18:227–244.

Qi, L. and Sun, J. (1993). A nonsmooth version of Newton's method. *Mathematical Programming*, 58:353–368.

Qiu, Y. and Magnanti, T. L. (1989). Sensitivity analysis for variational inequalities defined on polyhedral sets. *Mathematics of Operations Research*, 14:410–432.

Qiu, Y. and Magnanti, T. L. (1992). Sensitivity analysis for variational inequalities. *Mathematics of Operations Research*, 17:61–76.

Rademacher, H. (1919). Über partielle und totale Differenzierbarkeit von Funktionen mehrerer Variabeln und über die Transformation der Doppelintegrale. *Mathematische Annalen*, 79:340–359.

Ralph, D. and Dempe, S. (1995). Directional derivatives of the solution of a parametric nonlinear program. *Mathematical Programming*, 70:159–172.

Robinson, S. M. (1976). An implicit function theorem for generalized variational inequalities. Technical Summary Report 1672, Mathematics Research Center, University of Wisconsin–Madison.

Robinson, S. M. (1979). Generalized equations and their solutions, Part I: Basic theory. *Mathematical Programming Study*, 10:128–141.

Robinson, S. M. (1980). Strongly regular generalized equations. *Mathematics of Operations Research*, 5:43–62.

Robinson, S. M. (1981). Some continuity properties of polyhedral multifunctions. *Mathematical Programming Study*, 14:206–214.

Robinson, S. M. (1984). Local structure of feasible sets in nonlinear programming, Part II: Nondegeneracy. *Mathematical Programming Study*, 22:217–230.

Robinson, S. M. (1985). Implicit B-differentiability in generalized equations. Technical Summary Report 2854, Mathematics Research Center, University of Wisconsin–Madison.

Robinson, S. M. (1991). An implicit-function theorem for a class of nonsmooth functions. *Mathematics of Operations Research*, 16:292–309.

Rockafellar, R. T. (1970). *Convex Analysis*. Princeton University Press, Princeton.

Rockafellar, R. T. (1981). *The Theory of Subgradients and Its Applications to Problems of Optimization: Convex and Nonconvex Functions*. Heldermann Verlag, Berlin.

Rosen, J. B. (1965). Existence and uniqueness of equilibrium points for concave n-person games. *Econometrica*, 33:520–534.

Rudin, W. (1974). *Real and Complex Analysis*. McGraw-Hill, New York.

Scheel, H. and Scholtes, S. (preprint 1997). Mathematical programs with equilibrium constraints: Stationarity, optimality and sensitivity. *Mathematical Programming*. To appear.

Schittkowski, K. (1986). NLPQL: a FORTRAN subroutine solving constrained nonlinear programming problems. *Annals of Operation Research*, 5:485–500.

Scholtes, S. (1994). *Introduction to Piecewise Differential Equations*. Habilitation Thesis, Institut für Statistik und Mathematische Wirtschaftstheorie, Universität Karlsruhe, Germany.

Scholtes, S. and Stöhr, M. (preprint 1997). Exact penalisation of mathematical programs with equilibrium constraints. *SIAM J. on Control and Optimization*. To appear.

Schramm, H. (1989). Eine Kombination von Bundle- und Trust-Region-Verfahren zur Lösung nichtdifferenzierbarer Optimierungsprobleme. Bayreuther Mathematische Schriften, Heft 30, Universität Bayreuth.

Schramm, H. and Zowe, J. (1991). Bundle trust methods: Fortran codes for nondifferential optimization. User's guide. Report 269, Mathematisches Institut, Universität Bayreuth.

Schramm, H. and Zowe, J. (1992). A version of the bundle idea for minimizing a nonsmooth function: conceptual idea, convergence analysis, numerical results. *SIAM J. on Optimization*, 2:121–152.

Shapiro, A. (1990). On concepts of directional differentiability. *J. of Optim. Theory and Applications*, 66:477–487.

Shimizu, K. and Aiyoshi, E. (1981). A new computational method for Stackelberg and min-max problems by use of a penalty method. *IEEE Transactions on Automatic Control*, AC–26:460–466.

Shor, N. Z. (1985). *Minimization Methods for Nondifferentiable Functions*. Springer-Verlag, Berlin-Heidelberg.

Sokołowski, J. and Zolésio, J.-P. (1992). *Introduction to Shape Optimization: Shape Sensitivity Analysis*. Springer-Verlag, Berlin.

Vicente, L. N., Savard, G., and Judice, J. J. (1994). Descent approaches for quadratic bilevel programming. *J. of Optimization Theory and Applications*, 81:379–399.

von Neumann, J. and Morgenstern, O. (1944). *Theory of Games and Economic Behaviour*. Princeton University Press, Princeton.

Walras, L. (1954). *Elements of Pure Economics*. Allen and Unwin, London.

Wolfe, P. (1975). A method of conjugate subgradients for minimizing nondifferentiable convex functions. *Mathematical Programming Study*, 3:145–173.

Ye, J. J. and Zhu, D. L. (1995). Optimality conditions for bilevel programming problems. *Optimization*, 33:9–27.

Ye, J. J., Zhu, D. L., and Zhu, Q. J. (1997). Exact penalization and necessary optimality conditions for generalized bilevel programming problems. *SIAM J. on Optimization*, 7:481–507.

Yosida, K. (1968). *Functional Analysis*. Springer-Verlag, Berlin.

Zarantonello, E. H. (1971). Projections on convex sets in Hilbert space and spectral theory. In Zarantonello, E. H., editor, *Contributions to Nonlinear Functional Analysis*, pages 237–424, New York. Academic Press.

Zhang, R. (1994). Problems of hierarchical optimization in finite dimensions. *SIAM J. on Optimization*, 4:208–227.

Zolesio, J. P., editor (1983). *Shape Controlability for Free Boundaries in System Modelling and Optimization*. Lecture Notes in Control and Information Sciences 59. Springer-Verlag, New York.

Index